IET ENERGY ENGINEERING SERIES 198

# Battery Management Systems and Inductive Balancing

## Other volumes in this series:

| | |
|---|---|
| Volume 1 | **Power Circuit Breaker Theory and Design** C.H. Flurscheim (Editor) |
| Volume 4 | **Industrial Microwave Heating** A.C. Metaxas and R.J. Meredith |
| Volume 7 | **Insulators for High Voltages** J.S.T. Looms |
| Volume 8 | **Variable Frequency AC Motor Drive Systems** D. Finney |
| Volume 10 | **$SF_6$ Switchgear** H.M. Ryan and G.R. Jones |
| Volume 11 | **Conduction and Induction Heating** E.J. Davies |
| Volume 13 | **Statistical Techniques for High Voltage Engineering** W. Hauschild and W. Mosch |
| Volume 14 | **Uninterruptible Power Supplies** J. Platts and J.D. St Aubyn (Editors) |
| Volume 15 | **Digital Protection for Power Systems** A.T. Johns and S.K. Salman |
| Volume 16 | **Electricity Economics and Planning** T.W. Berrie |
| Volume 18 | **Vacuum Switchgear** A. Greenwood |
| Volume 19 | **Electrical Safety: A guide to causes and prevention of hazards** J. Maxwell Adams |
| Volume 21 | **Electricity Distribution Network Design, 2nd Edition** E. Lakervi and E.J. Holmes |
| Volume 22 | **Artificial Intelligence Techniques in Power Systems** K. Warwick, A.O. Ekwue and R. Aggarwal (Editors) |
| Volume 24 | **Power System Commissioning and Maintenance Practice** K. Harker |
| Volume 25 | **Engineers' Handbook of Industrial Microwave Heating** R.J. Meredith |
| Volume 26 | **Small Electric Motors** H. Moczala et al. |
| Volume 27 | **AC–DC Power System Analysis** J. Arrillaga and B.C. Smith |
| Volume 29 | **High Voltage Direct Current Transmission, 2nd Edition** J. Arrillaga |
| Volume 30 | **Flexible AC Transmission Systems (FACTS)** Y-H. Song (Editor) |
| Volume 31 | **Embedded Generation** N. Jenkins et al. |
| Volume 32 | **High Voltage Engineering and Testing, 2nd Edition** H.M. Ryan (Editor) |
| Volume 33 | **Overvoltage Protection of Low-Voltage Systems, Revised Edition** P. Hasse |
| Volume 36 | **Voltage Quality in Electrical Power Systems** J. Schlabbach *et al.* |
| Volume 37 | **Electrical Steels for Rotating Machines** P. Beckley |
| Volume 38 | **The Electric Car: Development and future of battery, hybrid and fuel-cell cars** M. Westbrook |
| Volume 39 | **Power Systems Electromagnetic Transients Simulation** J. Arrillaga and N. Watson |
| Volume 40 | **Advances in High Voltage Engineering** M. Haddad and D. Warne |
| Volume 41 | **Electrical Operation of Electrostatic Precipitators** K. Parker |
| Volume 43 | **Thermal Power Plant Simulation and Control** D. Flynn |
| Volume 44 | **Economic Evaluation of Projects in the Electricity Supply Industry** H. Khatib |
| Volume 45 | **Propulsion Systems for Hybrid Vehicles** J. Miller |
| Volume 46 | **Distribution Switchgear** S. Stewart |
| Volume 47 | **Protection of Electricity Distribution Networks, 2nd Edition** J. Gers and E. Holmes |
| Volume 48 | **Wood Pole Overhead Lines** B. Wareing |
| Volume 49 | **Electric Fuses, 3rd Edition** A. Wright and G. Newbery |
| Volume 50 | **Wind Power Integration: Connection and system operational aspects** B. Fox *et al.* |
| Volume 51 | **Short Circuit Currents** J. Schlabbach |
| Volume 52 | **Nuclear Power** J. Wood |
| Volume 53 | **Condition Assessment of High Voltage Insulation in Power System Equipment** R.E. James and Q. Su |
| Volume 55 | **Local Energy: Distributed generation of heat and power** J. Wood |
| Volume 56 | **Condition Monitoring of Rotating Electrical Machines** P. Tavner, L. Ran, J. Penman and H. Sedding |
| Volume 57 | **The Control Techniques Drives and Controls Handbook, 2nd Edition** B. Drury |
| Volume 58 | **Lightning Protection** V. Cooray (Editor) |
| Volume 59 | **Ultracapacitor Applications** J.M. Miller |
| Volume 62 | **Lightning Electromagnetics** V. Cooray |
| Volume 63 | **Energy Storage for Power Systems, 2nd Edition** A. Ter-Gazarian |
| Volume 65 | **Protection of Electricity Distribution Networks, 3rd Edition** J. Gers |
| Volume 66 | **High Voltage Engineering Testing, 3rd Edition** H. Ryan (Editor) |
| Volume 67 | **Multicore Simulation of Power System Transients** F.M. Uriate |

Volume 68 **Distribution System Analysis and Automation** J. Gers
Volume 69 **The Lightening Flash, 2nd Edition** V. Cooray (Editor)
Volume 70 **Economic Evaluation of Projects in the Electricity Supply Industry, 3rd Edition** H. Khatib
Volume 72 **Control Circuits in Power Electronics: Practical issues in design and implementation** M. Castilla (Editor)
Volume 73 **Wide Area Monitoring, Protection and Control Systems: The enabler for smarter grids** A. Vaccaro and A. Zobaa (Editors)
Volume 74 **Power Electronic Converters and Systems: Frontiers and applications** A. M. Trzynadlowski (Editor)
Volume 75 **Power Distribution Automation** B. Das (Editor)
Volume 76 **Power System Stability: Modelling, analysis and control** A.A. Sallam and B. Om P. Malik
Volume 78 **Numerical Analysis of Power System Transients and Dynamics** A. Ametani (Editor)
Volume 79 **Vehicle-to-Grid: Linking electric vehicles to the smart grid** J. Lu and J. Hossain (Editors)
Volume 81 **Cyber-Physical-Social Systems and Constructs in Electric Power Engineering** S. Suryanarayanan, R. Roche and T.M. Hansen (Editors)
Volume 82 **Periodic Control of Power Electronic Converters** F. Blaabjerg, K.Zhou, D. Wang and Y. Yang
Volume 86 **Advances in Power System Modelling, Control and Stability Analysis** F. Milano (Editor)
Volume 87 **Cogeneration: Technologies, optimisation and implentation** C. A. Frangopoulos (Editor)
Volume 88 **Smarter Energy: From smart metering to the smart grid** H. Sun, N. Hatziargyriou, H. V. Poor, L. Carpanini and M. A. Sánchez Fornié (Editors)
Volume 89 **Hydrogen Production, Separation and Purification for Energy** A. Basile, F. Dalena, J. Tong and T.N.Veziroglu (Editors)
Volume 90 **Clean Energy Microgrids** S. Obara and J. Morel (Editors)
Volume 91 **Fuzzy Logic Control in Energy Systems with Design Applications in Matlab®/Simulink®** I. H. Altas
Volume 92 **Power Quality in Future Electrical Power Systems** A. F. Zobaa and S. H. E. A. Aleem (Editors)
Volume 93 **Cogeneration and District Energy Systems: Modelling, analysis and optimization** M. A. Rosen and S. Koohi-Fayegh
Volume 94 **Introduction to the Smart Grid: Concepts, technologies and evolution** S.K. Salman
Volume 95 **Communication, Control and Security Challenges for the Smart Grid** S.M. Muyeen and S. Rahman (Editors)
Volume 96 **Industrial Power Systems with Distributed and Embedded Generation** R. Belu
Volume 97 **Synchronized Phasor Measurements for Smart Grids** M.J.B. Reddy and D.K. Mohanta (Editors)
Volume 98 **Large Scale Grid Integration of Renewable Energy Sources** A. Moreno-Munoz (Editor)
Volume 100 **Modeling and Dynamic Behaviour of Hydropower Plants** N. Kishor and J. Fraile-Ardanuy (Editors)
Volume 101 **Methane and Hydrogen for Energy Storage** R. Carriveau and D. S-K. Ting
Volume 104 **Power Transformer Condition Monitoring and Diagnosis** A. Abu-Siada (Editor)
Volume 106 **Surface Passivation of Industrial Crystalline Silicon Solar Cells** J. John (Editor)
Volume 107 **Bifacial Photovoltaics: Technology, applications and economics** J. Libal and R. Kopecek (Editors)
Volume 108 **Fault Diagnosis of Induction Motors** J. Faiz, V. Ghorbanian and G. Joksimović
Volume 110 **High Voltage Power Network Construction** K. Harker
Volume 111 **Energy Storage at Different Voltage Levels: Technology, integration, and market aspects** A.F. Zobaa, P.F. Ribeiro, S.H.A. Aleem and S.N. Afifi (Editors)
Volume 112 **Wireless Power Transfer: Theory, technology and application** N.Shinohara
Volume 114 **Lightning-Induced Effects in Electrical and Telecommunication Systems** Y. Baba and V. A. Rakov
Volume 115 **DC Distribution Systems and Microgrids** T. Dragičević, F. Blaabjerg and P. Wheeler
Volume 116 **Modelling and Simulation of HVDC Transmission** M. Han (Editor)

Volume 117 **Structural Control and Fault Detection of Wind Turbine Systems** H.R. Karimi

Volume 119 **Thermal Power Plant Control and Instrumentation: The control of boilers and HRSGs, 2nd Edition** D. Lindsley, J. Grist and D. Parker

Volume 120 **Fault Diagnosis for Robust Inverter Power Drives** A. Ginart (Editor)

Volume 121 **Monitoring and Control Using Synchrophasors in Power Systems with Renewables** I. Kamwa and C. Lu (Editors)

Volume 123 **Power Systems Electromagnetic Transients Simulation, 2nd Edition** N. Watson and J. Arrillaga

Volume 124 **Power Market Transformation** B. Murray

Volume 125 **Wind Energy Modeling and Simulation, Volume 1: Atmosphere and plant** P.Veers (Editor)

Volume 126 **Diagnosis and Fault Tolerance of Electrical Machines, Power Electronics and Drives** A.J. M. Cardoso

Volume 128 **Characterization of Wide Bandgap Power Semiconductor Devices** F. Wang, Z. Zhang and E.A. Jones

Volume 129 **Renewable Energy from the Oceans: From wave, tidal and gradient systems to offshore wind** and solar D. Coiro and T. Sant (Editors)

Volume 130 **Wind and Solar Based Energy Systems for Communities** R. Carriveau and D.S-K. Ting (Editors)

Volume 131 **Metaheuristic Optimization in Power Engineering** J. Radosavljević

Volume 132 **Power Line Communication Systems for Smart Grids** I.R.S. Casella and A. Anpalagan

Volume 134 **Hydrogen Passivation and Laser Doping for Silicon Solar Cells** Brett Hallam and Catherine Chan (Editors)

Volume 139 **Variability, Scalability and Stability of Microgrids** S. M. Muyeen, S. M. Islam and F. Blaabjerg (Editors)

Volume 145 **Condition Monitoring of Rotating Electrical Machines** P. Tavner, L. Ran, C. Crabtree

Volume 146 **Energy Storage for Power Systems, 3rd Edition** A.G. Ter-Gazarian

Volume 147 **Distribution Systems Analysis and Automation, 2nd Edition** J. Gers

Volume 151 **SiC Power Module Design: Performance, robustness and reliability** Alberto Castellazzi and Andrea Irace (Editors)

Volume 152 **Power Electronic Devices: Applications, failure mechanisms and reliability** F. Iannuzzo (Editor)

Volume 153 **Signal Processing for Fault Detection and Diagnosis in Electric Machines and Systems** Mohamed Benbouzid (Editor)

Volume 155 **Energy Generation and Efficiency Technologies for Green Residential Buildings** D. Ting and R. Carriveau (Editors)

Volume 156 **Lithium-ion Batteries Enabled by Silicon Anodes** Chunmei Ban and Kang Xu (Editors)

Volume 157 **Electrical Steels, 2 Volumes** A. Moses, K. Jenkins, Philip Anderson and H. Stanbury

Volume 158 **Advanced Dielectric Materials for Electrostatic Capacitors** Q Li (Editor)

Volume 159 **Transforming the Grid towards Fully Renewable Energy** O. Probst, S. Castellanos and R. Palacios (Editors)

Volume 160 **Microgrids for Rural Areas: Research and case studies** R.K. Chauhan, K. Chauhan and S.N. Singh (Editors)

Volume 166 **Advanced Characterization of Thin Film Solar Cells** N. Haegel and M. Al-Jassim (Editors)

Volume 167 **Power Grids with Renewable Energy Storage, Integration and Digitalization** A.A. Sallam and B. OM P. Malik

Volume 170 **Reliability of Power Electronics Converters for Solar Photovoltaic Applications** Frede Blaabjerg, Ahteshamul Haque, Huai Wang, Zainul Abdin Jaffery and Yongheng Yang (Editors)

Volume 171 **Utility-Scale Wind Turbines and Wind Farms** A. Vasel-Be-Hagh and D. S.-K. Ting

Volume 172 **Lighting Interaction with Power Systems, 2 volumes** A. Piantini (Editor)

Volume 174 **Silicon Solar Cell Metallization and Module Technology** T. Dullweber (Editor)

Volume 193 **Overhead Electric Power Lines: Theory and practice** S. Chattopadhyay and A. Das

Volume 199 **Model Predictive Control for Microgrids: From power electronic converters to energy management** J. Hu, J. M. Guerrero and S. Islam

Volume 905 **Power System Protection, 4 volumes**

# Battery Management Systems and Inductive Balancing

Alex Van den Bossche and Ali Farzan Moghaddam

The Institution of Engineering and Technology

Published by The Institution of Engineering and Technology, London, United Kingdom

The Institution of Engineering and Technology is registered as a Charity in England & Wales (no. 211014) and Scotland (no. SC038698).



First published 2021



The Institution of Engineering and Technology
Michael Faraday House
Six Hills Way, Stevenage
Herts, SG1 2AY, United Kingdom

www.theiet.org



**British Library Cataloguing in Publication Data**
A catalogue record for this product is available from the British Library

**ISBN 978-1-83953-357-0 (hardback)**
**ISBN 978-1-83953-358-7 (PDF)**

Typeset in India by MPS Limited

# Contents

# About the authors

**Alex Van den Bossche** received his M.Sc. and Ph.D. degrees in Electromechanical Engineering from Ghent University, Ghent, Belgium in 1980 and 1990, respectively. He worked at the university's Electrical Energy Laboratory. He is a professor at the same university in the same field. His research is in the field of electrical drives, power electronics on various inverter types and passive components, magnetic materials, and battery management system for electric vehicles. He started three spin-off companies active in power electronics and energy conversion. He is not only interested in renewable energy conversion, battery management, but also more direct use of solar PV. He is an author of the book entitled 'Inductors and Transformers for Power Electronics'.

**Ali Farzan Moghaddam** received his B.Sc. degree in Electrical Engineering from Azad University (South Branch), Tehran, Iran. In 2010, he graduated from Eastern Mediterranean University (EMU), Famagusta, North Cyprus and received his M.Sc. degree in Electrical and Electronic Engineering. He worked as a research assistant in EMU from 2008 to 2010. His employment experience included the electrical company (2010–15). In 2020, he received his Ph.D. degree in Department of Electrical Energy at Ghent University, Ghent, Belgium. Currently his research interests include battery management system for electric vehicles, active and passive cell balancing method for lithium-based batteries, power electronics, and DC–DC and DC–AC converters.

# Preface

The book is related to the main aspects of battery management systems. After some historical note, there is an overview of the previous and actual battery types. It helps to understand why in previous times battery management systems were not really needed and are necessary for each cell with the actual batteries. It is also not clear if Li types are always the best choices as some other types might have to be considered which may be superior some aspects (number of cycles, energy/weight, energy/volume) or cheaper in cost/stored kWh. Some batteries may get concurrency of the next generation candidates.

Most books describe the chemistry, the state of charge, state of health, and capacitive balancing of cells. But how are they connected to converters in an elegant way? There are also newer problem settings, such as what happens if solar panels are added to the vehicle and that charging is done ‘as and when the sun shines’? The inductive equalizing possibilities are more elaborated than usual. A basket of methods is given that to adapt depending on the needs of the application. The book investigates both active and passive balancing circuits. The proposed methods are tested with both simulation and practical set-up.

Active balancing circuits, for instance transformer, and coupled inductor-based balancing methods have fast balancing time compared to the capacitor-based balancing methods. A reason is less transistor drops in series. A difference is made between transformers that have a separate primary and coupled inductors with equal windings. So, the main objective of this book is to focus on topologies of converter balancing techniques using the transformers, or coupled inductors.

**Cover image:** Without battery assistance, climbing on mountains by bike is difficult, the purpose of a good battery technology is also to preserve nature. Without the right BMS, batteries last less long in a variety of applications, not only electric cars, also the BMS life expectancy counts.

Alex Van den Bossche
Ali Farzan Moghaddam
June 16, 2021

# Acknowledgements

Alex Van den Bossche thanks Ghent University for the given opportunities. I am also grateful to the master students that helped in their thesis to out concepts for the ultralight electric vehicle (F2E for two electric).

Dedicated to my wife Christine and three children Maxime, Nathan, and Laura.

Alex Van den Bossche

Writing a book is more fascinating than I thought. None of this would have been possible without my family. My sincere gratitude goes to my family who provided me an opportunity to continue my studies, who provided me financial support and motivation to write this book. Without their precious support, it would not be possible to conduct this book. Everything we are today and everything we may become tomorrow are all for the sacrifices our parents made for us. Thanks to my father, 'Hossein', for his continuous support. You are a true example of what it means to be a good person, Dad. My special thanks goes to my mother, 'Shahin', who unfortunately passed away during editing this book. You made me happy even when I had no reason to be. Thanks to my lovely sister 'Sara' for her beautiful heart, and your merciful temper.

'The way in the world is one, and that is the truth' (Zoroaster 1500 BC-Persia).

Ali Farzan Moghaddam

# Summary

Presently, the world is facing an increasing amount of $CO_2$ in the environment. One of the alternatives is the use of renewable energy to power vehicles with electricity. But it is only a part of this revolution, today there is a long list of appliances that use battery storage. A drawback is that the production of batteries releases a quite large amount of $CO_2$ – in fact, not only $CO_2$ but also other pollutants. This is normally recovered in the battery life compared to the actual solutions. However, good battery management systems (BMSs) may contribute to extending the average life of the batteries. Not to forget that BMSs may themselves have a limited life. As today, e.g. the performance of electric vehicles (EVs) depends greatly on the batteries; thus BMSs also play an important role. Indirectly, the BMS has also an important influence on the safety, efficiency, and autonomy. Manufacturing tolerances result in variations of the battery capacity and internal resistance from cell to cell. These unavoidable differences are due to the chemical and electrical characteristics, such as asymmetrical degradation with ageing, temperature difference, production tolerance, and non-uniform internal impedance.

The problems of the capacity and resistance variation are increased by differences in ageing among the cells in one pack or between packs. This imbalance results in a reduction in the charge storage potential and the lifetime of the battery pack. Furthermore, with repeated charging and discharging cycles, a charge imbalance is generated among cells, visible as a voltage imbalance, mainly at full and no charge. This charge imbalance is a major factor to deteriorate the performance and reliability of the battery pack because it decreases the available capacity due to a low battery cell voltage and the explosion risk due to overcharging. To overcome the problem of imbalances, a BMS is introduced to monitor battery modules and to provide an early warning protection.

A BMS has many features. It may observe the battery voltage, battery charge, battery temperature, discharge current and also over-voltage protection, under-voltage protection, over current protection, over-temperature protection, and short circuit protection. During the past decades, many battery equalizing techniques have been proposed in the literature. The balancing processes can be divided into passive and active balancing methods. The passive balancing technique dissipates extra energy as heat while shunting a resistor at each cell. The advantages are low cost and easy implementation; however, the energy loss and the heat problem are disadvantages. The active balancing technique is used to overcome the problem of the energy loss and equalizes voltages of battery cells by transferring charge from higher voltage battery cells to lower voltage battery cells.

In this book, the types of batteries, as well as the existing battery balancing methods, are described. The book contains 14 chapters, which was intentionally done to improve clarity, so to avoid that the reader mixes up between different principles.

In this book, several types of cell active balancing methods for lithium-ion batteries in EVs and other application are investigated. Still in Chapter 5, a passive balancing circuit is shown but where the standby losses are minimized. Moreover, a cell voltage monitoring circuit was implemented to measure the individual cell voltage of a battery module. The book is mainly focused on active balancing methods which are especially based on single inductors, coupled inductor, and transformer balancing techniques. In the introduction, lithium-based batteries are described and the different topologies of cell balancing methods are described.

Chapter 1 studies first, the history of the battery, introduction to the battery, types and shapes of batteries. Next, it investigates battery mistreatments, battery charging and discharging process. It also explains the definitions as nominal voltage, cells, battery, and pack. Then it will explain charging methods and battery characteristics. In the end, a wide variety of battery applications are listed.

Chapter 2 is a discussion about batteries, 'lead acid batteries, lithium-ion batteries'. It also talks about life and reuse of batteries. Then it studies the effect of $CO_2$ and batteries. Even batteries are not the 'perfect' solution for every storage problem. In the end it gives ideas of avoiding batteries and concurrent storage technologies.

In Chapter 3, an overview of actual types is presented. Some system is found looking first at batteries using water-based electrolytes. Then, organic electrolyte based batteries are investigated, i.e. sodium-, potassium-, magnesium-, calcium-, zinc-, and aluminium-based batteries. A next category are the molten salt batteries. Then fuel cells are explained, and also flow battery systems. Some chemical basics of the Nernst equation and the electrochemical potential are tackled as well.

Chapter 4 investigates first the state of charge and state of health. Next, an introduction of BMSs is given. Why is it needed and what are the types? At the end, several cell balancing topologies are investigated, and advantages and disadvantages are given.

In Chapter 5, Section 1, a passive equalizing and high-voltage cell detection circuit is proposed. Also, a monitoring system with a liquid crystal display is implemented to determine if one cell is high or all cells are at a high voltage level. A Bluetooth module is added to command the charger to be turned ON/OFF if desired. A major feature of the system is to draw a very low current in standby. Therefore, the system does not contribute significantly to the self-discharge of the battery. The experimental results were provided to verify the validity of the system. Section 2 proposes a smart cell voltage monitoring system via a mobile application. It is implemented to display the cell voltage on cell phone. The chapter aimed to design a low-cost individual cell voltage measurement circuit and a new Android smartphone application to monitor the measuring voltage from a safe distance. The system can be split into three parts: measurement circuit, wireless communication system, and Android smartphone monitoring application. First, the individual cell voltage will be measured by an operational amplifier which is used as a differential amplifier. Following, the measured data will be transferred via Bluetooth to the smartphone. Finally, the information which is the cell voltages will be displayed on the smartphone screen. Bluetooth was used as a wireless communication means between Arduino and smartphone. The smartphone monitoring application was a new application designed by the open-source developed program (MIT App Inventor 2). The main advantages of the system are low cost and safety. The proposed transistorized voltage detection circuit had some negative temperature coefficient, where the Li-ion batteries have very little. An adaptation of the circuit is proposed, where the temperature coefficient has been compensated.

In Chapter 6, a resonant switched-capacitor converter balancing method is examined. A switched-capacitor converter (SCC) is worth considering as it can be implemented easily. Low volume, low cost, avoiding magnetic components are its features. However, a slow balancing speed is an important problem of the SCC. In this chapter, a resonant switching capacitor is proposed to reduce the balancing time and reduce energy loss.

Chapter 7 introduces the separate inductor balancing circuit. The conventional inductor based balancing needs a long balancing time because the energy transfer from the first cell to the last cell is long and resulting in long equalization time. In Section 1, the proposed circuit can achieve fast equalization compared to the conventional one by using a path between middle cells. The proposed inductor balancing has faster balancing time. In Section 2, a Ćuk converter balancing method is proposed, which has a high efficiency with non-dissipative currents and a bidirectional energy transferring capability. As compared with the conventional Ćuk balancing method, which requires $2N-2$ switches for $N$ cells, the proposed method uses $N$ switches and $N-1$ inductors for $N$ number of cells, which results in fewer components and losses.

In Chapter 8, Section 1, a coupled inductor balancing method is developed. For the eight cells in series example, the developed balance circuit has four inductors, one magnetic circuit with one winding per two cells, and one control switch per cell as compared to the traditional inductor based equalizer that needs $N-1$ inductors and magnetic circuits for $N$ number of cells and more switches. Finally, a more efficient, cost-effective circuit and a low bill of materials are built up. The inductors are well coupled using a single magnetic core with an air gap, and care is taken to have the same electromotive force (EMF). In Section 2, a Ćuk converter balancing method by using a coupled inductor for lithium-based batteries is investigated. The proposed circuit is an active balancing circuit that will equalize eight cells in series. The conventional Ćuk converter balancing circuit requires $2N-2$ switches in order to equalize $N$ cells, but the proposed circuit requires only $N$ switches for $N$ cells in a battery pack. A conventional Ćuk converter uses one switch and one diode, but the proposed circuit uses two MOSFETs as switch per cell, therefore, resulting in less voltage drop. Furthermore, the proposed circuit uses coupled inductors instead of uncoupled inductors, in the conventional type. The fact of using coupled inductors not only needs a fraction of the number of magnetic cores, but also fewer windings compared to separate inductors, moreover the symmetry is maintained and equalizes faster.

In Chapter 9, Section 1, the concept of active balancing technique by using multi-winding transformer balancing circuit (with forward structure) is presented. It equalizes eight cells in series battery pack. As compared to the conventional method with a flyback structure, which uses a diode in each secondary side, with an air-gapped transformer, this scheme with forward structure uses switches thus reduces the effect of unequal voltage of the secondary windings. At the primary side, it has full-bridge inverter, and no air gap is needed. Section 2 investigates the concept of active cell balancing technique, by using a multiple outputs double forward converter. It controls two times more cells than secondaries, and it equalizes eight cells in series. In this method, four secondaries can reasonably be wound with the same back EMF, 'four in one hand'. Traditional schemes employing a multi-winding transformer require as much secondaries as cells. This design requires one secondary for two neighbouring cells, consequently, the number of secondaries is reduced by a factor of 2. Also, the redistribution of a charge from a high cell to a low cell does not require many switching components rather needs some little intelligence to determine low cell voltage detection. In Section 3, a flyback balancing method has been investigated. The flyback stores energy when the transistor is on and then gives to the secondary loaded with the lowest voltage. A flyback converter is widely used in power supplies, which combines low cost, simple design, and reasonable efficiency. The primary side and secondary side are isolated and it can provide multi-output. In the multiple-output version loaded by cells, the primary will not get an over-voltage, unless none of the secondaries make contact. If only a few are not connected, or a diode missing, the other still clamps the voltage. This is not true in single-input–single-output flyback converters. If the secondary is not connected and no feedback is present, the transistor goes in over-voltage and is expected to fail.

In Chapter 10, a special multiple-output forward converter equalizing circuit is proposed. It is fed by a buck converter from lithium-based batteries in ultralight electrical vehicles. The special arrangement makes it possible to equalizing eight cells or more in series in a battery pack. It can be repeated for the other battery packs in the full battery. A common 40-V buck converter is feeding them, so it is a solution for a full battery equalization. Each battery pack consists of eight cells in series and the total battery bank consists of four battery packs in series. The proposed method combines a simple structure with fast balancing. This method applies a single-transistor zero-voltage switching technique, which also avoids a demagnetizing winding. It needs only one transistor and has lower switching losses and does not require a demagnetizing winding. The proposed forward type uses only one diode per cell and one transistor making it cost-effective.

Chapter 11 studies trickle charging and a vehicle start-up system. The design of a charger with four battery packs, each consisting of eight cells in series is investigated. A current controller is proposed to limit the current from the charger, so that the equalizer can handle that current. Also a starting resistor is needed to avoid high starting currents; thus the start-up system is required and designed to overcome this problem.

In Chapter 12, an intermediate voltage is made using current driver, and its working principles are studied. It permits to distribute the imposed voltage over different packs. The power-loss calculation and heat sink design are investigated. The rationale of the proposed method is that when all signals are absent, and thus all 32 cells remain below their threshold voltage, the charging current is not limited and the charger can deliver its full current to the battery. When one cell of the 32 cells goes above the threshold voltage, the charging current is reduced to 300 mA, so that the balancing circuit is capable of balancing the voltages of the affected module. Once all four modules indicate that all their eight cells are high, the battery is disconnected automatically from the charger.

Chapter 13 proposes a push–pull converter battery charger fed by photovoltaic (PV) for EVs. The aim of this chapter is to charge the batteries with PV panel separate from the grid or using panels on the vehicle. The maximum power voltage of this PV panel is typically much lower than the battery packs; thus a DC–DC converter is needed to overcome this problem. Therefore, a push–pull converter is designed and implemented to be connected to the solar panel. First, the advantages of PV panels are explained, especially combined with vehicles, and advantages and disadvantages are given. A push–pull converter is designed, and power loss and all design parameters are studied.

In Chapter 14, an auxiliary DC/DC converter is proposed. The main purpose of doing this 'two-stage DC/DC converter' is to equalize all battery cells from the total battery packs. In the first stage, the first converter, which is a buck converter, will be fed by the total battery pack consisting of four packs in series. Also, each pack includes 8 cells in series, thus a total of 32 cells in series. The nominal voltage of each cell is 3.2 V in this case; therefore, the total pack results in 102.4 V. The first converter 'buck converter' will decrease the total battery packs voltage to a lower voltage, for example 40 V, which is suitable to feed the second stage converter to equalize eight cells in series. The second converter could be, a forward converter, a flyback converter, or a multi-secondary transformer balancing circuit.

# Abbreviations

| | |
|---|---|
| AGM | absorbent glass matt |
| BJT | bipolar junction transistor |
| BLDC | brushless DC motor |
| BMS | battery management system |
| BOM | bill of materials |
| CC | constant current |
| CV | constant voltage |
| DOD | depth of discharge |
| DMFC | direct methanol fuel cell |
| DT | dead time |
| EMF | electromotive force |
| EMI | electromagnetic interference |
| ESR | equivalent series resistance |
| ES | electric scooter |
| EV | electric vehicle |
| FEC | fluoroethylene carbonate |
| HEV | hybrid electric vehicle |
| HTFC | high-temperature fuel cell |
| HVED | high-voltage equalizing detection |
| IC | integrated circuit |
| IEEE | Institute of Electrical and Electronics Engineers |
| IoT | Internet of Things |
| KERS | kinetic energy recovery system |
| KOH | potassium hydroxide |
| KVL | Kirchhoff's voltage law |
| LCD | liquid crystal display |
| LED | light-emitting diode |
| LCO | lithium–cobalt oxide |
| LCR meter | inductance, capacitance, and resistance meter |

| | |
|---|---|
| LFP | lithium–iron phosphate |
| LMO | lithium–manganese oxide |
| LTFC | low-temperature fuel cell |
| LTO | lithium–titanate oxide |
| NiCd | nickel–cadmium |
| NiMH | nickel metal hydride |
| NiFe | nickel–iron |
| NiZn | nickel–zinc |
| NMC | lithium–nickel–manganese–cobalt oxide |
| NTC | negative temperature coefficient |
| MATLAB | matrix laboratory |
| MLT | mean length of turns |
| MOSFET | metal–oxide–semiconductor field-effect transistor |
| MPPT | maximum power point tracking |
| OCV | open-circuit voltage |
| PCB | printed circuit board |
| PCM | phase-change material |
| PoE | power over Ethernet |
| PPTC | polymeric positive temperature coefficient |
| PTC | positive temperature coefficient |
| PV | photovoltaic |
| PWM | pulse-width modulation |
| RCD | resistor–capacitor–diode |
| Rdson | drain–source on-resistance |
| RoHS | restriction of hazardous substances |
| RSCC | resonant switched capacitor converter |
| SCC | switched-capacitor converter |
| SEI | solid electrolyte interphase |
| SMD | surface-mount device |
| SMPS | switched-mode power supply |
| SoC | state of charge |
| SOFC | solid oxide fuel cell |
| SoH | State of Health |
| SoP | State of Power |
| UPS | uninterruptible power supplies |

| | |
|---|---|
| VRFB | vanadium flow redox battery |
| ZVS | zero voltage switch |
| Anode-Cathode | Positive-Negative |

To avoid confusion it is good to mention also the way how some words are used:

- while charging, the positive pole is an anode, the negative is a cathode;
- while discharging the positive pole is a cathode, the negative is an anode;
- chemists like the use of the words anode (the way to-) and cathode (the way back-) of positive charges;
- if the battery is charging, the positive pole is anode;
- if the battery is loaded, the positive pole is cathode.

From the electric point of view, the terms positive and negative electrodes are easier to work with!

*Part I*

# Introduction to batteries and balancing

# Chapter 1
# Introduction to batteries

## 1.1 Preface

The area of the batteries is a continuously changing world. Although it is always 'chemical storage', they may react in a very different way on the electric side. In this introduction, a view on battery types is given. So, why suddenly is a battery management system (BMS) needed in the newer batteries, where the older ones did without? The fact is that the (rechargeable) batteries got more energy-dense and affordable and triggered a lot of applications. In fact mobile phones contain only one cell and do not contain a 'battery' in the narrow definition of terms. When cells are put in series, it is possible that by diverse reasons, in some cells less charge remains than other. In the older types (lead acid and Nickel Metal Hydride), the internal leakage increased with voltage, so they were self-stabilizing. The newer types do not have this property and may get easily damaged while overcharged or undercharged. Something has to keep them within limits, the task of a battery management system. A 'battery' in language means in fact an assembly of units, it could be an assembly of drums in music or an artillery battery in military language as well. People think of *electric cars*, but laptops got much more mobile than desktops by their batteries. In between, we have electric bikes (pedelec) or speed elec (speed pedelec) and a number of mobility devices such as Segway, Hoverboard, step, monowheel and also (indoor) karts. Also, garden equipment is subjected to a revolution in cordless or battery version: lawn mower, robot mower, riding mower, lawn trimmer, weed sweeper, leaf blower, chainsaw, hedge trimmer, electric pruning shears, rainwater pump, and wheelbarrow, also household tools such as, toothbrush, lamps, fans, insect killers, various gadgets, such as a flour sifter. Then we have the hobby and professional tools, such as forklifts, tractors, drones, small aeroplanes, screwdriver, drill, light-emitting diode (LED) lamps, angle grinder, drones, battery body warmer, battery cooler, wearable electronics, and exoskeleton. Also stationary batteries for a home battery, no-break emergency supplies, safety purposes, and so on.

A similar list could be given for medical equipment, even inside the body. The batteries for electric vehicles (EV) mainly include lead–acid, nickel–metal hydride, and lithium-ion batteries. Formerly, the battery applications were limited mainly to heavy and bulky lead–acid batteries [1]. The lead–acid battery was widely used because of its mature technology, stable performance, and low price [2,3]. However, it has some disadvantages such as a low energy density, long charging time, short life, and lead contamination [4]. The nickel–cadmium battery has been used for its large charge–discharge rate. However, a memory effect and toxicity are its disadvantages [5]. NiCd is banned from household and similar use by the regulation on hazardous substances RoHS, and nickel metal hydride is replacing them. nickel–metal hydride batteries have been widely applied in hybrid cars for their large charge–discharge rate and being environmentally friendly [6–8]. However, their single-cell voltage is low, and the losses are still high. The appearance of lithium-ion batteries brought a remarkable improvement in the area of battery systems [9]. Nowadays, lithium-based batteries extend the driving range for electric and hybrid vehicles as a competitive transportation means. The lithium-ion batteries are widely accepted because of their high cell voltage, high energy density, good cycling performance, and low self-discharge [7,8]. They are regarded as a reasonable choice for the actual needs of power batteries. The increasing use of lithium-based battery cells in electric and hybrid vehicles presented many challenges in terms of system design and a safe use of the cells. Li-ion batteries are getting a bad reputation by the important risk of fire. Recalls of vehicles can cost much more than the right battery monitoring and equalizing system. But maybe the accessibility should also be OK to avoid big costs in man-hours if something has to be replaced. The new battery chemistry requires new technologies for monitoring and maintaining the battery cells. Real-time decision-making based on measured parameters and optimizing the overall system is required.

A properly maintained battery system should keep cells balanced. Each cell in the battery system should be at the same state of charge (SoC). The levelling is often based on the voltage of the cell. The unbalancing of a lithium-based battery system is the result of numerous discharging and charging cycles during which the SoC of each battery begins to vary slightly. Some cells may age faster and get less capacity. Many variables that can lead to this unbalancing include temperature, individual cell chemistry, leakage currents, and the use of a BMS [10].

The balancing of a lithium-based battery system has one very large benefit. As the batteries are charged and discharged, their capacity is reduced. The cell with the lowest charge capacity may determine the total capacity of the battery but not worse than that. If cells are shifted in charge, the useful charge may be much lower. It is also possible to increase the charge beyond the worst cell characteristic if powerful bidirectional converters are used in the BMS. On the other hand, a longer lifetime (of both battery and BMS) might be obtained if the BMS is carefully designed for a long life. The end user and the environment may benefit form a good BMS.

## 1.2 History of the battery

The application of electricity moves back to the ancient period. First kind of battery, also known as the Parthian Battery, was invented about 2000 years ago [11]. It refers to the Parthian Empire (150 BC–223 AD) in ancient Persia.

The battery is a set of three artefacts that were found together: a ceramic pot, a tube of copper, and a rod of iron. The clay jar was probably filled with a vinegar

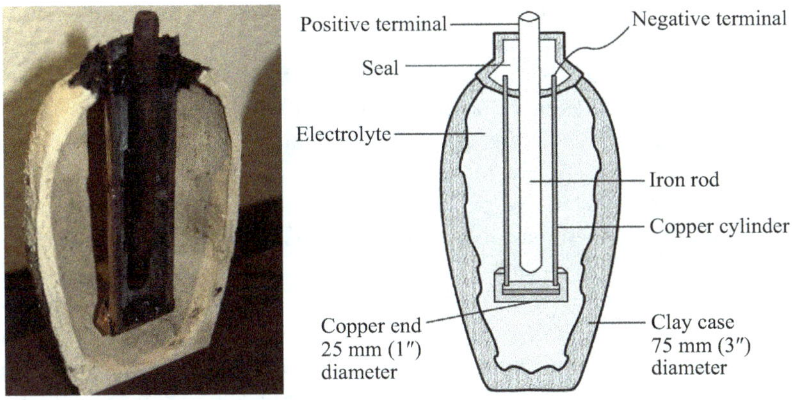


*Figure 1.1 Parthian Battery invented during the Parthian Empire (150 BC–223 AD)*

*Table 1.1 Open-circuit voltage for copper and iron in different materials*

| Substance | OCV [mV] |
|---|---|
| Mandarin orange | 549 |
| Lemon | 485 |
| Apple | 561 |
| Vinegar 13° | 496 |
| Salt 20% | 320 |
| Wood ash | 255 |
| Tongue-tip potential | 314 |

solution into which an iron rod surrounded by a copper cylinder was inserted. This device produced probably around 0.5 V–0.6 V open-circuit voltage (OCV) if indeed iron was used for the negative electrode. Zinc-plated iron could give more, but zinc (or derived products) seemed not to be present in the jar. Figure 1.1 illustrates the Parthian Battery. Several of them have been found, so not one cell, but a 'battery' could have been made. The artefacts consist of a pot approximately 130 mm (5 in.) tall (with a 1.5-in. mouth) containing a cylinder made of a rolled copper sheet, which houses a single iron rod. At the top, the iron rod is isolated from the copper by bitumen, with plugs or stoppers, and both rod and cylinder fit snugly inside the opening of the jar. The copper cylinder is not watertight, so the jar being filled with a liquid, this would surround the iron rod as well.

The artefact had been exposed to the weather and had suffered corrosion. We tried with iron nails and copper wire using several liquids as vinegar, grape fruit, salt water, and other electrolytes that could have been used, the repeatability is some 10% for the OCV (Appendix A; Table 1.1). Moving, shaking, or mixing help to increase the OCV, which indicates the usefulness of the jar. The copper side is positive.

One has to imagine that they had no 'volt meter' at that time and even did not know what voltage and current were. Normally, the tongue tip voltage subtracts, and the effect of voltage is hardly noticeable (the difference between connected and disconnected), so one has to be attentive and not stressed. But if the metals are crossed, the tongue-tip EMF adds to the original voltage, and some effect can more easily be detected, so sufficient for a demonstration with people around it.

If it was not detected by one cell, how to think using several in series? Anyhow, a number of them have been found, so it is likely that they might have put them in series, and we can talk about a 'battery'. Such a cell is called today a 'primary' cell as it cannot be recharged. Anyhow, how would they 'recharge'?

Although some people suggest technical applications, they are not sure, as then the principle probably would have continued. But a strange tongue sensation would have been enough to take it as a demonstration anyhow and to experiment with it. It is even not sure that they could generate a spark. We will not enumerate the hypotheses on internet of hypothetical gold plating and eternal burning electric lamps. The old dream could work now if batteries and LEDs last very long. Today some LEDs are well visible at 5 μA, so if a battery of 88 Ah had no auto-discharge, it could theoretically work 2000 years.

Alessandro Volta repeated almost two millennia later probably about the same: a silver coin and a tin strip. Afterwards he used zinc. Zinc was only reasonably available in the eighteenth century, it was known in the ancient Persian time but very rare, and not clear if the Persian people had used zinc-plated iron.

Then Volta got to copper and zinc to make a pile of it in 1799, about 1 V/cell [12]. The previous types were 'primary cells', so non-rechargeable. When DC generators were discovered, recharging became possible and there was an advantage of having rechargeable or 'secondary cells'. The lead–acid battery was invented in 1859, by Gaston Planté [13].

## 1.3 Introduction of the battery

The term 'cell' is the smallest individual electrochemical unit. It delivers a voltage that depends on the mixture of chemicals and compounds chosen to make the cell. The electrochemical unit is characterized by an anode 'negative electrode' (attracts anions, negative ions, metal ions move away from metal electrode) and a cathode 'positive electrode' (attracts cations, positive ions, collects metal ions). They are used to receive, store, and deliver electrical energy. Single-use cells are called primary cells and rechargeable cells are called secondary cells. Batteries or battery packs are made up of groups of cells in series or parallel or both. Technically, a cell is different from a battery, so, in this book, the term 'cell' refers to an individual electrochemical unit, and the terms 'battery' and 'pack' refer to an electrically connected group of cells. The 'nominal voltage' is the value specified to a cell or battery; normally, it is the average discharge voltage. The cell or battery can operate above or below the nominal voltage. Cells deliver a range of voltage depending on conditions such as temperature and how much charge they are holding.

The battery cells with lead–acid (PbA) chemistry have a nominal voltage of 2.0 V [14]. Other types, for instance, nickel–cadmium (NiCd) cells, have a nominal voltage of 1.2 V [15], and most lithium-based cells have nominal voltages over 3 V [16]. Associated with the cell's charge capacity, the C-rate is a relevant measure of cell current. It is the constant current charge or discharge rate that the cell can maintain for 1 h. It is the nominal ampere-hour (Ah) rating of the cell multiplied by 1 $h^{-1}$. A fully charged 20-Ah cell should be able to deliver 20 A (at '1 C' rate) for 1 h. The cell nominal energy capacity is the quantity of electrical energy in watt-hours (Wh) that a cell is rated to handle and is calculated as the cell's nominal voltage multiplied by its nominal charge capacity.

## 1.4 Lithium-based batteries

Lithium-ion batteries possess very low self-discharge rates [17] compared to other battery types. They have no memory effects, which means it can be charged and discharged more flexibly than other types. For instance, in lead–acid cells, when the fully charged state voltage is reached, it provides some inherent balancing function as the leakage current increases. In contrast, lithium-ion batteries cannot be operated in this way. They lead to overcharging, causing internal corrosion and unsafe operating conditions. An example of a lithium-ion battery is shown in Figure 1.2. The battery is a prismatic cell type and was introduced in the early 1990s. A lithium-ion cell includes the following parts:

- **positive electrode**
- **electrolyte**
- **separator**
- **electrolyte**
- **negative electrode**
- **enclosure**

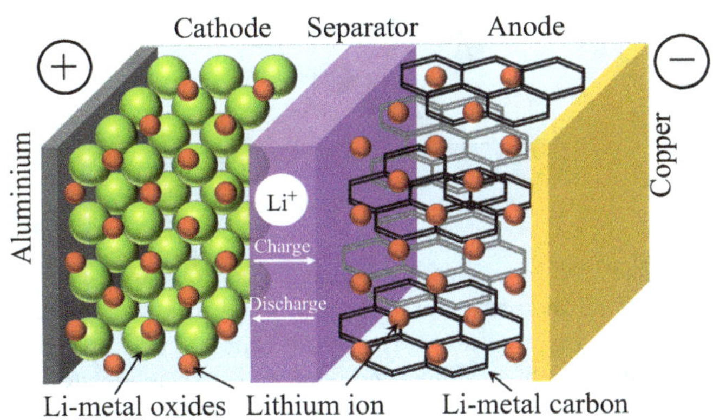


*Figure 1.2 A lithium-ion battery in detail*

There are some popular shapes of lithium-ion batteries as described next.

***Cylindrical*:** The cylindrical cell is still one of the most widely used packaging styles. The advantages are easy to manufacture and good mechanical stability. The tubular cylinder can withstand high internal pressures without deforming. Lithium-based cylindrical cells may include a positive thermal coefficient (PTC) thermistor [18]. When an excessive current occurs, the normally conductive polymer heats up and becomes resistive, which stops the current to flow. When this condition is terminated and removed, the PTC cools down and returns to the conductive state. A cylindrical cell is shown in Figure 1.3.

*Figure 1.3 Cylindrical cell*

***Prismatic cell*:** The prismatic cell satisfies the requirements for slim sizes. These cells are generally found in mobile phones, tablets, and laptops ranging from 800 mAh to 4000 mAh [19]. In addition, prismatic cells are also available in large formats that have a 20–50-Ah [20] rating and are used for EV. They reduce the required space in the application and permit a flexible design; however, it can be more expensive to manufacture. They are less efficient in thermal management and have a shorter cycle life than the cylindrical design. A pouch cell is shown in Figure 1.4.

*Figure 1.4 Prismatic cell*

***Pouch cell*:** They are enveloped in a pouch, made from two pieces of plastic film melted together around the perimeter. The pouch cell offers a simple, flexible, and lightweight solution to battery design. The pouch cells can deliver high load currents but perform best under light loading conditions and with moderate charging. They produce the most efficient use of space and achieve 90%–95% packaging efficiency (ratio of cell mass, compared to total mass). The pouch packs are used in consumer, military, and automotive applications. No standardized pouch cells exist; each manufacturer designs its own. Pouch packs are common for Li-polymer or $LiFePO_4$. Small cells are popular for portable applications requiring high load currents, such as

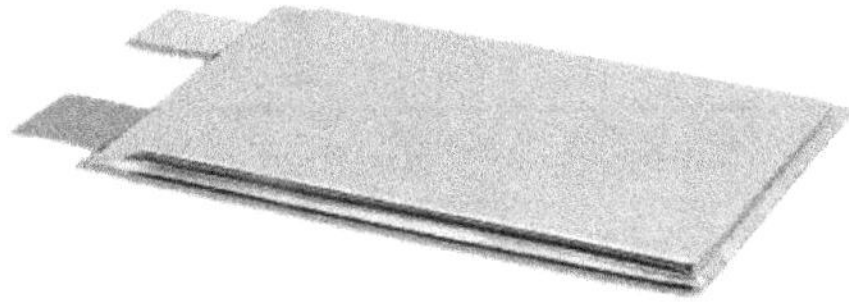

*Figure 1.5 Pouch cell*

drones and hobby gadgets. The larger cells [21] in the 40 Ah range serve in energy storage systems and vehicles. A prismatic cell is shown in Figure 1.5.

***Button cell*:** The button (coin) cell allowed a versatile design in portable devices. To have higher voltages, cells are stacked into a tube. Applications are cordless telephones, medical devices, and security devices at airports. Figure 1.6 shows a button cell. The battery chemistry can be divided into some categories as listed next:

*Figure 1.6 Button cell*

**Lithium–cobalt oxide:** These cells provide high energy density and its high specific energy enables Li–Co the popular choice for mobile phones, laptops, and digital cameras. Li–Co should not be charged and discharged at a current higher than its C-rating. The drawbacks of Li–Co are a relatively short life span, low thermal stability, and limited load capabilities.

**Li–Mn oxide:** It is a lithium-ion cell that uses manganese dioxide and shapes a three-dimensional spinel structure at the cathode. It improves the ion flow on the electrode, resulting in lower internal resistance. A low internal cell resistance allows and enables faster charging and high-current discharging. Furthermore, the advantage of a spinel crystal structure is a high thermal stability and enhanced safety, compared to pure metal.

**Lithium–nickel–manganese–cobalt oxide:** This is one of the most prosperous Li-ion systems using a cathode combination of nickel–manganese–cobalt. This type is the battery of choice for power tools, e-bikes, and other electric powertrains and has the lowest self-heating rate. The cathode combination is typically one-third nickel, one-third manganese, and one-third cobalt.

**Lithium–iron–phosphate (LFP):** $LiFeOP_4$ presents great electrochemical performance with low resistance. Moreover, a high current rating and long cycle life are also advantages. It does not contain cobalt, which is a scarce and expensive material. It is more tolerant to full charge conditions. LFP is often used to replace the lead–acid starter battery.

**Lithium–titanate (LTO):** LTO can be charged fast and delivers a high discharge current. The cycle count is higher than that of a regular Li-ion. LTO is safe and has excellent low-temperature discharge characteristics. The thermal stability under high temperature is also better than other Li-ion systems; however, the battery is expensive. Typical uses are electric power trains, UPS, electric buses, and

*Table 1.2 Characteristics of some batteries*

| Batteries name | Values | Nominal voltage [V] | Operating range [V] |
|---|---|---|---|
| Lithium–cobalt oxide | $LiCoO_2$ | 3.6 | 3.0–4.2 |
| Lithium–manganese oxide | $LiMn_2O_4$ | 3.70 | 3.0–4.2 |
| Lithium–nickel–manganese–cobalt oxide | $LiNiMnCoO_2$ | 3.60, 3.70 | 3.0–4.2 |
| Lithium–iron–phosphate | $LiFePO_4$ | 3.20, 3.30 | 2.5–3.65 |
| Lithium–titanate | $Li_2TiO_3$ | 2.4 | 1.8–2.85 |

solar-powered street lighting. In Table 1.2, characteristics of some lithium-based batteries are presented.

Regardless of the many advantages and features of lithium-ion batteries, they could present danger if the battery is used outside of its safe operating area, especially the large batteries. A summary of battery mistreatments are listed next.

**Overcharge:** The overcharge condition happens when a cell is charged beyond the allowed limitation, which is an SoC greater than 100%. It causes deterioration performance inside the cell. It can lead to thermal runaway, swelling, and other problems.

**Overdischarge:** This condition occurs when a cell is discharged beyond the permitted limit. If this condition occurs, the cell voltage falls rapidly. It can result in internal cell damage. The cell, which is repeatedly overdischarged, can lead to safety risk when recharging.

**High–low temperature:** Batteries function best at room temperature. Operating a battery at upraised temperatures improves performance but a prolonged exposure will shorten life. It was already true in lead–acid batteries. A warm battery starts the car engine better than a cold one but shortens life. Cold temperature increases the internal resistance and lowers the capacity but does not destroy as such. High temperature leads to cell degradation and thermal runaway. It will occur when a cell is under high ambient temperature. So, low temperature reduces the performance but the high temperature may result in breakdown.

**Overcurrent:** It occurs when Li-ion cell or battery pack is charged or discharged at a much higher current than allowed by the cell manufacturer. A high current leads to internal heating, which results in an over-temperature situation. Overcurrent may lead to swelling or rupturing of cells and the risk of damaging the battery.

## 1.5 Charging and discharging

The lithium-ion battery utilizes organic solvents and lithium salt electrolytes, which serve as a transport layer to move lithium ions from the negative to the positive electrode during discharge. The direction is merely reversed as the battery is charged. The state of health is dependent on the operational current and the deep depth of discharge (DOD). High operational currents generate excessive heat, which results in microfracturing of the active materials. Through active thermal control, these impacts can be reduced.

Figure 1.7 shows a battery-equivalent circuit for the lithium-ion cell where resistance parameters change on the basis of charging, $R_{xc}$, or discharging, $R_{xd}$, respectively. Low drop diodes are present to show that charging and discharging may occur at a different chemical potential. $R_{2c}$ and $R_{2d}$ represent the internal resistance during charging and discharging, while $R_{1c}$ and $R_{1d}$ represent the over voltage resistance during charging and discharging. Over-voltage conditions account for the increased voltage required to inject energy into the battery during charging or the voltage drop, which occurs as a result of loading during discharging. This portion governs the impulse response of the battery where $C_1$ is the over-voltage capacitance. The transfer function depicting the equivalent impedance is

$$\mathrm{Zeq}(s) = R_{2c} + \frac{R_{1c}}{R_{1c}C_1 s + 1} \quad \text{(charging)} \tag{1.1}$$

$$\mathrm{Zeq}(s) = R_{2d} + \frac{R_{1d}}{R_{1d}C_1 s + 1} \quad \text{(discharging)} \tag{1.2}$$

A list of definitions through this book is given, the list will not be exhaustive.

- **Cell:** The term ‘cell’ is a small unit that contains an anode (+), cathode (−), an electrolyte, and possibly a separator. A cell is shown in Figure 1.8.
- **Battery:** It is a collection of cells connected in series and/or parallel. A battery is shown in Figure 1.9. It is also called ‘module’.
- **Pack:** A battery pack is an assembly of multiple batteries connected in series and/or in parallel. A battery pack is shown in Figure 1.10.
- **Primary cells:** They cannot be recharged, e.g. zinc carbon (Leclanché), alkaline cells (can be recharged to some extend but are usually not designed for it).

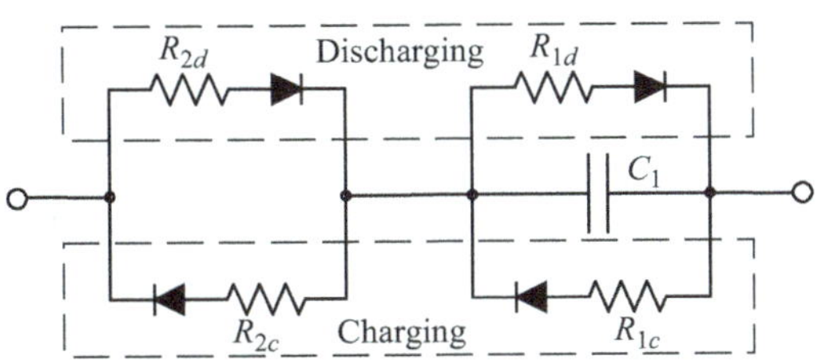


*Figure 1.7 Charging and discharging of a battery*

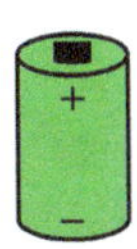


*Figure 1.8 Cell*

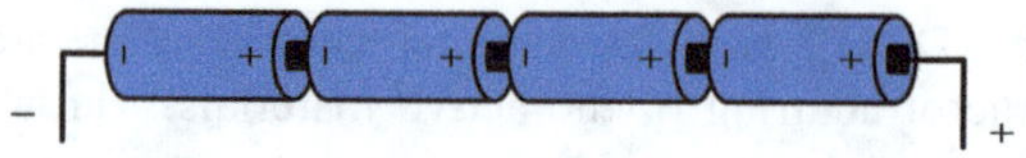

*Figure 1.9 Battery*

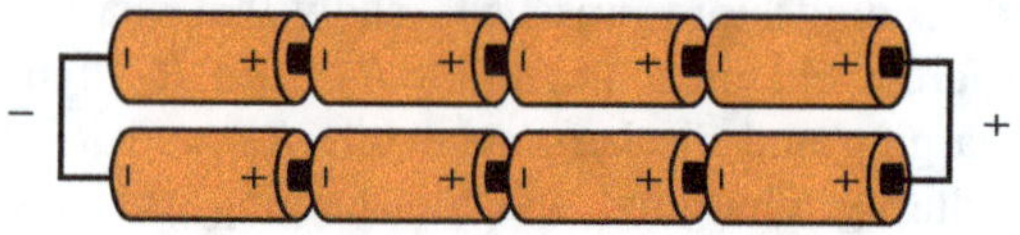

*Figure 1.10 Pack*

- **Secondary cells:** Secondary cells can be recharged a sufficient number of times, e.g. >200.
- **Nominal voltage of a battery:** The average voltage during a (slow) discharge at, for example, 0.1 C, this is discharging during 10% of the capacity in Ah.
- **Open-circuit voltage:** Voltage at rest after 4 h (chemically at rest).
- **SoC:** SoC, often stated in % of full charge.
- **Discharge curve:** A diagram with voltage vertical and charge (Ah) horizontal.
- **Charge curve:** The same but while charging.
- **Charge efficiency = Coulombic efficiency = Faraday efficiency:** The output Ah to the load/the input Ah from the charger (a high internal leakage current reduces the charge efficiency).
- **Voltage efficiency:** The average voltage during discharge/the voltage during charging.
- **Battery efficiency:** The output energy/input energy, a high battery implies a high charge and voltage efficiency, the battery efficiency is close to charge efficiency × voltage efficiency.

## 1.6 Chemistry

The cathode and anode have their own reactions and potential compared to a reference electrode. If only one reaction is dominant at each electrode, a constant voltage will appear during discharge and charge. The cells with low impedance and a flat charge and discharge curve may have the best efficiency. Different metals, oxides, and other reactant products not only have their specific potential, which depends also on the charge number. Above a defined potential, the electrodes are not stable compared to the electrolyte so that decomposition can occur (gassing, corrosion of internal conductors). In lead–acid, it can corrode the positive conductor in lead. In Li types, very thin Al and Cu foils are used that can also corrode. That is why batteries are often stored at 50% of SoC. Corrosion of electrodes makes the batteries more mechanically fragile and can disconnect parts or complete electrodes, resulting in a high impedance. Too much discharging can make substances inactive. But milliseconds of voltage dips caused by high currents are voltage drops in the conductors so do not affect the internal potential and could be allowed.

## 1.7 Charging methods

***CC/CV***: The most common charging method is constant-current–constant-voltage (CC/CV); constant current at low voltage and voltage limited.

***Topping charge***: A temporary higher voltage.

***Float charging***: A continuous but not extreme high voltage to keep it charged.

***Trickle charge***: To maintain the accumulator in a good condition and to compensate auto-discharge; this is rather a small current source, where float charging is rather an imposed voltage.

***Reflex charging***: For example, each minute a negative pulse is given to avoid a build-up of gas bubbles, or to influence the granularity of the lattice of the electrodes.

***Fast charging***: In the middle of the charge characteristic, the batteries usually withstand a much higher current. The limitation is not only a maximal voltage but also heating. Batteries can be made with less active material and more conductor material allowing higher currents. Fast charging is usually in compromise with a long life. Fast discharging has less compromises.

***Solar charging***: So it may be operated 'as and when the sun shines'. To avoid overloading the grid, it makes sense to minimize unnecessary exchange with the grid, to charge, depending on the availability of the sun, and also to reduce grid and conversion losses: DC–DC converters have lower losses than DC–AC.

***Smart charging***: If the renewable energy is at some distance (wind, solar), and the charging is not urgent, charging can be improved depending on the price or availability, using Internet of Things.

## 1.8 Characteristics

Batteries are widely used in these days and the main issue is how to reach a higher lifetime of the batteries. In EV, up to 60% of the cost of the car depends on the batteries. So, to solve the problem of unbalancing, a BMS is applied to charge and discharge the batteries in such a way that all the cells in the battery will be equalized at the same voltage, in order to prolong the lifetime. Charging and discharging is a chemical reaction, but Li-ion is claimed to be the exception, where lithium is stored in a lattice. Battery specialists divide the chemistry in an anode and in a separate cathode part and consider the $Li^{+}$ion movement in each part and consider the intercalation (so not a real chemical reaction) in the electrode lattice.

The capacity fade often results from a reduction of the possibility to trap ions in the lattice, but as with all battery systems, internal corrosion and other degenerative effects also play a role. The Li-ion charger is a voltage-limiting device that has similarities to the lead–acid system. The differences with Li-ion lie in a higher voltage per cell, tighter voltage tolerances, and the absence of trickle or float charge at full charge. While lead–acid offers some flexibility in terms of voltage cut-off, manufacturers of Li-ion cells are very strict on the correct setting because Li-ion cannot accept overcharge. Care should be taken to lower the charging current if one cell gets too much voltage. In practice, it means that the charging philosophy should be matched with the BMS capability: if one cell is too high, the charge rate should be reduced, so that the BMS can still manage it.

The OCV that results as part of an electrochemical reaction varies with the metals and electrolyte used. Charging raises the voltage and discharging lowers it. The voltage behaviour under a load and charge is ruled by the current flow and the internal battery resistance. A low resistance provides low voltage variation under load or charge. On the other hand, a high internal resistance causes the voltage to swing excessively.

A battery cell has an OCV. If a voltmeter is connected across the terminals of a cell, it will display some value. Therefore, the model of the cell starts as an ideal voltage source. The schematic for this model is drawn in Figure 1.11. In this model, the open cell voltage $v(t)$ is not a function of the charge, the integral $i(t)$. The model can be indeed improved by including a relationship between the OCV and the charge status of the cell. To do so, we must first define what we mean by this 'charge status'. Hence, the SoC of a cell can be defined 100% when the cell is fully charged and 0% when the cell is fully discharged. The SoC is a unitless quantity and is denoted by the symbol $h$ in this chapter. The total capacity is usually measured in Ah or milliampere-hours (mAh) and is denoted by the symbol $Q$. The model equation can be expressed as

$$\frac{\partial h(t)}{\partial t} = -\frac{\eta(t)i(t)}{Q} \tag{1.3}$$

The sign of $i(t)$ is positive on discharge. Hence, positive (discharge) current lowers the cell's SoC, and negative (charge) current increases the cell's SoC. The term $i(t)$ is measured in amperes, and to be compatible, $Q$ needs to be converted to ampere-seconds (i.e. Coulombs). Both terms $h(t)$ and $\eta(t)$ are unitless. It can be noted that $\eta(t)$ is the coulombic efficiency or charge efficiency of the cell. The coulombic efficiency is $\eta(t) = 1$ at time instants when the sign of current is positive (discharging), but $\eta(t) \leq 1$ at time instants when the sign of current is negative (charging). The coulombic efficiency is different from the energy efficiency of a cell. The coulombic efficiency in a typical lithium-ion cell is around 99% or higher and is equal to (charge out)/(charge in). The energy efficiency is closer to 95% at

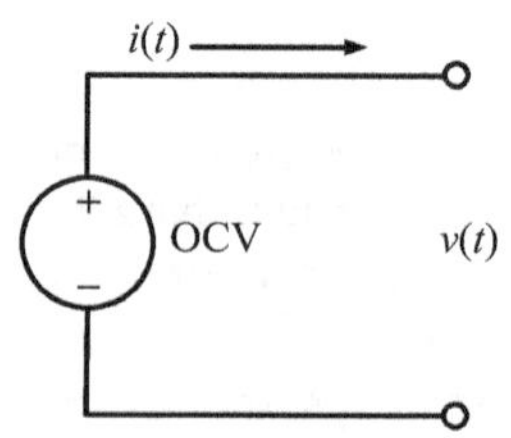


*Figure 1.11 The schematic of the cell model*

low current and is equal to (energy out)/(energy in). A change in the SoC over some time interval $t < t_0$ can be expressed as

$$h(t) = h(t_0) - \frac{1}{Q}\int_{t_0}^{1} \eta(\lambda)\, i(\lambda)\mathrm{d}\lambda \tag{1.4}$$

To complete the model, a discrete-time model should be considered as well. In discrete modelling, the cell input and output are sampled at a routine rate with the period $\Delta t$. The reason to model the system discretely is that it can be used by the microcontroller. Therefore, (1.3) can be rewritten as

$$h((k+1)\Delta t) = h(k\Delta t) - \frac{\Delta t}{Q}\eta(k\Delta t)\, i(k\Delta t) \tag{1.5}$$

where $t_0 = k\Delta t$ and $t = (k + 1)\,\Delta t$. The term $\Delta t$ can be changed to a square bracket [] to present the discrete time. Therefore, (1.5) can be rewritten as

$$h[k+1] = h[k] - \frac{\Delta t}{Q}\eta[k]\, i[k] \tag{1.6}$$

Equation (1.6) represents the mathematical model for SoC. The cell open-circuit is a function of the SoC. The improved cell modelling is shown in Figure 1.12. It includes OCV dependence on the cell's SoC. The ideal voltage source is replaced by a controlled voltage source with the value of OCV $s(t)$. There is another term that can be investigated as DOD. The DOD is an alternative way to indicate a battery's SoC. The DOD is the complement of SoC: as one increases, the other decreases. It is expressed as a fraction or in Ah. So, the DOD $= 1 - h(t)$. If the DOD is expressed in Ah, it can be rewritten as DOD $= Q(1 - h(t))$. The model can further be more enhanced by adding dynamic behaviour to the model in order to define what will occur when the cell is subjected to the time-varying input current. Some current dependency can be introduced by adding a resistor in series with the voltage source. Therefore, the equation for the new model can be expressed as

$$v(t) = OCV(h(t)) - i(t)\, R_0 \tag{1.7}$$

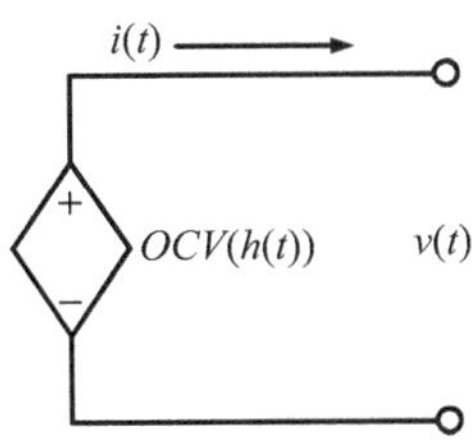


*Figure 1.12 The schematic of the improved cell model*

It can be noted that $v(t) > OCV(h(t))$ when $i(t) < 0$ or in other words, when it is charging and $v(t) < OCV(h(t))$ when $i(t) > 0$ or when it is discharging. Equation (1.7) can be expressed in discrete time; therefore, it can be rewritten as

$$v[k] = OCV(h[k]) - i[k]\,R_0 \tag{1.8}$$

It can be noted that the series resistance in the model also indicates that the power is dissipated by the cell internal resistance as heat. Hence, the power dissipated by the series resistance can be calculated as $i^2(t)\,R_0$. The schematic of the revised model with the equivalent series resistor is shown in Figure 1.13.

That model is sufficient for many uncomplicated electronic circuit configurations, but it is not adaptable for large-scale battery packs, for instance, in EV and grid storage systems. There is another term that should be considered to the model, which is called diffusion voltage. In battery cells, polarization indicates any passage of the cell's terminal voltage away from OCV due to passage of current within the cell. The previous model modelled instantaneous polarization via the $i(t)\,R_0$ expression. But real cells have more complicated behaviour, where the voltage polarization slowly increases over time as the current is required from the cell and then slowly decays over time when the cell is allowed to rest. This phenomenon is caused by slow diffusion processes of lithium in a lithium-ion cell, so a slowly changing voltage will refer to the diffusion voltage. Its effect can be approximated closely in a circuit using one or more parallel resistor–capacitor subcircuits.

In Figure 1.14, the combination of resistor $R_1$ and capacitor $C_1$ performs this function. In the model, the SoC equation remains the same as before, but the voltage equation for continuous and discrete turns to

$$\begin{aligned} v(t) &= OCV(h(t)) - v_{C_1}(t) - v_{R_0}(t) \\ v[k] &= OCV(h[k]) - v_{C_1}[k] - v_{R_0}[k] \end{aligned} \tag{1.9}$$

However, when employing data to classify model parameters, it turns out to be more simplistic if these expressions can be written in terms of element currents instead. Therefore, (1.9) can be rewritten as

$$\begin{aligned} v(t) &= OCV(h(t)) - R_1 i_{R_1}(t) - R_0 i(t) \\ v[k] &= OCV(h[k]) - R_1 i_{R_1}[k] - R_0 i[k] \end{aligned} \tag{1.10}$$

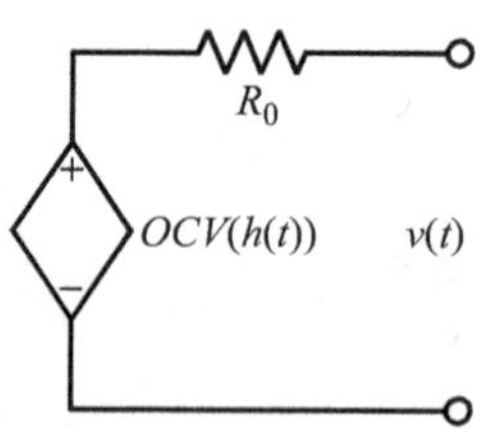


*Figure 1.13 The schematic of the revised cell model with resistor*

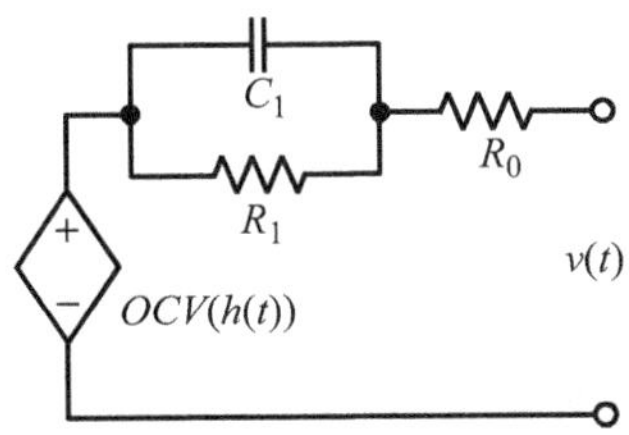


*Figure 1.14 The schematic of the revised cell model with diffusion voltages*

Now an expression for the resistor current $i_{R_1}(t)$ can be derived as shown later. First, it should be recognized that the current through $R_1$ plus the current through $C_1$ must equal $i(t)$. The capacitor current $i_{C_1}$ can be calculated as

$$\begin{aligned} i_{C_1} &= C_1 \frac{\partial v_{C_1}(t)}{\partial t} \\ i(t) &= i_{R_1}(t) + C_1 \frac{\partial v_{C_1}(t)}{\partial t} \end{aligned} \tag{1.11}$$

The equation can be rewritten according to the expression $v_{C_1}(t) = R_1 i_{R_1}(t)$

$$\begin{aligned} & i_{R_1}(t) + R_1 C_1 \frac{\partial i_{R_1}(t)}{\partial t} = i(t) \\ & \frac{\partial i_{R_1}(t)}{\partial t} = -\frac{1}{R_1 C_1} i_{R_1} = \frac{1}{R_1 C_1} i(t) \end{aligned} \tag{1.12}$$

The obtained differential equation can be solved to determine $i_{R_1}(t)$. It can also be converted to discrete time from. Now, transferring the generic ordinary differential equation (ODE) from continuous time to discrete time as

$$\dot{x}(t) = a\,x(t) + b\,u(t) \tag{1.13}$$

It is desired to estimate $x(t)$ at discrete time $t = k\Delta t$, where $x[k] = x(k\Delta t)$. The input term $u(t)$ is constant over the sampling interval. The general solution to (1.13) can be expressed as following. First, we rearrange the equation, then multiply both sides by $e^{-at}$

$$e^{-at}(\dot{x}(t) - a\,x(t)) = e^{-at} b\,u(t) \tag{1.14}$$

It can be noted that the left side of the equation can be written as $(d/dt)[e^{-at}\,x(t)]$, thus

$$\frac{d}{dt}[e^{-at} x(t)] = e^{-at} b\,u(t) \tag{1.15}$$

Presently, by integrating both sides of this equation between times 0 and $t$ interval, and replacing $t$ inside the integral with $\lambda$, the new equation can be expressed as

$$\int_0^t \frac{\mathrm{d}}{\mathrm{d}\lambda}\left[\mathrm{e}^{-a\lambda}x(\lambda)\right]\mathrm{d}\lambda = \int_0^t \mathrm{e}^{-a\lambda} b\, u(\lambda)\mathrm{d}\lambda$$

$$\int_0^t \mathrm{d}\left[\mathrm{e}^{-a\lambda}x(\lambda)\right] = \int_0^t \mathrm{e}^{-a\lambda} b\, u(\lambda)\mathrm{d}\lambda \tag{1.16}$$

$$e^{-at}x(t) - x(0) = \int_0^t \mathrm{e}^{-a\lambda} b\, u(\lambda)\mathrm{d}\lambda$$

Therefore, the general solution to (1.13) can be expressed as

$$x(t) = \mathrm{e}^{at}x(0) + \int_0^t \mathrm{e}^{a(t-\lambda)} b\, u(\lambda)\mathrm{d}\lambda \tag{1.17}$$

In order to drive the discrete-time equation, substitute $x[k+1] = x((k+1)\Delta t)$

$$k[k+1] = \mathrm{e}^{a(k+1)\Delta t}x(0) + \int_0^{(k+1)\Delta t} \mathrm{e}^{a((k+1)\Delta t-\lambda)} b\, u(\lambda)\mathrm{d}\lambda$$

$$k[k+1] = \mathrm{e}^{a(k+1)\Delta t}x(0) + \int_{k\Delta t}^{(k+1)\Delta t} \mathrm{e}^{a((k+1)\Delta t-\lambda)} b\, u(\lambda)\mathrm{d}\lambda \tag{1.18}$$

$$+ \int_{k\Delta t}^{(k+1)\Delta t} \mathrm{e}^{a((k+1)\Delta t-\lambda)} b\, u(\lambda)\mathrm{d}\lambda$$

Now, the term '$\mathrm{e}^{a\Delta t}$' can be factored of the first two terms then

$$x[k+1] = \mathrm{e}^{a\Delta t}\mathrm{e}^{ak\Delta t}x(0) + \mathrm{e}^{a\Delta t}\int_0^{k\Delta t} \mathrm{e}^{a(k\Delta t-\lambda)} b\, u(\lambda)\mathrm{d}\lambda$$

$$+ \int_{k\Delta t}^{(k+1)\Delta t} \mathrm{e}^{a((k+1)\Delta t-\lambda)} b\, u(\lambda)\mathrm{d}\lambda$$

$$x[k+1] = \mathrm{e}^{a\Delta t}\left(\mathrm{e}^{ak\Delta t}x(0) + \int_0^{k\Delta t} \mathrm{e}^{a(k\Delta t-\lambda)} b\, u(\lambda)\mathrm{d}\lambda\right) \tag{1.19}$$

$$+ \int_{k\Delta t}^{(k+1)\Delta t} \mathrm{e}^{a((k+1)\Delta t-\lambda)} b\, u(\lambda)\mathrm{d}\lambda$$

$$= \mathrm{e}^{a\Delta t}x(k\Delta t) + \int_{k\Delta t}^{(k+1)\Delta t} \mathrm{e}^{a((k+1)\Delta t-\lambda)} b\, u(\lambda)\mathrm{d}\lambda$$

Eventually, the result using discrete-time equation can be expressed as

$$x[k+1] = \mathrm{e}^{a\Delta t}x[k] + \mathrm{e}^{a(k+1)\Delta t}\int_{k\Delta t}^{(k+1)\Delta t} \mathrm{e}^{-a\lambda} b\, u(\lambda)\mathrm{d}\lambda \tag{1.20}$$

Assume that $u(t)$ is constant from $k\Delta t$ to $(k + 1)\Delta t$ and equals $u(k\Delta t)$, then

$$\begin{aligned} x[k+1] &= \mathrm{e}^{a\Delta t}x[k] + \mathrm{e}^{a(k+1)\Delta t}\left(\int_{k\Delta t}^{(k+1)\Delta t} \mathrm{e}^{-a\lambda}\mathrm{d}\lambda\right) b\, u[k] \\ &= \mathrm{e}^{a\Delta t}x[k] + \mathrm{e}^{a(k+1)\Delta t}\left(-\frac{1}{a}\mathrm{e}^{-a\lambda}\Big|_{k\Delta t}^{(k+1)\Delta t}\right) b\, u[k] \\ &= \mathrm{e}^{a\Delta t}x[k] + \frac{1}{a}\mathrm{e}^{a(k+1)\Delta t}\left(\mathrm{e}^{-ak\Delta t} - \mathrm{e}^{-a(k+1)\Delta t}\right) b\, u[k] \\ &= \mathrm{e}^{a\Delta t}x[k] + \frac{1}{a}\left(\mathrm{e}^{a\Delta t} - t\right) b\, u[k] \end{aligned} \tag{1.21}$$

So, that is the discrete-time ODE that is equivalent to the continuous-time ODE of (1.13). In order to apply this result in the cell model, the correspondences between constants in (1.13) and (1.12) should be noted

$$a = -\frac{1}{R_1C_1} \quad \text{and} \quad b = \frac{1}{R_1C_1} \tag{1.22}$$

Now, by substituting these values into the generic result, the following discrete-time equation for the resistor will be obtained as

$$\begin{aligned} i_{R_1}[k+1] &= \mathrm{e}^{(-(\Delta t/R_1C_1))} i_{R_1}[k] \\ &\quad + (-R_1C_1)\left(\mathrm{e}^{(-(\Delta t/R_1C_1))-1}\right)\left(\frac{1}{R_1C_1}\right) i[k] \\ &= \mathrm{e}^{(-(\Delta t/R_1C_1))} i_{R_1}[k] + \left(1 - \mathrm{e}^{(-(\Delta t/R_1C_1))}\right) i[k] \end{aligned} \tag{1.23}$$

Finally, it can be observed that a diffusion-voltage response of a cell is a function of cell's SoC.

## 1.9 Application areas

As discussed before, Li-ion batteries are available in all shapes and sizes, letting them be the perfect option for power needs regarding how large is the size of the system. There is a list of real-world applications, usage of lithium-ion batteries beside your phone, for example medical equipment to garden equipment and toys, which are the top uses for rechargeable lithium batteries. A few years ago, people would not be aware of such a change or revolution, even the following list will not be exhaustive:

- **Automotive:** Cars, bike, truck, railway, and aeroplanes.
- **Computers:** Laptop and tablet.
- **Tools:** Screw, cutting disc, circular saw, and reciprocal saw.
- **Portable garden equipment:** (Robot) lawnmower, riding mower, hedge trimmer, lawn trimmer, chainsaw, pruning shears, leaf blower, sprayer, wheelbarrow, rain water pump, cultivator, and electric tractors.
- **Utility:** Street lamps, traffic lights, monitoring systems, and lamps and powering construction sites.
- **Household equipment:** Mixer, (robot) cleaner, disinfecting UV-led robots or the combination, portable refrigerator using Peltier devices, wheelchairs, diverse headlamps, and many gadgets.
- **Industrial:** Forklifts, automatic guided vehicles, and labelling tools.
- **Grid-connected:** Power backup/UPS and energy storage systems.
- **Stand-alone:** Traffic lights, solar park meter, and automatic road radar.
- **Marine traction:** Short-distance ferries, battery backup not only for the safety of on-board systems, but also for the reliability and robustness of the equipment. Combination with PV panels in yachts, and standby of vessels.
- **Medical:** Defibrillators, implantable devices, infusion pumps, and dispensers.

Most of the application areas did have a gasoline, grid-electric, or a pneumatic version before. Some battery-freaks do convert existing equipment, even selling the original engine and transmission. No belts no pulleys, no hydraulics, less noise. No engines to start while pulling a chord.

## Summary

Although prime discoveries about primary batteries were made almost two millennia ago, the breakthrough of the secondary (rechargeable) battery came in the late nineteenth century when DC generators could recharge them. Gradual improvements were done during one century, where the lead–acid was still dominant, and to some extent Ni based types. A revolution occurred after 1990 when non-water-based Li cells were developed, although the early cells were rather dangerous. Some explanation about commonly used terms is given. This chapter also provides an introduction to the types of batteries, their application, and their characteristics, and also different charging methods are introduced. First, the history of the battery is introduced, and an introduction to the lithium-ion battery is given. Next, a number of shapes and chemistries of lithium-based batteries are investigated. Moreover, the battery characteristics and modelling of a single lithium battery cell are presented. By the fact that much lighter batteries with more available cycles were available, the amount of application areas increased enormously; an overview is given.

## Appendix A: Open-circuit voltage (OCV)

Open-circuit voltage (OCV) for copper and iron in different materials are presented: (a) OCV 'mandarin', (b) OCV 'lemon', (c) OCV 'apple', (d) OCV 'Vinegar', (e) OCV 'salt', (f) OCV 'wood ash'.

(a) OCV 'mandarin'

(b) OCV 'lemon'

(c) OCV 'apple'

(d) OCV 'Vinegar'

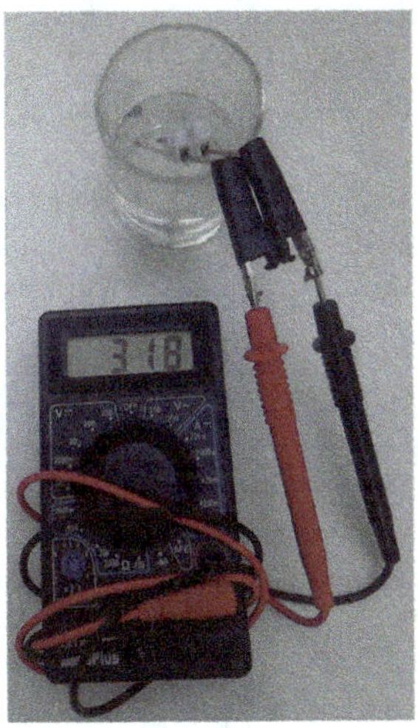


(e) OCV 'salt'

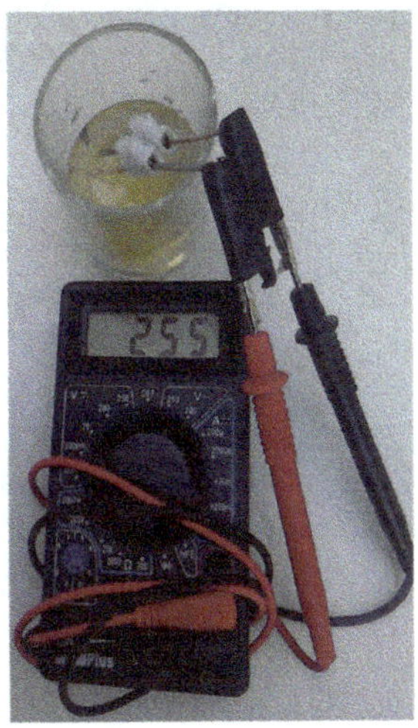


(f) OCV 'wood ash'

## References


[1] Fernandez M and Trinidad F. Charging strategies for valve-regulated lead/acid batteries in electric-vehicle applications. *Journal of Power Sources*. 1997;67(1–2):125–133.

[2] Chang Y, Mao X, Zhao Y, *et al.* Lead-acid battery use in the development of renewable energy systems in China. *Journal of Power Sources*. 2009;191(1):176–183.

[3] Lailler P, Sarrau JF, and Sarrazin C. Comparative study for "36 V" vehicle applications: advantages of lead-acid batteries. *Journal of Power Sources*. 2001;95(1-2):58–67.

[4] Boden D. Comparison of methods for adding expander to lead-acid battery plates–advantages and disadvantages. *Journal of Power Sources*. 2004;133(1):47–51.

[5] Herritty RM and Midolo J. Nickel cadmium batteries for photovoltaic applications. In: Thirteenth Annual Battery Conference on Applications and Advances. Proceedings of the Conference. IEEE; 1998. p. 255–257.

[6] Sakai T, Yuasa A, Ishikawa H, *et al.* Nickel-metal hydride battery using microencapsulated alloys. *Journal of the Less Common Metals*. 1991;172:1194–1204.

[7] Frost DF and Howey DA. Completely decentralized active balancing battery management system. *IEEE Transactions on Power Electronics*. 2017;33(1):729–738.

[8] Affanni A, Bellini A, Franceschini G, *et al.* Battery choice and management for new-generation electric vehicles. *IEEE Transactions on Industrial Electronics*. 2005;52(5):1343–1349.

[9] Pistoia G. *Lithium-Ion Batteries: Advances and Applications*. Oxford: Elsevier; 2013.

[10] Emadi A, Lee YJ, and Rajashekara K. Power electronics and motor drives in electric, hybrid electric, and plug-in hybrid electric vehicles. *IEEE Transactions on Industrial Electronics*. 2008;55(6):2237–2245.

[11] Warner JT. *The Handbook of Lithium-Ion Battery Pack Design: Chemistry, Components, Types and Terminology*. Amsterdam: Elsevier; 2015.

[12] Alessandro Volta [database on the Internet]. USA: www.wikipedia.org. 2021. Available from: https://en.wikipedia.org/wiki/Alessandro_Volta.

[13] Lead–acid battery [database on the Internet]. USA: www.wikipedia.org. 2021. Available from: https://en.wikipedia.org/wiki/Lead/acid_battery.

[14] West S and Krein PT. Equalization of valve-regulated lead-acid batteries: issues and life test results. In: INTELEC. Twenty-Second International Telecommunications Energy Conference (Cat. No. 00CH37131). IEEE; 2000. p. 439–446.

[15] Lynch WA and Salameh ZM. Electrical component model for a nickel-cadmium electric vehicle traction battery. In: 2006 IEEE Power Engineering Society General Meeting. IEEE; 2006.

[16] Kim IS. The novel state of charge estimation method for lithium battery using sliding mode observer. *Journal of Power Sources*. 2006;163(1):584–590.

[17] Zimmerman AH. Self-discharge losses in lithium-ion cells. *IEEE Aerospace and Electronic Systems Magazine*. 2004;19(2):19–24.

[18] Wakihara M. Recent developments in lithium ion batteries. *Materials Science and Engineering*: R: Reports. 2001;33(4):109–134.

[19] Maleki H and Howard JN. Internal short circuit in Li-ion cells. *Journal of Power Sources*. 2009;191(2):568–574.

[20] Gitzendanner R, Russell P, Marsh C, *et al.* Design and development of A 20 Ah Li-ion prismatic cell. *Journal of Power Sources*. 1999;81:847–852.

[21] Ren D, Feng X, Lu L, *et al.* An electrochemical-thermal coupled overcharge-to-thermal-runaway model for lithium ion battery. *Journal of Power Sources*. 2017;364:328–340.

# Chapter 2
# Discussion about batteries

## 2.1 Introduction

In general, only mentioning the positive aspects may be a way to write a book. However, most technologies have disadvantages or some weakness aspects, so also batteries. Being open to a discussion may even a way to improve the battery development and the battery management system (BMS). What is a problem today could be solved a few years after, each day discoveries are made, but only a fraction of 1% gets to the market. Most of the battery news is a small increment in a big entity. We have seen a similar evolution in PV panels, gradually less (fossil) energy was needed to manufacture, and less critical materials for the same power. About nine elements can be enumerated in a battery: cathode conductor, cathode materials, cathode electrolyte, separator (membrane), anode electrolyte, anode materials, anode conductor, housing, and BMS. So if even for lithium, for each, ten different techniques are considered, the theoretical total count is $10^9$ possibilities. They cannot all be combined, but the effort and skills of many engineers and designers will be needed to land on really effective technologies for all aspects.

## 2.2 Are minerals available?

Batteries may require more or less kg of materials to produce them. Often minerals do not contain only one element or oxide, but a mix of several elements. Sometimes these elements are very useful, such as indium, which is a 'by-product' of zinc. However, mostly the other elements are a part of the pollution associated with mining, where heavy metals and other toxic products have to be disposed.

### *2.2.1 Lead–acid batteries*

As an example, it was easier to look first at lead–acid batteries as much less materials are involved. Lead as such is toxic, but it is also one of the examples of best recycled metals. The world consumption of refined lead is 11.7 million metric tons of lead [1]. The lead reserves as of 2019 are 90.36 million tons. The biggest reserves are in Australia 36 million, second China 18 million (Figure 2.1). The consumption is still rising and is about 12 million tons/year. In fact, in mining activities, more prospection is started if a shortage is likely, but if a shortage really appears prices may rise in an uncontrolled way; it happened to magnet materials a

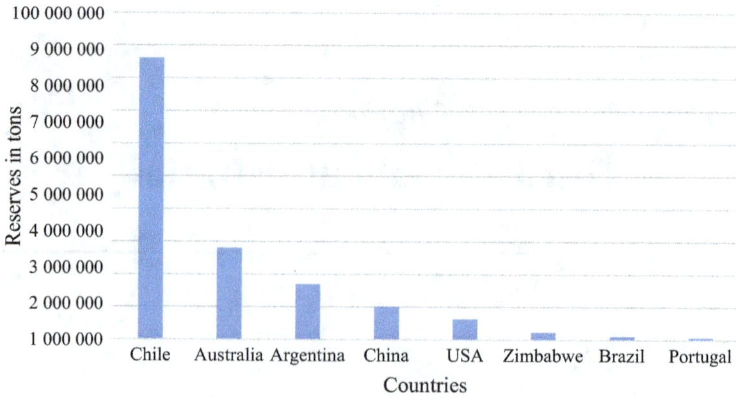


*Figure 2.1 Largest reserves worldwide*

decade ago. As in Europe very little lead mineral is found, most comes from other continents; it is a long distance for a heavy material such as lead. This also shows the importance of recycling, or what they call 'urban mining'. Lead is not used anymore for water conducts, also almost not anymore in roof flashing, and tetraethyl lead is fortunately not used anymore as an octane booster. In fact ethanol does that job now and replaces also other toxic products as benzene, toluene, and xylene. We hope also to remove the need of gasoline and its associated $CO_2$ by using battery technology [2]. The sulphur to make sulphuric acid is in fact a waste product now of producing oil products out of tar sands. Huge yellow sulphur waste pyramids emerge as in the Alberta province of Canada. The plastic components in lead batteries represent little weight; in previous times, glass was used as a container.

### *2.2.2 Lithium-ion batteries*

In comparison, lithium-ion batteries contain a large variety of materials. In fact, much more materials are involved than lithium. It originally started with a lithium metal anode, but that was quite dangerous in practice. A newer 'generation 3' seems to make it possible again, but most use electrodes where ions are intercalated, but this means also other materials [3]. So, the use of ions and not the metal results in the actual name. Lithium as such, is in fact not the biggest problem, it appears mainly in salts in desert regions in Chile and Australia, mostly underground. Countries such as Bolivia and Algeria may also have a lot, but the prospection hardly started. It is expected that by 2025, the production of lithium will be around between 400 000 and 500 000 t/year.

Lithium was originally found in rocks, by the name of stone Lithos, $\lambda\iota\theta o\varsigma$, in Greek. A brine is made where lithium salts dissolve less than other salts, so it uses crystallization when the water is evaporated and has a 'low tech' refining possibility. However, a process using a selective membrane can extract more from the same brine and so pollute less by its brine residue disposal [4]. Sodium, being a bigger ion than lithium, is stopped by the membrane. In fact the mined lithium is much cheaper than the recycled, as the recycled lithium is a part of the metallic slag of the metallurgic process in the form of oxides and more difficult to extract (might need strong acids) than a chloride in the form of a salt. So there might be a waste disposal issue. The anode of Li-ion batteries contains porous carbon: graphite or variants. The porous carbon such as graphite has a poor conductivity, so it might need copper or aluminium to be used as a current collector (spine). Other carbon forms as graphene have also a good conductivity, in the order of $10^5$ S/m, but are produced at the expense of an energetic chemical process by chemical vapour deposition. Carbon fibre can show conductivities similar to metal alloys: in the order of $10^6$ S/m, the manufacturing starts from a polymer that is oxidized and carbonized, so also quite energy demanding. The cathode contains nickel and often cobalt; in fact they need a lattice to store the Li ions by intercalation, and cobalt helps to do this in a larger number of cycles without deforming the electrodes too much. Batteries of laptops and cell phones use quite a lot of cobalt; in automotive, efforts are made to reduce or avoid it as much as possible. The conditions at which the cobalt mineral is dug and the fact that more than 50% of world is in the region of Katanga, south east of Congo, is not a comfortable situation. EV makers push to improve social conditions in mining. A part of cobalt is mined together with copper. Nickel is not very abundant and is already used in a lot of applications, such as stainless steel, but also some magnetic materials, some catalysts and as a moulding surface finish to avoid adherence. It is predicted that EVs will double the nickel price by 2040 [5]. However, the most troublesome is cobalt, where the demand may double by 2030. If the supply cannot follow, speculation may start and the price may skyrocket for some time, as we have seen such an effect with neodymium for magnets in the past. The technology is called LMC; it is a bit confusing, as people confuse ‘C’ with ‘carbon’, but it is lithium–manganese–cobalt. The intercalation materials such as Mn and Co are a bigger fraction in weight and price than the active material Li. In fact lithium–iron–phosphate ($LiFePO_4$) could take over, at some less energy density, but has a larger number of cycles, degrades less. So, finally, the real remaining range after some 10 years could be similar for $LiFePO_4$ and LMC. Each month, discoveries are made, promising higher energy density or more cycles, or being more cost-effective. As there are many details in a battery, it often concerns a detail, and the final development takes effort and time. The problem of active material could be solved with Na (sodium) a good candidate as well, but the truth will also be in the weight and cost of intercalation materials and the number of cycles. In fact only a part of the weight is really active, so similar or better final energy densities could still be obtained. Magnesium has two valence electrons, so potentially good for energy density, but it also means two different voltage levels, which probably means a lower efficiency. Aluminium has three valence electrons, which is in principle still 30% more heavy/electron, but if the intercalation is more effective, it is also a candidate. In fact we have not to forget the electrolyte, which is not water based but, for example, fluoroethylene-carbonate based. So the question also arises about the fluorine availability. It seems reasonably available, and some other uses were cancelled due to the risk of ozone depletion when used for refrigeration gases. Heavy molecules containing fluorine have much less ozone depletion potential, as they are less volatile. The BMS has much less mass and is not expected to be a bottleneck for minerals.

## 2.3 Life extension

If the battery is well maintained in its good operating area (for each cell), means not too deep discharged, not to high charged, in a mild temperature range, it may be thankful to you and have a longer life. But we have to be honest: a limited life also applies to the BMS. True, the modern batteries need a BMS as without, their life would be very short, or would require a continuous manual check. But the question also arises about the life of the BMS. To start, the operation principle and quality control must be all right. However, it is not so clear that good batteries will really fail before the end-of-life of a BMS. However, a failure of the BMS might destroy the battery. So we can look a bit deeper. Some components are known to have a limited life, such as electrolytic capacitors. They have a bathtub-type of curve: burn-in and end-of-life. Hydrogen disappears, the aluminium oxide gets thicker, the electrolyte gets dry and more resistive, and there is also the deterioration of polymers. A typical life expectancy of good electrolytic capacitors is 10 000 h at high temperature and 100 000 h at low temperature. A high number of capacitors (cells) increase the risk. There are several reasons for it: the electrolyte composition, corrosion, even rubber [6]. They argue that a 2000 h/85 °C capacitor operating at 40 °C roughly lasts for 6 years. It can be more than 10 years for a 1000 h/105 °C capacitor operating at 40 °C. To prolong the life of a BMS, one of the elements is to be avoided, that is electrolytic capacitors in the design [7]. Electrolytic capacitors are not the only reason; in general, it is not sure that a BMS lasts longer than a well-designed battery, but once the problem is well formulated, the technology may gradually improve.

Electronics and contacts do not like moisture, so it is one possible reason. In extreme dry conditions, electrostatic discharges and mobile electrostatic charges might also disturb sensitive ICs. Salt, sand, and pollution also shorten the life. Of course, a good engineering design is essential, but the finishing touch may also be important. Is all soldering perfect? Welding of the housing? Will cables not pull on their contacts? The right coating? Personally, I do not prefer potting as it might generate internal stresses. The classic maintenance of lead–acid was cleaning contacts (but not with a brush as acid may get into the eyes) and applying technical Vaseline. Smaller contacts may need contact cleaner and applying silicone oil to avoid oxidation. Should a BMS be replaced as a preventive action? Should the circuits, the battery be cleaned? The learning school of reality will probably help us in improving the technology. The question of the ability to repair is also relevant. Can a bad cell be identified and removed or over-bridged? The application could be designed to run using a few cells less. The shelf-life of Li batteries can be several years, as far as no current drawing circuit is connected. In fact data should be available about the standby current of BMSs and other connected equipment. If you do nothing special, Tesla indicates an auto-discharge of 100 days, without grid connection; this is mainly due to a continuous standby computer of about 30 W. After that, forums call it a ‘brick’. In lead–acid batteries, a long shelf-life is obtained by storing them without acid, but once filled, it is better to compensate the auto-discharge using a very small current. In conventional cars, the connected equipment may reduce it to 2–3 weeks. So most of new combustion engine cars are shipped without battery. NiMH batteries can be stored in a discharged state, they are still popular in some household appliances.

## 2.4 Reuse or second life

**Lead–acid:** Some batteries get a second life in less-used vehicles, old tractors, lawn mowers, where the reliability is less essential. Batteries from forklifts may have a second life in home solar or less critical UPS systems. Older engines need much less peak crank power to start, compared to brand new ones, that still need to reduce their friction. End-of-life or accident-faced cars (electrical or fuel) may still have a fairly good battery. Reuse is better than recycling, but often it is the last step before recycling. Question is if the existing BMS can be reused: probably not, as it may be highly software-dependent, using a CAN-bus; how can the correct information be retrieved in a new application? This book can also propose a system which could be rather independent from what was used before, for reusing in other vehicles, or grid-coupled batteries, or stand-alone circuits. If one looks at using a battery to cross a night in a household even a part of a car battery is relevant, in the order of 5 kWh. A car of 40 kWh at 400 V is 100 Ah battery, 16 cells used at 3.5 V = 56 V, 5.6 kWh, say 5 kWh useful (efficiency of inverter), even more in DC use. Could batteries be used instead of generators in a construction site? One puts a set of PV on a trailer + some batteries, unfold panels, so start without waiting for a grid connection. It can be used for operating drills, cutting discs, concrete mixer, crane. To reuse a cell, a visual test and an open-circuit voltage test can be done. An impedance check around 1 kHz allows comparing cells if their total electrode surface is active (no internal cracks). After that, a controlled charge and discharge is used to detect the remaining SoH. Using 1 kHz impedance is already a common practice for lead–acid batteries [8]. Care should be taken to reduce the inductance of the leads (a four-wire test, Kelvin contacts). A frequency around 1 kHz will test if the internal conductors are still good, and, hence, is a relevant frequency in the electrochemical impedance spectroscopy IES frequency range. If a part of the electrodes is cracked, it will have a higher internal resistance, a lower capacity, and overheat easily. Much lower frequencies test the mobility of ions, but this is rather related to the battery concept. The frequency of 1 kHz allows testing without very heavy equipment as a converter and transformer with capacitor coupling is realistic. **Lithium:** 20%–50% less capacity/cell is still acceptable for home battery but less good for a car.

## 2.5 Recycling

Lead can be almost fully recycled. This is about 98% of the lead, and 92% of the battery components. But as some batteries may take a decade or more before recycling, so still new lead is needed. More than 60% of global production comes from recycling, and that % increases [9]. A car battery lasts from 5 to 7 years [10]. However, my own experience is that some AGM, Absorbed Glass Mat (often used in start-stop batteries) last up to 10 years. So, if the technology improves, the battery takes more time to get recycled, which is not negative. Another effect in recycling lead batteries is the reduction of antimony, Sb. One decade ago, 4%–10% of antimony was added to the lead to improve the alloy strength. Nowadays, AGM-like batteries contain about 1%–2% antimony, as the electrode deformation is rather controlled by the glass mat. Recently, the product antimony oxide got more expensive, it is used in other applications such as fire retardant, removing halogens in gas phase. Intensively used forklift lead acid batteries typically last 2 years and still 1 year after refurbishment. A typical way for reconditioning is using high frequency (ultrasonic) bursts for desulfating the battery. It intends to dissolve the lead sulfate crystals which tend to accumulate on the electrodes which block the free movement of current and electrolyte. We will not go in detail as lead acid is not the aim of the book. Recycling of Li-ion started first for mobile phones, laptops, and other IT equipment. The company Umicore in Belgium started quite early, as well as Glencore International AG (Switzerland), International Metals Reclamation Company, LLC/INMETCO (the United States), Retriev Technologies (the United States), and Raw Materials Company (Canada), which is the principle of 'urban mining' [11]. Current battery recycling operations are predominantly pyrometallurgical. High-temperature processes have limited applicability for the recycling of lithium ion batteries without valuable metals (i.e. cobalt) because lithium ends up in the smelter slag, rendering the process uneconomic. So the full recycling of Li-ion is not that simple. Moreover, many variants exist in the Li-ion technology.

## 2.6 $CO_2$ emission to produce a kWh of battery storage

If we look at papers and reports, we find very large variations for $CO_2$ emissions, especially for the lithium types. One element is that small series need more energy to manufacture than well-controlled series productions. But on the other hand, manufacturers do not like to show that a lot of energy is needed so may also hide information. Some people take a worst case and generalize, so other do the same but with the best case, showing a minimal value. For cars, sometimes a longer lifetime or more travelled distance is taken into account to show some advantages. If an average car in Belgium runs 15 000 km/year for 10 years, it covers 150 000 km. The warrantee is often not longer than 10 years and the question is whether the BMS will also last much longer if it is not designed on that purpose. But it means that a lot of cars do less than 100 000 km or even 50 000 km in their life as well. The 50 000 km user could better share a vehicle or use a second hand, but the $CO_2$ should be at acceptable level for 100 000 km life. Then some countries use more nuclear energy or more hydro to produce electricity, whereas other countries more coal. The learning school of the power electronic area teaches us that 'summation of losses' always underestimates losses: capacitor losses are forgotten, eddy current losses, hidden temperature effects in switching losses of diodes, PCB conduction losses, losses of the connection wires.

In a similar way, the $CO_2$ for battery manufacturing can be underestimated: are all thermal and drying processes taken into account? Diverse transport? Heating, ventilation, air-conditioning of the factories? There are always unknowns. The above example mentions 100 000 km as an example, some cars do 500 000 km, and it would give better figures for electric, but a question could be if it will be reached with the same battery? There might be also a risk that the BMS might have a shorter life. If bad results are obtained far below 100 000 km life, light renewable methanol vehicles could have a lower carbon footprint, more $CO_2$/km, but less in manufacturing. Finally, the price may be some indication: if it is cheap, not much fossil $CO_2$ can have been emitted, as fossil fuel has a cost. But in some countries, energy is subsidized. But nothing says that a cheap product may not generate a lot of pollution during production, use, or end-of-life or have a shorter life span. If something is very expensive, some hidden $CO_2$ emissions may be somewhere present.

### *2.6.1 Lead–acid and $CO_2$*

The case of lead–acid is simpler as less variety of materials are involved. Some papers compare the performance with flywheels in the case of UPS for data centres, concerning VRLA, valve-regulated lead–acid. According to [12], 98% of the lead of a battery is recycled, but it does not mean that no new lead is needed. Lead–acid batteries have a life between 3 and 15 years, so some take (fortunately) a long time to appear in recycling; if it is a growing market of 5% year, one needs to add $1.05^{10} - 1 = 62\%$ new led to produce it. That one needs new minerals does not mean that a big part is not recycled. In the life cycle, one can consider ‘cradle to grave’, or ‘cradle to cradle’. Manufacturing, packaging, and installation have been excluded in that paper (as they only compare). They take 200 miles, 322 km as an example, but in Europe the mines are not in the centre: Sweden, Poland, and Ireland; so it easily reaches 1500 km [13].

Indeed some 40 L of fuel/100 km can be adopted for large truck transport of 20 t load. If 1500 km is travelled (materials and finished product), it means $1500 \times 40/100/20\,000 = 0.03$ L/kg at 2.64 kg $CO_2$/L is about 0.08 kg $CO_2$/kg battery. Manufacturing the lead 1.33 kg $CO_2$/kg lead, the lead is 72% of the battery weight (Table 2.1). The fabrication can be evaluated by another article where the total energy consumption of a factory was observed and related to the kWh of batteries produced. It is the industrial case of a production plant in Colombia [14]. Figure 2.2 shows the electricity consumed and the equivalent battery production. It seems to be about 5 kWhe/kWhb (Figure 2.2). At an average European 300 g $CO_2$/kWh, it adds 1.5 kg $CO_2$/kWhb. Table 2.1 shows different items with emission and percentage.

*Table 2.1 Emissions of different items, for 30 kg/kWhb*

| Item | Emission factor (kg $CO_2$/kg) | Percentage |
|---|---|---|
| Lead | 1.33 | 72 |
| Fiberglass separator | 1.35 | 2 |
| Plastic container | 1.14 | 5 |
| Transport | 0.08 | 100 |

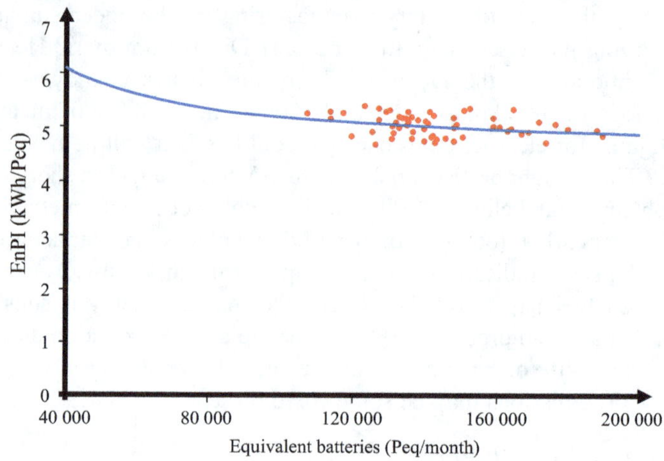


*Figure 2.2 Electricity used vs lead–acid battery production*

It results in $1.33 \times 0.72 + 1.35 \times 0.02 + 1.14 \times 0.05 + 0.08 = 1.12$ kg$CO_2$/kg. If a lead–acid battery of 1 kWh weighs 30 kg, it is about 33.6 kg $CO_2$/kWhb. The battery manufacturing seems to add about 5 kWhe/kWhb, and at an average European 300 g $CO_2$/kWh, it adds 1.5 kg $CO_2$/kWhb. The sum gets 35.1 kg $CO_2$/kWhb. This makes the lead–acid battery still interesting for stationary use where the number of cycles is low and not fully discharged. If renewable energy would generate no $CO_2$, about 150 full equivalent cycles with a cycle efficiency of 80% are needed to be at the same level of the European grid CO2 emission. At that level, the manufacturing of batteries would generate as much, so in fact double the CO2 than what is expected from the grid. So, it could still be a good solution to store weekly fluctuations. Let us take the example of the importance of weight if is used as a starting battery in a car that makes 100 000 km in its life (it may not be the same battery), an efficient diesel car of 4 L of fuel/t curb. Some reports found out that the vehicle consumption is almost proportional to its weight and also some more (battery) weight need other weight to support it. About 1/1000 of 4000 L is 4 L/kg gives $4 \times 2.64 = 10.56$ kg $CO_2$/kg. So the weight of a lead–acid battery in a conventional car may generate indirectly about eight times more $CO_2$ than its production. Note that if you start from 105 g $CO_2$/km, you get 10.5 kg $CO_2$/kg battery, about the same. A lead–acid battery may have a typical 30 kg/kWh, where it is not good to discharge it fully. If an electric vehicle uses lead–acid batteries, it could need 1000 kg of batteries, compared to 1000 kg curb, and then still not a long range. If the vehicle needs 18 kWh/100 km, 50% could be attributed to the battery weight, 9 kWh/100 km. At an European average of 300 g $CO_2$/kWh, and 100 000 km, it means $9000 \times 0.3 = 270$ kg $CO_2$ for 30 kWh is 9 kg $CO_2$/kWhb due to the battery weight during the car life + 30 kg $CO_2$ of manufacturing. Probably the battery will not last that long, three batteries could be needed in the vehicle life. The total is then $35.1 \times 3+9 = 114.3$ kg $CO_2$/kWhb directly and indirectly emitted over the considered

*Table 2.2 Total kWh lost and the battery throughput*

| Technology | kWh lost over lifetime | kWh throughput | Ratio% |
|---|---|---|---|
| PbA fast | 84.919 | 575.2 | 14.8 |
| PbA slow | 150.538 | 1200 | 12.5 |
| NiZn | 338.71 | 2700 | 12.5 |
| MnO2 | 430.108 | 1875 | 22.9 |

distance. If one considers that the vehicle typically transports 100 kg of driver with some luggage, the vehicle mainly transports its own battery [15]. Even the losses in the battery may influence the total life cycle $CO_2$. Table 2.2 shows the results of the total kWh lost and the battery throughput, and the ratio computed.

Note that the charger and inverter losses are not included in the table. A discussion is also where the lost kWh comes from renewable or production with electricity from coal. In general, articles will show each time different results depending on the considered case.

### *2.6.2 Lithium ion and $CO_2$*

As many substances are involved, the calculation will not be that simple; it is easy to hide some part or have the error to extrapolate a bad example. In fact, the producers may know but may not be ready to share all details, especially if it appears that a lot of energy is used, and that they may be sooner or later get some kind of $CO_2$ taxation. For this reason, the results are also widely different with a climate impact ranging from 39 kg $CO_2$e/kWh to 196 kg $CO_2$e/kWh. If here, a deep analysis is made, it will be just one more in the list of more than 100 articles [16]. They get to a final result of 73 kg $CO_2$e/kWh. An example is given that the transport from Korea to the United States is 4.1 kg $CO_2$e/kWh. Sea transport has about two times less $CO_2$/km/t than road, but often much more distance is made. So depending on the authors, very different results are proposed. Also the comparison with fuel vehicles is subject to a lot of difference and intrigues. There is also a habit to hide things. The fact that the required energy is often listed in MJ/kWhb does not specify if the MJ comes from electricity or heat, and which fuel is used. Some materials may travel around the world before being used: lithium from Chile to China, cobalt from Congo, RDC = DRC, to the United States. Only a small fraction comes from recycling, a recycling that needs real metallurgic temperatures. The question is if the consumption of the whole factory is taken into account or based on reports that Chinese companies like to communicate. Another overview is given by [17]. They find out that some articles exaggerate with 175 kg $CO_2$/kWhb and but found some 40–100 kWhb more realistic, where they take a typical value of 75 kg/kWhb. However, they say that some studies under- or overestimate the life of a car; indeed, in Europe it is about 150 000 km, but it means that 100 000 km and 200 000 km should also be $CO_2$-effective. In the Netherlands, a car does some 13 000 km/year in Belgium 15 000 km. So the question is whether a number of 10 years will be the end-of-life of the battery. Gradually less distance is made per year, this tendency might be confirmed by EVs as making long distances makes more charging stops. A side effect of the Covid 19 pandemic is that people learn to work from home and more meetings are done remote, so the average distance/ year might reduce even more. The tendency is to increase the kWh for the battery and increase vehicle weight. So 1500 kg, corresponding with 16 kWh/100 km, and a battery of 48 kWh, gives a range of 300 km. In 100 000 km, 16 000 kWh electricity is used, as, for example, 300 g $CO_2$e/kWh is 48 g $CO_2$/km, 48 kWh at 75 kg $CO_2$e/kWh over 100 000 km is 36 g $CO_2$e/km. Together it is 84 g $CO_2$e/km. This is indeed better than the major part of combustion engines but not enough to really have a big effect on climate change. However, if we take a lower budget solution we get a different view. The Dacia Spring is proposed with 230 km range in WLTP combined cycle, the 'secret' is the lower curb weight of 921 kg, so it can drive with a smaller motor of 33 kW and a battery of 26.8 kWh at about 12 kWh/100 km. If we take 75 kg $CO_2$/kWhb and 300 g $CO_2$/kWh grid, we get $75 \times 26.8/10^5 + 0.12 \times 0.3 = 56$ g $CO_2$/km, if the car is mainly charged on PV during more than 50% time of the year, we could get lower than 40 g $CO_2$/km. So the 'secret' is a lighter vehicle, not too much battery, and much more renewable energy than the European grid average.

The attribute 'ZEV' 'zero-emission vehicle' for electric is not true, rather they are only ZTEV 'zero tailpipe emission vehicle'. The manufacturing of the battery, the electricity, and the tire wear have still non-negligible emissions. Another remark is that one should avoid to store solar first in a home battery to charge a car battery with it, as then the $CO_2$/kWh of the battery appears two times as two batteries are involved for the same kWh. Swapping batteries would be a better option, but it is not easy to standardize, and it involves a container that increases weight. A lower weight is important, and there are three reasons why weight influences consumption: more weight requires more rolling consumption, more weight requires more energy to accelerate (be happy if 50%–60% is recuperated), more weight requires more energy for altitude changes. The real progress is to drastically reduce the vehicle weight and its frontal area. A lot of vehicles are used for commuting: rather low speed and a low range need. It might be an option for the 'second car' or shared with family or neighbours. Electric bikes weigh some 20% of people. A fully covered two-person vehicle seems feasible in 160 kg–200 kg, maximal speed 90 km/h. The consumption is in the range of 2–4 kWh/100 km [18]. A fast recharge would be possible on a normal 16 A–230 V wall plug. If airplane-like technology (reinforced fibre and high strength aluminium) is combined with laminated wood, for crash energy absorption, a low weight can be obtained. Already around 1920, some electric vehicles were made needing about 120 kg for two persons, known as 'Smith Flyer' or 'Auto Red Bug'. Their trick was using steel instead of cast iron and a flexible chassis in ash wood. It was open, the speed was not high and safety not perfect, but today materials pretend also to be much better, so such concepts could have been closed, faster and safer for a similar curb weight. A discussion can be found in [18]. Another example: the Formula E cars weigh about 800 kg, driver included, 200 kg battery, peak powers of 250 kW [19]. However, a five-time lower power is enough for a normal vehicle, so the weight can be reduced accordingly. The canopy of a traditional Formula 1 is some 40 kg. Reducing weight is just a matter of good engineering and mentality change [20].

## 2.7 Avoiding battery storage

It seems strange to include such a paragraph in a book about batteries. However, it can make sense. Is it possible without a battery? Some possibilities can be suggested as listed next.

### 2.7.1 *Water heaters*

It is possible that the energy of a battery is finally used for heating water. This may be for sanitary water or sink water. A hot water heater could be considered a 'thermal battery'. It is typically 'charged' with a heating resistor and fed by cold water and discharged by tapping water. The efficiency can be increased by a heat pump. The storage is much cheaper than batteries, for example, 150 L is often sold around 200–300 euro to heat water from 15 °C up to 75 °C, 150 × (75–15) × 4.19/3.6 = 10.5 kWh, so rather 20–30 euro/kWh storage, which is about ten times cheaper than batteries. However, a typical 150 L boiler has a heat loss of 1.5 kWh/day = 62.5 W losses, so a time constant of 7 days, which is poor. The reason is the thin polyurethane insulation of about 50 mm, where in buildings often 100 mm is used for much less temperature difference: about 8 K compared to 55 K. Probably not more insulation is used for a water heater, as it would not pass a small door width. However, the thermal insulation can be improved while mounting it on wooden blocks to the wall, and some 200 mm rock wool. Tape well the gaps, as air can also circulate in rock wool. Not to forget the in- and outgoing water pipes as they are also leaking paths of about 5 W each. New techniques are coming up using Vacuum Insulation Panels, VIP. It obtains the same insulation of 200 mm rock wool in 40 mm thickness but cannot be cut or adjusted. It is good to store heat but this is not reasonably convertible to electricity again. The reason is the Carnot law. The maximal convertible energy fraction is $(T1-T2)/T1$, where the temperatures are in degrees Kelvin. $T1$ is the hottest point, $T2$ is the cold one. So between, for example, 75 °C and 15 °C (75 − 15)/(273 + 75) = 0.172, thermal machines lower this by about a factor of 2. If reverse Peltier elements are used, rather 3%.

### 2.7.2 *Storing cold*

If storing heat is possible, it is logic to think about storing cold as well. Most devices are internally based on heat pump technology. An example is a deep freezer. If solar energy is available, it could operate during the day and freeze deeper (using 'normal' or 'max', and not 'eco') and then use some thermal inertia to cross the night. Starting up of a conventional freezer or refrigerator needs about 1500 W during 0.5 s, where afterwards only 40 W–80 W is needed to continue. A battery can deliver such a starting power, and the grid also, but a PV system would not be able to start in the early morning. Classic air-conditioning systems have the

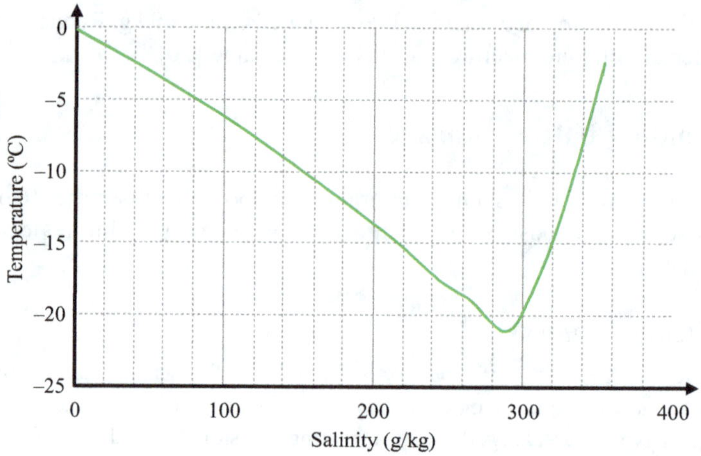


*Figure 2.3 Freezing point of salt water*

same starting problem. More expensive devices contain a permanent magnet motor to drive the compressor and use a power electronic converter for it. The efficiency is somewhat higher, and it would reduce seriously the starting power. It contains a rectifier at the input, so some manufacturers may propose systems working directly on PV without battery. The cold could be stored in ice at 0 °C for air conditioners. Large systems do this already. The latent heat of the ice to water phase change is 334 kJ/kg. So it means about 10.8 kg of ice is needed to store 1 kWh of cold. Also refrigerators in dairy farms use ice to cool milk down to 3 °C in a short time, without having to draw a sudden large power from the grid. In the case of a deep freezer, it is good using phase changing materials (PCM) that absorb and release heat at just below −18 °C. It has been tried out in larger installations [21]. According to estimates from ZAE Bayern, around 1 GW–2 GW of load could be shifted flexibly if just half of Germany's fridges deployed thermal storage. Solutions that contain several substances in a non-eutectic way will get a segregation while freezing. A low-concentration salt brine is cheap but when freezing rather pure ice is formed, leaving a more concentrated brine behind. Some trajectory instead of a single temperature is formed [22]. Some 295 g salt/kg could do the job, close to eutectic (Figure 2.3). The same happens with substances as glycol but are not so aggressive to metals as salt brine. But a question about food contamination remains. The best class of freezers has some 330 Wh consumption/day at a COP of, for example, 3, it means that about 0.990 kWh cold is produced. If it is stored for three-fourth of the day in winter, it needs some 0.7466 kWh = 2673 kJ cold storage. Some vegetable oils freeze mainly at −20°C such as sunflower oil. This is the freezing temperature of the unsaturated fraction, saturated fractions may give clouds around 0 °C. They call the oil 'winterized' if those fractions are removed [23]. A latent melting heat of 149.2 kJ/kg is mentioned for bio-based PCM, in the example 17.9 kg would be needed, 19.5 L. Note that if 5 °C change is

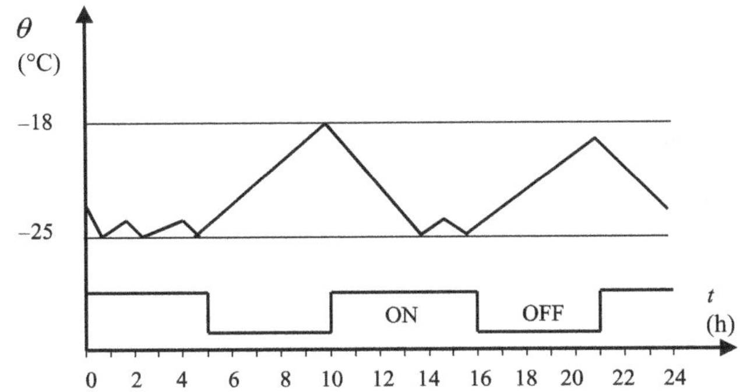


*Figure 2.4 Deep-freeze supply on/off and temperature change when set at −25 °C (max. setting)*

allowed in ice at 2.04 kJ/kg almost 15 times more would be needed, almost filling the useful volume. However, almost 20 L oil is still quite a lot. A test has been made using 8 L, to see what is possible. The recipients are put close to the wall where typically ice is formed, so where the cooling tubes are. Transparent recipients help to see cloud forming, at − 25 °C, the oil gets harder but still malleable. Even in winter, there is sufficient light between 10 h and 16 h and the tariff valley-hours are, for example, between 21 h and 05 h to keep it cold, which means 5 h to hold in between. To start, the PCM bottles are not inserted at once but gradually and not touching the food. It needs a few days to get fully in steady state. An inside temperature seems to rise about 1–1.5 °C/h when the cooling is off and reduce 2–3 °C/h when the compressor is working. Figure 2.4 shows a simplified temperature evolution of the inside air. In practice, the inside temperature is not homogenous, as the different foods have different time constants but still on the safe side if applied well, a few litres of oil and a timing clock helps to avoid peak, reduce the peak, and increase valley demands in the grid when a lot of solar PV injection is used. In the literature, it is known as the 'duck curve' effect (put the keywords in your browser). The utility may reflect this in the tariff while billing more the peak hours in the morning and the evening. The case of **refrigerators** is less easy as a control between 1 °C and 5 °C is expected. But if the concept is changed, ice could be used for storage. **Air conditioners** may have excess capacity in the morning to make ice and to use it in the evening after sunset. In the afternoon, the system would typically not need storage. Ice has a too low storing temperature, for a high efficiency, but it is not easy to find a better non-toxic and cheap material. Oversizing PV might be more cost-effective. A PCM in the range of 5 °C–10 °C would be preferred but result in more expensive, flammable, or toxic substances (e.g. paraffin waxes $C14$–$C18$) [24].

### *2.7.3 Direct use of solar PV*

**Direct DC:** Our modern appliances are often internally supplied by Switched Mode Power Supplies, SMPS. They are often designed for a worldwide use [25]. This means, e.g. in a part of Japan at 100 V, with − 15% tolerance, means 85 Vac. The peak value is $\sqrt{2}$ times higher, but the internal capacitor has a typical 10% discharge in a half period, $85 \times 2^{0.5} \times 0.9 = 108.2$ Vdc. In practice, it means that a lot of SMPS work down to 110 Vdc. If it comes from solar PV or from a battery is of secondary importance. Normal switches seem to be compatible with DC and SMPS; the reason is that an SMPS contains a capacitor inside, so that when a switch opens, the voltage on the switch rises only slowly, and no arc is formed. Some LED lamps have a buck converter inside: it allows an operation using less LEDs in series. They are listed as 'non-dimmable'. Some of them even work down to 70 Vdc even if only 230 Vac is listed on their label. A Japanese DAB+ radio seemed to work down to 40 Vdc, which is also not listed on its label but may be nice to know for people who have a 48 V battery system, so that it can operate without an inverter. The maximal allowable voltage is rather determined by the Australian grid: 240 V+10%, so $240 \times 1.1 \times 2^{0.5} = 373.4$ Vdc. A side remark is that some upcoming DC grids use 350, 375, or 400 Vdc. At +10% tolerance, 350 V can reach 385 Vdc and hence is not compatible with existing appliances without adding a step-down converter. If a grid is made of 300 Vdc ±10% and appliances 300 Vdc ±20%, a lot of designs are possible. For example, 80 Li-ion cells of 3.7 V (10 banks of 8 cells) is 296 V nominal. A quantity of 330 V corresponds with 4.125 V/cell: more than a normal charge. A quantity of 270 V corresponds with 3.375 V/cell: close to discharge. There is some room for voltage drop in both directions, +20% at the level of PV panels and −20% if appliances draw a high current and the voltage drops by that current (5% in the battery and 5% in de conductors of the installation). Most of the time, a too low voltage does not damage the equipment (a 'brown out' as they say), but the normal functioning is not warranted. However, a too high voltage (that can deliver some current) is often lethal for the equipment, even below 1 ms. The fact of using a PV voltage or battery voltage directly (so without intermediate converter) does not mean that no protection is needed. DC can generate arcs even if the current is limited (PV panels) and batteries may deliver a dangerous peak current when short circuited. In Figure 2.5,

*Figure 2.5 A switch made in 1905, compatible with DC voltage, brand 'Lancot'*

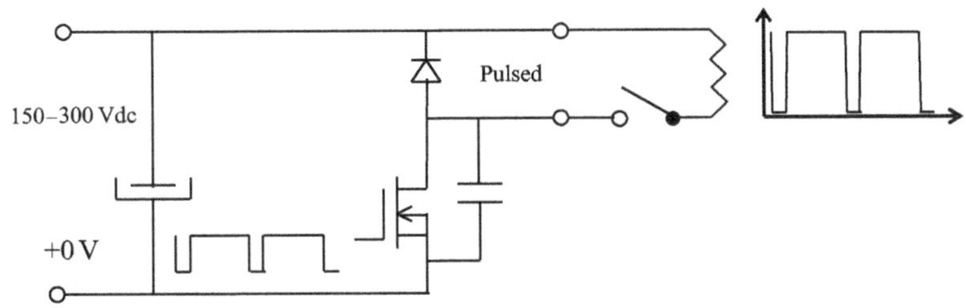


*Figure 2.6 Pulsed DC for water heaters with thermostat*

a switch is shown, made in 1905, compatible with DC. The switch was tried out at 180 V and 40 Ω, where it switches some little sparking. At a bit more voltage, the sparks became too serious. It was probably compatible with 110 Vdc grids that were common at that time, at max. 6 A DC, fuses of that time. The today electrical installation switches and thermostat switches are not made for DC anymore but most of plugs withstand.

**Pulsed DC:** A typical problem is that conventional switches tend to arc in DC, typically for resistive and resistive–inductive loads. Very old porcelain switches were made to cope with DC: two interruptions and spring-loaded contacts. Modern switches have often only small apertures between contacts and do not manage to interrupt DC; my experience is that even down to 40 Vdc may be a problem for a thermostat controlling a resistor. Where is the problem? In power electronics, we switch easily 100 kHz at 1000 V now. Indeed, a thermostat switch can be replaced by a transistor. However, a thermostat switch in an electric boiler has also a backup function. If the temperature gets really too high, a second bimetal cuts both wires, and a manual reset using some pin in a hole is needed to operate it. It happens typically if the plumber forgets to put water inside. A transistor will not be accepted as a galvanic separation, even if it switches better. And for a galvanic separation, some 1500 or 3000 V insulation would be desired. But there is a solution. The arc in AC extinguishes as there is some time to deionize the arc. If DC is pulsed, we can simulate the same effect. A 0.5–2 ms pause is sufficient. A chopper is in fact the easiest power electronic circuit. If no battery is used, an electrolytic capacitor is needed to flatten the voltage, by absorbing the current ripple.

A circuit is shown in Figure 2.6. The capacitor parallel to the MOSFET is used to reduce the EMC, and inductive filters can be added as well. The advantage is that the electric water heater remains compatible with an AC grid use in case the DC is not available (winter, snow, or more demand). Without battery, the circuit can be controlled by a voltage feedback: more voltage is increasing the duty ratio. It stabilizes the voltage for other applications. The voltage needs to be a bit lower in summer. A real maximum power point tracker, is difficult to apply to control a heating resistance, as it will reduce the power of other appliances on the same PV panels. Above 250 Vdc, the maximal duty ratio has to be limited not to overload the heating resistor.

## 2.7.4 Flywheel storage

### 2.7.4.1 Principle

The advantage is that the instantaneous power is not limited by chemical diffusion as in batteries. A disadvantage is that the energy is limited by the used strain limit: $(\text{strain})^2/\text{kg}$. The energy in a ring is $m/2 \times c_t^2$, where $c_t$: tangential speed [units kg $\text{m}^2/\text{s}^2 = \text{kg m/s}^2 \times \text{m} = \text{J}$]. If the ring is cut in the middle, the force on a half ring is (half mass × distance of the mass centre × centrifugal acceleration) (Figure 2.7):

$$m/2\times(2/\pi) \times c_t^2/r = 2/(\pi \times r) \times m/2 \times c_t^2 = \sigma \times (2 \times S)$$

where $\sigma$ is the stress in MPa and $S$ is the section of the ring.

So the maximal kinetic energy for a thin ring is:

$$\sigma\times S \times \pi \times r = \text{volume}/2\times\sigma \text{ [units: N/m}^2 \times \text{m}^2 \times \text{m} = \text{N m} = \text{J]}$$

Note that this simplified calculation has some similarities with a pressured tank calculation. It is also remarkable that the density does not appear. What happens is that less dense materials will have to rotate faster. Another remark is that a thin ring is not the optimal shape, and optimal shapes have no hole in the middle. This will not be discussed here.

### 2.7.4.2 History

The idea is not new as such, early attempts such as in 1950 had been made such as a 'gyrobus' by a Swiss company and some use in Ghent Belgium [26]. A flywheel of 1500 kg at 3000 rpm was used, and the enclosure was filled with hydrogen to reduce aerodynamic friction and still allow convection heat transfer. The gyro effect made taking corners difficult, and it got in shortage of energy in traffic jams. Figure 2.8 shows the interior of a gyrobus, in which only the motor is visible and the flywheel is below it. The flywheel is in the middle of the bus.

It charged about 3 kWh at a stop. We guess that it has a ring with an average radius diameter of 1.2 m, a weight of 1500 kg, at 50 revolutions/s. It has in theory at 1500 rpm an energy of about $(1.5\times\pi\times50/2)^2 \times 1500/3.6/10^6 = 5.68$ kWh. However, often only 75% of the energy is used (between half and full speed). One

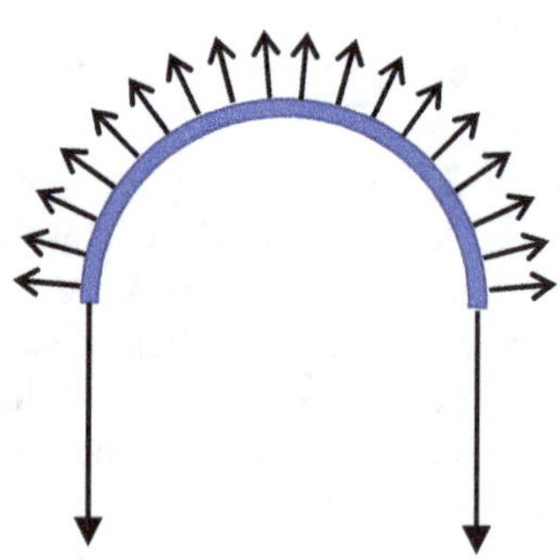

Figure 2.7 Stresses in a thin ring

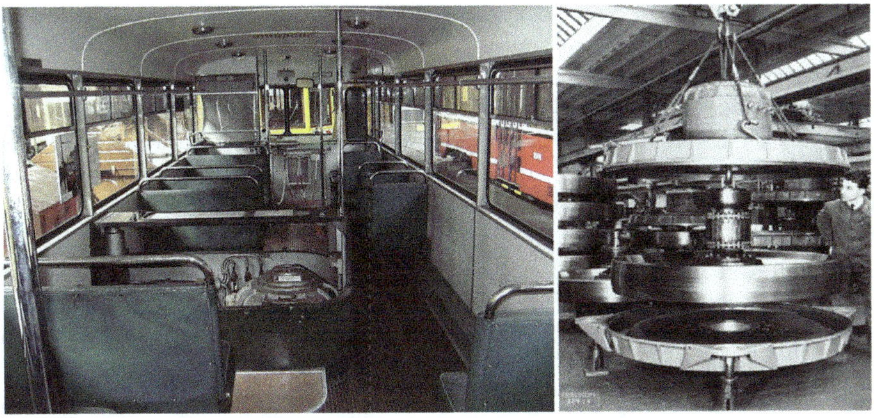

*Figure 2.8 Interior of a gyrobus, flywheel in the middle*

acceleration of a 20 t bus (passengers included) up to 15 m/s (54 km/h) at 100% efficiency is $0.5 \times 20\,000 \times 15^2/(3.6 \times 10^6) = 0.625$ kWh. Note that the flywheel was not pushed to the limits, in which 1500 rpm was convenient for charging the flywheel, starting from a 50 Hz grid at that time. It was running in hydrogen atmosphere to reduce friction, but still not using vacuum to avoid a too easy flashover. Older people told that it sometimes stuck in a traffic jam and had to be towed. If we look at it today, the power electronics, some emergency battery and some PV panels, would be worth redesigning such concepts.

#### 2.7.4.3 Today state of the art

Lower losses can be obtained when the flywheel rotates in vacuum. In vacuum, electric motors without brushes are needed, as brushes do have a short life in vacuum. Also the produced heat has to be removed by radiation and conduction, as convection is not present in vacuum. It can also be interesting for temporary grid storage or frequency stabilization. However, in the last three decades, a few accidents occurred with flywheels, which rise a lot of questions. Are the minimal properties well known? Some possible defects? Is non-destructive testing necessary and/or sufficient? Number of cycles: they are very high but still limited? What about the case for confinement? Bearings? A good overview is given in [27]. Flywheels got a revival in Formula 1 races where it is known as KERS, Kinetic Energy Recovery System. The very high revolutions/min are used sometimes with a shaft feedthrough to vacuum. Mechanic and electric variants exist. The long-term vacuum is not at big item in that case.

### *2.7.5 Gravity storage*

People do not realize that it was used during centuries to operate clocks. But the breakthrough was using hydro energy. The potential is simple $W = mgh$. On a small scale, it does not store much, as an example we take 1 t, 10 m high, which is

98.1 kJ, which corresponds to 0.02725 kWh at 100% efficiency. The human intuition feels more, but the reason is that we compare with our own ability. Biking slow at 100 W during 10 h is 1 kWh. A faster ride at 250 W during 4 h is the same. If a cyclist+bike weighs 100 kg, he has to climb 3670 m to get 1 kWh if air and rolling resistance is ignored. The previous sentences may create some respect for '1 kWh' of energy which is in fact already quite large. If we take 1000 m altitude difference, $10^5$ $m^3$ (a lake of 10 ha, 1 m difference), it corresponds to 272.5 MWh. However, few places on earth fit well for that. Traditionally, sites with rivers and lakes are considered where artificial lakes with dams are added. However, also sites without river could be considered as they might have a better topography. Water has the tendency to evaporate, freeze, and to let grow (possibly toxic) algae. For the algae, today's tendency is to cover the lake at least partially with floating PV panels. A reservoir could also be below ground. To eliminate the water completely, bags with gravel and cranes are considered and also railway on hills. Some people think in underwater gravity storage, as no tower is needed. Try to put this in a search engine: gravity energy storage. I realized one thing: there is no energy storage without danger. Hydraulic power plants may show good statistics as they often do not include the dam breaking disasters in China in the past, so the statistics look better.

### 2.7.6 Compressed air

Especially if it is done in cavities under the ground, large energies can be stored. The compressors and turbines have a lower efficiency than batteries, but the main problem resides in the fact that air heats a lot when compressed and cools down when expanded. In the ideal *isothermal* compression and expansion (at constant temperature), without friction, without leaks, 100% energy is recovered, but this means a big heat transfer during both processes. In the other extreme, *adiabatic* compression and expansion (without heat transfer to and from the air) and also a high efficiency are obtained; however, the air should then be stored at high temperature, which is not realistic in large volumes underground. Several solutions can be done, but at the increase of cost.

- Storing the compression heat to warm the air before expansion.
- Using the generated heat in a process.
- Using the cold air from expansion to cool some products, food, or goods or to make ice.
- Using solar heat to heat the air before expansion to improve the mechanical energy yield at expansion.

One of the own experiences is that air from 20 bar cools enough in a good piston expander to freeze normal oil. The formation of water during cooling down after compression and the formation of ice during expansion are all things that may influence the system. In warm countries, it may be a solution to create a reserve for electricity and create at the same time some cold for air con in the evening. Air is compressed and stored underground where it cools partly down. During the expansion, mechanical energy is generated for light, computer, refrigerator, and other appliances. The cold air from the expansion is used for cooling. For an adiabatic process $PV^{\gamma} = K1 =$ constant where $\gamma = C_p/C_v$, for air it is 1.4.

However, $PV = nRT$. So $V^{(\gamma - 1)}/T = K2$
Air has a $C_p/C_v = \gamma = 1.4$
$V^{0.4}/T = K2$.

If we take a factor of 2 in volume, the absolute temperature changes with a factor of 1.32, which is a lot. So, if from 1→2 at 290 K (17 °C), we end 92.8 K higher at 110 °C; if we start from 17 °C for expansion, we end 70 K lower or −53 °C, a normal oil could freeze at that temperature. A volume reduction of a factor 2 corresponds with a pressure rise of $2^{1.4} = 2.64$ bar. Figure 2.9 is about to scale. The shaded area is the reduction in recovered energy due to temperature change. So, quite high-temperature changes occur for rather low volume or pressure ratios. A reduction of the absolute temperature reduces the possible energy recuperation proportionally. One way to reduce the effect is to split the compression into a number of steps with intercooling, and also heat exchanges at expansion. Too much temperature differences in the heat exchanger reduce the efficiency. Figure 2.10 shows such an effect.

The figure shows that a higher compression and expansion can be made without too much energy loss, but at the expense of an intercooler in compression and interheater in the expansion. The dashed line in figure shows the energy loss that would occur if the same volume ratio was reached in one stage. The large temperature change effect is interesting for cryogenic purposes. However, for normal refrigeration, cycles using butane or propane result in a much smaller

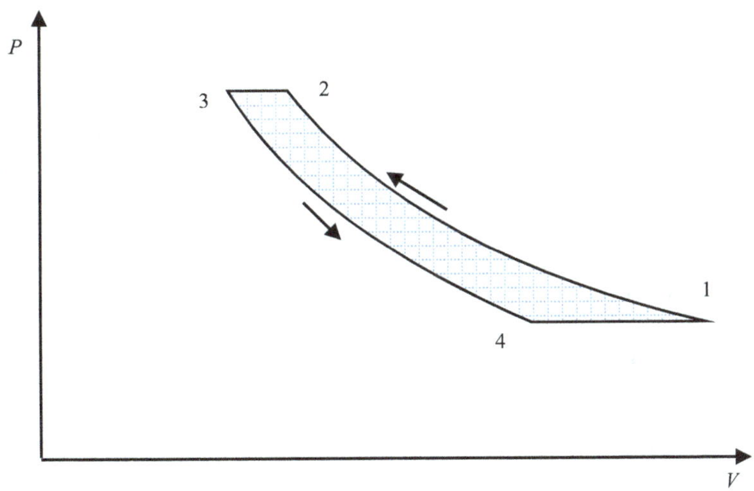


*Figure 2.9 Adiabatic process*

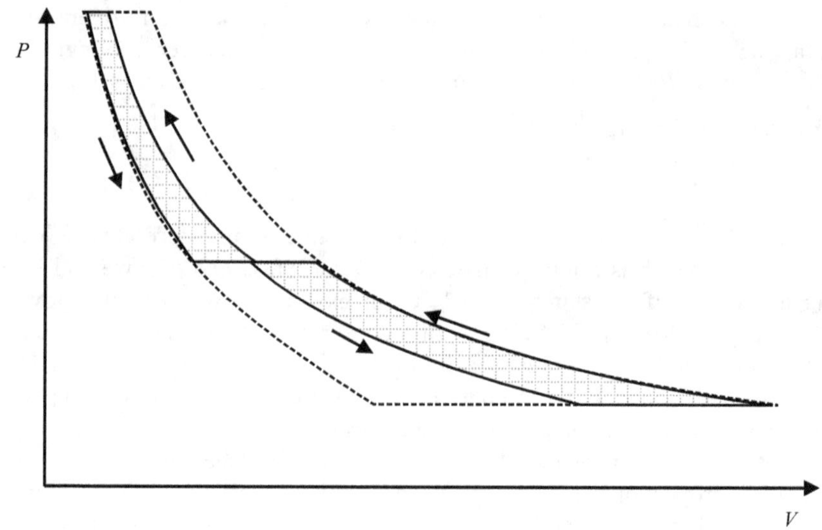


*Figure 2.10 Two-stage adiabatic compression with intercooling and interheating*

temperature difference, and another advantage is that boiling and condensation have a superior heat exchange for given temperature difference and surface.

### 2.7.7 Seasonal storage

In the northern latitudes, there is a lot of PV electricity in summer and little in winter. In Belgium, 52°N, the light hours in summer are two times more than in winter whereas the PV electricity in December–January is five times less than in June–July. Moreover, any energy could be used in winter for room heating. Wind energy is more available in winter, but it will be difficult to install, for example, three times more wind turbines than today. In solar, a potential of a factor of 10 is still realistic (not only roofs to south but also east and west). So the 'Holy Grail' would be seasonal summer–winter storage. If a battery equipment has a lifetime of 20 years, the manufacturing $CO_2$ counts each year for 5%. Also the cost/kWh would be prohibitive. In fact, battery systems fail to propose a solution for seasonal storage. Flow batteries could be the closest to it, as the materials could have a longer life than the cells. Also hydrogen storage is expensive for seasonal storage. However, for heating, large, well-insulated water tanks could do the job, in the order of 100 m$^3$/house [28]. A part of the winter energy need could be filled in by biomass (wood logs, chips and pellets, and other energy crops). But this is often in concurrency with food production. In fact, PV panels produce in the order of 30 times more energy/m$^2$ than biomass (some 1.3 kg/year/m$^2$ dry biomass, 6.5 kWh heat, compared to 200 kWh/m$^2$/year PV panel electricity some 1000 kWh/kWp, and 5 m$^2$ panel for 1 kW). A possible way is producing energy fuels as ammonia $NH_3$ or methanol MeOH. A whole economy is emerging around it: the power to ammonia and power to methanol. The round trip efficiency to make electricity from electricity is around 25%, but also heat could be harvested in CHP-combined heat power and a heat pump to generate more heat than electricity. If it is well done, more heat in kWh can be produced than the kWh PV in summer where it started from. Another solution is that southern countries may have excess PV in winter as less air con is needed and could sell electricity to north, or to avoid excessive high voltage line costs to produce hydrogen and transport it by pipeline, sacrificing some efficiency to reduce transport costs and have some short-term storage.

## Summary

The discussion about batteries fits in the more general debate about energy storage. In principle, batteries could solve the energy storage, but they have their own problem of materials, production energy, recycling, and waste management. However, gradually the battery technology still improves. If the final purpose is to store for heat or cold, maybe other alternatives could be considered. This might be more cost-effective than an extensive use of batteries, for electric water heaters, freezers, and air conditioners. The recycling of Li types is more involved than, for example, lead–acid, or other stationary battery solutions. The transport of materials for batteries is not negligible but not the main aspect. However, for mobile applications, carrying its own battery is big part of the energy consumption of the vehicle. For handheld equipment, reducing weight is a part of the comfort of the user, so lithium gets a big part of the market. Gradually the process of Li-ion batteries gets more energy lean. A big part of the material is not the active product, so still some improvement may be expected. Also sodium, potassium, magnesium, and aluminium may be competitors if they show better intercalation efficiencies. One way to reduce the energy and $CO_2$ effectiveness of battery storage is to prolong its life. This is a part of the duty of the battery management system. But on the other hand, the life expectancy of the battery management itself should be considered. If life expectancies of 15 years are projected, the BMS should also at least reach it. For very short periods, double-layer capacitors could be considered, but also fast-charging battery types emerge, the best known is LTO. There they are winning in concurrency with flywheel storage. For seasonal storage, batteries are not a solution as the number of cycles is too low compared to the energy to produce the battery. Energy carriers as ammonia or methanol could be more suited for that. In between, we have underground or underwater compressed air storage. Even if efficiencies are lower, there is a market for excess solar energy in summer to store it for winter. Anyhow, still a diversity of engineering will be needed for the challenges of the next decades.

## References

[1] Garside M. Lead reserves worldwide as of 2019, by country [database on the Internet]. USA: www.statista.com. 2019. Available from: https://www.statista.com/statistics/273652/global-lead-reserves-by-selected-countries.

[2] Ethanol and octane for beginners [database on the Internet]. USA: www.mnbiofuels.org. 2016. Available from: https://mnbiofuels.org/media-mba/blog/item/1511-octane-and-ethanol-for-beginners.

[3] Happich J. MIT spinoff to release high-capacity lithium metal battery [database on the Internet]. Belgium: www.smart2zero.com. 2016. Available from: https://www.smart2zero.com/news/mit-spinoff-release-high-capacity-lithium-metal-battery.

[4] Ounissi T, Dammak L, Larchet C, *et at.* Novel lithium selective composite membranes: synthesis, characterization and validation tests in dialysis. Journal of Materials Science. 2020;55(34):16111–16128.

[5] Els F. Electric vehicle demand will double nickel price as soon as 2022 [database on the Internet]. USA: www.mining.com. 2018. Available from: http://www.mining.com/electric-vehicle-demand-will-double-nickel-price-soon-2022.

[6] Capacitor plague [database on the Internet]. USA: https://nl.wikipedia.org. Available from: https://en.wikipedia.org/wiki/Capacitor_plague.

[7] Van den Bossche A, Haddad S, and Mordjaoui M. Power electronic converters without electrolytic capacitors. International Journal of Renewable Energy Technology. 2019;10(1–2):83–92.

[8] Barai A, Uddin K, Widanage W, *et at.* A study of the influence of measurement timescale on internal resistance characterisation methodologies for lithium-ion cells. Scientific Reports. 2018;8(1):1–13.

[9] Recylex, European expert in lead recycling [database on the Internet]. France: www.recylex.eu. 2018. Available from: https://recylex.eu/en/lead.

[10] When does a car battery need to be replaced? [database on the Internet]. www.testingautos.com. 2020. Available from: https://www.testingautos.com/car_care/when-replace-car-battery.html.

[11] Lithium-ion battery recycling market [database on the Internet]. www.marketsandmarkets.com. 2020. Available from: https://www.marketsandmarkets.com/Market-Reports/lithium-ion-battery-recycling-market-153488928.html.

[12] Torell W. Lifecycle carbon footprint analysis of batteries vs. flywheels. Schneider Electr–Data Cent Sci Cent. 2015.

[13] Metal ore mining in Europe [database on the Internet]. Germany: www.at-minerals.com. 2018. Available from: https://www.at-minerals.com/en/artikel/at_Metal_ore_mining_in_Europe_3257608.html.

[14] Balbis Morejón M, Cabello Eras JJ, Sagastume Gutierrez A, *et at.* Factors affecting the electricity consumption and productivity of the lead acid battery formation process. The case of a battery plant in Colombia. International Journal of Energy Economics and Policy. 2019;9. https://doi.org/10.32479/ijeep.8021.

[15] Spanos C, Turney DE, and Fthenakis V. Life-cycle analysis of flow-assisted nickel zinc-, manganese dioxide-, and valve-regulated lead-acid batteries

designed for demand-charge reduction. Renewable and Sustainable Energy Reviews. 2015;43:478–494.

[16] Melin HE and Circular Energy Storage. Analysis of the climate impact of lithium-ion batteries and how to measure it [database on the Internet]. www.transportenvironment.org. 2019. Available from: https://www.transportenvironment.org/sites/te/files/publications/2019_11_Analysis_CO2_footprint_lithium-ion_batteries.pdfl.

[17] Hoekstra A and Steinbuch M. Comparing the lifetime green house gas emissions of electric cars with the emissions of cars using gasoline or diesel. Eindhoven University of Technology. 2020.

[18] Van den Bossche A. How to reduce the energy needs of electrical and conventional vehicles. In: Workshop on Electric Mobility (WELMO-2015). www.welmo.org. 2015. p. 1–32.

[19] Formula E car [database on the Internet]. USA: https://en.wikipedia.org. 2019. Available from: https://en.wikipedia.org/wiki/Formula_E_car.

[20] Edwards R, Rejtharova J, Padella M, and O'Connell A. Draft methodology for calculation of GHG emission avoidance first call for proposals under the Innovation Fund [database on the Internet]. https://ec.europa.eu. 2020. Available from: https://ec.europa.eu/clima/sites/clima/files/innovation-fund/20200205_ghg_en.pdf.

[21] Storage highlights countdown 6; load shifting fridges ZAE Bayern [database on the Internet]. www.pv-magazine.com. 2019. Available from: https://www.pv-magazine.com/2019/02/20/storage-highlights-countdown-6-load-shifting-fridges-zae-bayern.

[22] Freezing-point depression [database on the Internet]. USA: https://en.wikipedia.org. 2019. Available from: https://en.wikipedia.org/wiki/Freezing-point_depression.

[23] Rasta I, Wardana I, Hamidi N, *et at.* The role of Soya oil ester in water-based PCM for low temperature cool energy storage. Journal of Thermodynamics. 2016;2016.

[24] Abduljalil A, Mat S, Sopian K, *et at.* Review of thermal energy storage for air conditioning systems. Renewable and Sustainable Energy Reviews. 2012;16:5802–5819.

[25] Van den Bossche A. Direct use of PV panels [database on the Internet]. www.researchgate.net. 2019. Available from: https://www.researchgate.net/publication/331907430_Direct_use_of_PV_panels.

[26] Gyrobus [database on the Internet]. USA: https://nl.wikipedia.org. Available from: https://nl.wikipedia.org/wiki/Gyrobus.

[27] vor dem Esche R. Safety of flywheel storage systems [database on the Internet]. www.researchgate.net. 2016. Available from: https://www.researchgate.net/publication/309394878_Safety_of_Flywheel_Storage_Systems.

[28] Van den Bossche A. Safety of flywheel storage systems [database on the Internet]. www.researchgate.net. 2019. Available from: https://www.researchgate.net/publication/336315152_Is_the_2050_target_realistic_or_is_it_dreaming.

# Chapter 3
# Overview of battery types

## 3.1 Introduction

Since the invention of the older battery types, during one century mainly improvements happened to existing types. However, in the last decades we can say that some revolution occurred. This is mainly as types were not only developed with non-water-based electrolytes, but also with the right use of membranes to separate the anode and cathode chemistry. This not only allowed increasing the cell voltage but also the use of lighter ions as active material. In the meantime, another thing happened that the ion-based batteries did not show any increased leakage current at high voltage but instead deteriorated gradually or had safety problems. This initiated the need for battery management systems (BMSs) that had to monitor each cell to keep it between limits in voltage. Further, new materials were developed and gradually become affordable, such as ceramic materials, carbon fibre, graphene, silicon, a choice of organic electrolytes, and also more (power-) electronic control. The result is a technology that reaches much more energy and power density than before. Research also improved the understanding of ageing, so that the number of cycles increased. All these things opened new application areas; first in smaller appliances such as mobile phones, then computer, tools, all kinds of vehicles, grid storage.

## 3.2 Overview of actual types

The graph shown in Figure 3.1 gives an overview of actual types [1]. Note that the limit of the theoretical energy density for Li-ion is not reached yet. At high power, Li types can compete with super capacitors (double-layer capacitors): high power density Li-ion types are used in electric model planes and drones, where 100 C discharge is allowed (= a discharge in 36 s) up to 130 C, 139 g 850 mAh at about 2.5 V $0.85\times130/0.139\times2.5 = 2$ kW/kg. If it is discharged slowly: 3.8 V 0.85 Ah/ 0.139 kg = 22.6 Wh/kg, which is low, but the battery was optimized for high power. Another source for a similar graph [2] is shown in Figure 3.2. Note that the continuous power density of the classic induction motors is about 0.1 kW/kg and 0.25 kW/kg peak, respectively. Optimized electric new-generation outer rotor PM motors are 2 kW/kg continuous and 5 kW/kg peak. A comparison of batteries and fuel storage [3] is shown in Figure 3.3. Note that the efficiency of batteries is much

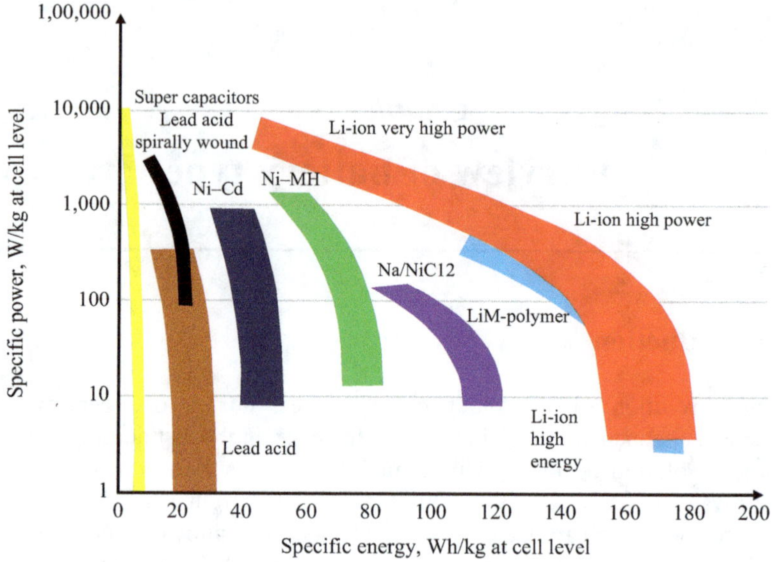


*Figure 3.1 Energy density against power density*

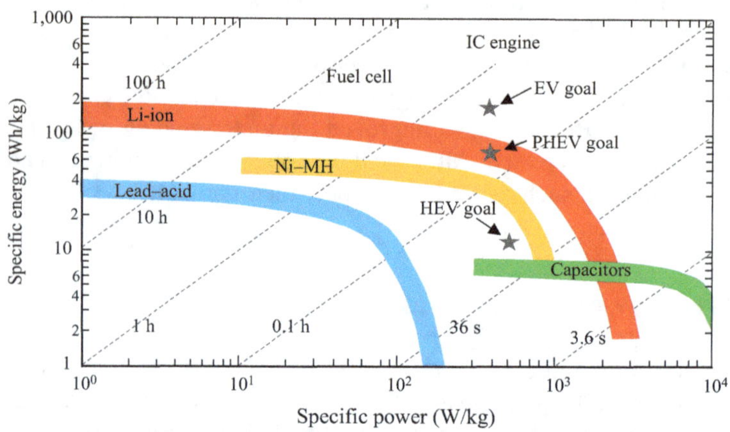


*Figure 3.2 Energy density against power density 'different viewpoint'*

higher than fuel cells, and fuel cells are significantly higher than internal combustion engines.

Note that with batteries much less weight and cost of auxiliary circuitry is needed compared to fuel cells, which might compensate for the lower specific energy. Note also that, due to the internal and external connections and the housing,

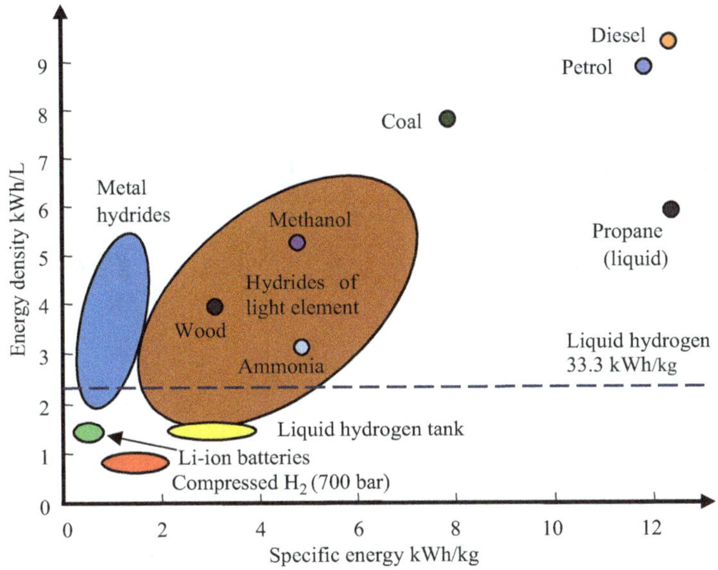


*Figure 3.3 Energy density against specific energy*

the weight/cell typically increases the weight between 10% and 100% when used in a battery unit. A 'beautiful extruded aluminium housing' might be a significant part of the weight, but it might be a containment if it is in fire, or the battery explodes. The performance of a battery in energy or power to weight is not the same as for an individual cell.

All batteries need in some way a protection and/or a BMS, but some are much easier than other ones, for example if the internal battery leakage current increases with voltage, the cell voltages tend to equalize in a natural way. Not only short circuits in batteries can be dangerous, but also batteries can explode (even lead–acid ones). As terminal wires may be integrated into the same connector, or pass through the same hole, a short circuit is already possible at that level. A way is putting one or more fuses between parts of the battery, not only outside the terminals.

## 3.3 Batteries with water-based electrolytes

Acid or base batteries were the first discovered types. The lead–acid batteries are not only the most commonly known ones, but NiMH, NiFe, NiZn, and rechargeable alkaline batteries do exist as well. The voltage is limited by the potential of oxygen and hydrogen. This potential can be influenced by a base, acid or salty liquid.

### *3.3.1 Lead–acid*

The positive electrode (anode) is lead oxide PbO, 'brown', and the negative is metallic lead: Pb, 'grey'. A number of partial reactions are possible, resulting in a trajectory in voltage. These partial reactions can be divided into negative plate, positive plate reaction, and finally total reaction. So, for the negative, positive, and total reactions, we have the following expression:

Negative plate reaction

$$Pb(s) + HSO_4^-(aq) \rightarrow PbSO_4(s) + H^+(aq) + 2e^-$$

The positive plate reaction

$$PbO_2(s) + HSO_4^-(aq) + 3H^+(aq) + 2e^- \rightarrow PbSO_4(s) + 2H_2O(l)$$

The total reaction can be written as

$$Pb(s) + PbO_2(s) + 2H_2SO_4(aq) \rightarrow 2PbSO_4(s) + 2H_2O \tag{1}$$

$$E^{O}_{cell} = 2.05\ V$$

During discharge, lead–sulfate is formed on both positive and negative plates. Water is bonded when the cell is more charged and the acid gets denser. It permits to check open cell types if they are charged. A diffusion of sulphuric acid is necessary for the operation, so liquid batteries can do this faster by convection than gel batteries. In other battery types, the electrolyte does not participate like this in the previous reaction. The nominal voltage is 2.0 V (= average discharge voltage), (12 V, for 6). At 1.8 V and lower, white deposits, lead sulfate crystals, appear that do not participate in the reaction, which is called hard sulfation, and it deform the plates as lead–sulfate takes more volume than Pb and PbO. It also makes dendrites at the cathode that can make short circuits between anode and cathode. The fully charged voltage (topping) is 2.30 V per element (13.8 V), and above that voltage, there is a tendency to corrode the positive electrode, which is not problematic for a short time but should not be prolonged, mainly at high temperature. Above 2.3 V, the battery has a tendency to generate $H_2$ and $O_2$ and hence loses water. An addition of calcium in the electrodes allows the recombination of gas to water and reduces water loss but increases its leakage current at higher voltage. The voltage of 2.5 V (or more) is only allowed as a peak voltage (15 V on a 12 V battery), and a lot of gassing occurs and electrodes corrode. A car battery is floated at 13.8 V. Considering 0.2 V drop in the cable, it also corresponds to 14 V at the generator. Appliances such as headlamps are rated at 13 V. Heavy loads such as glow plugs are rated at 10 V, illustrating a quite high voltage drop in 12 V systems. A charged cell in rest has 2.12 V (12.7 V). The reason for the low nominal battery voltage of 12 V was that filament bulbs under vibration need a thick filament, which is not needed anymore with today's LED lamps. Also arcing of relays beyond 30 V is no reason as MOSFETs (or solid-state relays) are cheaper now than relays.

There is a natural equalizing between the cells: the cell with the highest voltage has a higher leakage current [4]. A normal 'float voltage' for lead–acid is about 2.25 V/cell, so for longer time. The temperature coefficient: −2 to −3.0 mV/K. Another way is a feedforward of the estimated leakage current, in trickle charge a current of about 0.2 mA/Ah (for a battery of 50 Ah, it is 10 mA). It corresponds to a discharge in 5000 h. For batteries used in cars, the 'parasitic drain' can be 50 mA, together with the auto-discharge a normal battery a battery is discharged between 1 and 2 months [5]. Lead–acid batteries are still used a lot:

- In conventional cars: a battery life of 5–10 years if everything goes normal; they are made for 200–500 full cycles, but the performance is rather the starting current.
- In stationary systems: up to 20 years if well maintained and not often cycled (Planté battery, as invented in 1859)
- In forklifts: cycled two times/day at about 40%, two years, means about 1200 partial cycles and then reconditioned by reflex charging and a few slow charging cycles again for one year.
- Although lead is toxic, and its use is accepted as >95% can be recycled [5].

**Advice:**

- Do not store a battery on a concrete floor: the acid evaporates internally and falls down in lower concentration, floats at the surface, it discharges the battery.
- A trickle charger rated at 0.2 mA/Ah compensates for self-discharge, 5000 h self-discharge, more current results in water loss (most of trickle chargers deliver too much current).
- Older engines need less crank torque and may use a smaller rated battery.
- Use thick oil (such as SAE80W90) or technical Vaseline on the battery contacts, silicone oil on small contacts.
- Use safety glasses and possibly a bottle of water in the neighbourhood to rinse eyes, do not use steel brushes to clean, it can project dust with acid.

### *3.3.2 Nickel–metal hydride NiMH*

NiMH replaces the NiCd types of previous times, and NiCd are not allowed in consumer applications in Europe. There is a worldwide pressure to ban those batteries due to the high toxicity of cadmium where a fraction could penetrate in nature by the waste cycle. They do not satisfy the RoHS 'Regulation on Hazardous Substances'. They have to be removed, so that the ashes of household waste can be disposed as non-chemical waste. NiMH has similar characteristics as NiCd. The electrolyte is KOH in water. In consumer applications, NiCd and NiMH are rated at 1.20 V/cell; industrial, aviation, and military batteries adhere to the original 1.25 V. There is no difference between the 1.20 V and 1.25 V cells, it just depends how deep you define the discharge. The NiMH battery acts like a 'zener diode' at 1.50 V. An internal recombination occurs, causing a high leakage current that warms the battery and if the current is too high, it sets the cell under pressure. At normal temperature, a trickle charge current up to 0.1 C can be allowed (room temperature) and around 0 °C, only 0.05 C. If more current is given at full charge, gas is generated; the cells get under pressure and may leak.

**Special remarks:**

- The self-discharge can be quite high, some only hold a week, other versions are sold as 'stay charged' that keep 1 year.

- At repetitive low current discharge, a 'memory effect' happens: the capacity gets lower after a number of cycles.
- At high current charging, 'wires' can be formed that may short circuit anode to cathode. With a high peak current, those links can be blown (it allows repairing with an electrolytic capacitor discharge).
- It is almost the only battery that accepts a complete discharge. However, a reverse voltage is not good (it can happen in series connections).
- The charge efficiency can be very low, as at 1.5 V, current can be injected for a long time without a charging effect.
- If the charge is continued at full charge, the voltage drops 3 mV, to the author's opinion an unreliable criterion to detect full charge.
- If the charge is continued at its full, the cells heat up; this might be detected and is reliable.
- At lower cell temperature, the capacity is largely reduced.

### *3.3.3 Salt water battery*

The designers claimed that they had induced the cell to form a Solid–Electrolyte Interphase, SEI, a novelty for an aqueous electrolyte [7]. The SEI allows the battery to operate at higher voltages and a slower self-discharge. The high-salt concentration allows the interphase to form. It raised the maximum voltage for such a battery from 1.23 V to around 3 V. At 2.4 V, the battery's specific energy was approximately 100 Wh/kg and it displayed consistent performance over 1000 charge/discharge cycles. A typical salt water battery [8] is shown in Figure 3.4.

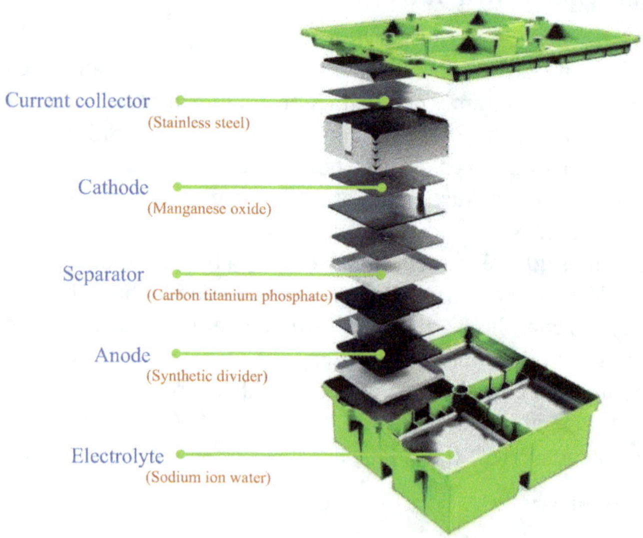


*Figure 3.4 Salt water battery*

### *3.3.4 Rechargeable alkaline battery*

Alkaline batteries use KOH–water electrolytes, a Zn cathode, and Zn/$MnO_2$ anode [6]. Low-ripple direct current is not suitable for charging disposable alkaline batteries; more suitable is a current pulsed at a rate of 40–200 pulses per second, with an 80% duty cycle. Pulsed charging appears to reduce the risk of electrolytes, usually potassium hydroxide (KOH) leakage. If they are discharged by less than 25%, they can be recharged for hundreds of cycles to about 1.42 V. If they are discharged by less than 50%, they can be almost fully recharged for a few dozen cycles, to about 1.32 V. After a deep discharge, they can be brought to their original high-capacity charge only after a few charge–discharge cycles. Without a charge monitoring system, it is difficult to operate them.

## 3.4 Organic electrolyte-based batteries

### *3.4.1 Lithium family*

Typical members of the 'lithium family' are lithium polymer (LiPo), lithium–iron–phosphate (LFP or $LiFePO_4$), lithium–titanate (LTO), and lithium–sulphur (Li–S). For light batteries, it is good to have electrode potentials that are far away in the potential table, also using a light element and also more than one electron exchanged. Lithium is in the top of the table but much more electropositive than hydrogen in the table of potentials; so it cannot use a water-based electrolyte, as it simply reacts with water. Early Li types used Li as a metal. Lithium is dangerous as it easily burns in contact with air and water and also melts at 180.50 °C, density 534 kg/$m^3$; so it floats in oil. For the anode, a porous carbon electrode is used. The cathode is a porous crystalline solid of cobalt, nickel, and other elements. Lithium salts are considered in practice, which are called 'Li-ion', and the metal state is avoided. The typical Li-ion uses cobalt in the anode which is also an expensive material. It has a high energy/weight ratio and is preferred for mobile phones, laptops, and tablets (and expensive electric cars). Lithium is more abundant than the actual exploited mines now. The lithium metal production costs about 5000–6000 $/t (production: Chile; Argentina; US, NV; Bolivia; China; Australia; reserves also in Algeria). The chemistry of the lithium ion is very complex. It concerns anode, cathode, separating membranes, and electrolytes. Electrolytes have to be chosen to have a good temperature range: the melting point (mp), boiling point (bp), flash point (fp), (FW) chemical formula weight. Table 3.1 shows properties of common solvents [9].

$LiCoO_2$ is most used in mobile phones and laptops, about 200 Wh/kg, about 1000 charges of 80% at 0.2 C. $LiFePO_4$ has a lower cost, a bit lower density and is more reliable: about 150 Wh/kg, 2000–5000 charges of 80%, also at low current. The peak power/kg depends on other parameters and is not so different. The internal leakage and the auto-discharge of Li-battery types are very low, so a battery can be stored during about 2 years if nothing is connected (also, no BMS !!). Unfortunately very little is known about the internal current drain of BMS systems, except if you test them.

*Table 3.1 The physicochemical properties of some common solvents*

| Solvent | FW | d, g/cm³ (25 °C) | $\varepsilon_r$ (25 °C) | $\eta$, mPa s (25 °C) | $E_{homo}$, eV | $E_{lomo}$, eV | mp, °C | bp, °C | fp, °C |
|---|---|---|---|---|---|---|---|---|---|
| Ethylene carbonate (EC) | 88 | 1.32 (40 °C) | 90 (40 °C) | 1.9 (40 °C) | −12.86 | 1.51 | 36 | 238 | 143 |
| Propylene carbonate (PC) | 102 | 1.2 | 65 | 2.5 | −12.72 | 1.52 | −49 | 242 | 138 |
| Dimethyl carbonate (DMC) | 90 | 1.06 | 3.1 | 0.59 | −12.85 | 1.88 | 5 | 90 | 17 |
| Ethyl methyl carbonate (EMC) | 104 | 1.01 | 3 | 0.65 | −12.71 | 1.91 | −53 | 108 | 23 |
| Diethyl carbonate (DEC) | 118 | 0.97 | 2.8 | 0.75 | −12.59 | 1.93 | −74 | 127 | 25 |

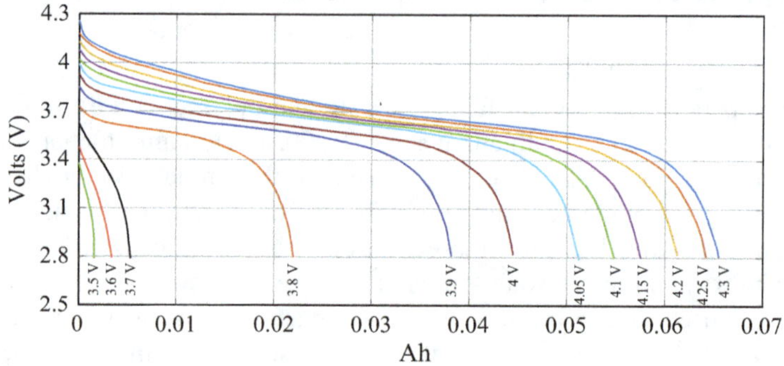


*Figure 3.5 Lithium-polymer battery discharge curves*

### *3.4.2 Li-ion battery characteristic*

The discharge curves for a 60 mAh lithium-polymer battery charged at voltage between 3.5 V and 4.3 V [10] are shown in Figure 3.5. The 1500 mAh pouch cells for mobile phones were first charged at a current of 1500 mA (1 C) to 4.20 V/cell and then allowed to saturate to 0.05 C (75 mA) as part of the full charge saturation. Eleven new Li-ion were tested [11] on a Cadex C7400 battery analyser shown in Figure 3.6.

### *3.4.3 Lithium–iron–phosphate characteristic $LiFePO_4$*

China is present: BYD is the biggest $LiFePO_4$ battery manufacturer in the world. Lithium–iron–phosphate battery refers to the use of lithium–iron–phosphate as the anode material of lithium-ion battery. The characteristic of $LiFePO_4$ in discharging process [10] is shown in Figure 3.7.

In Table 3.2, the cycle life as a function of depth of discharge is presented [11]. Partial cycling prolongs the battery life. A factor 2 in DoD change could result in about $2^{1.5} = 2.8$ times more cycles. It is a matter of disforming electrodes. Elevated

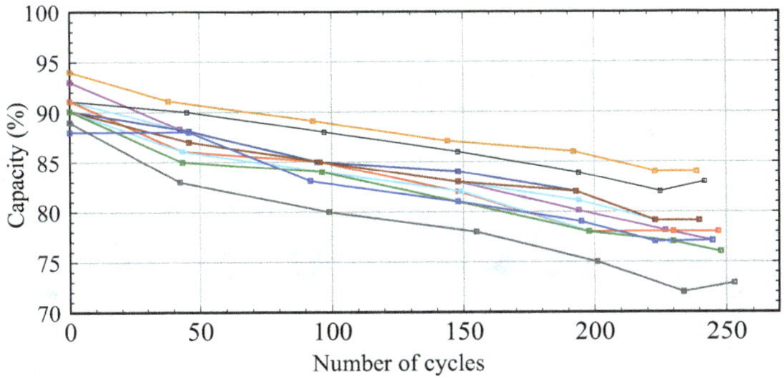


*Figure 3.6 Number of cycles against capacity for Li-ion '11 new batteries'*

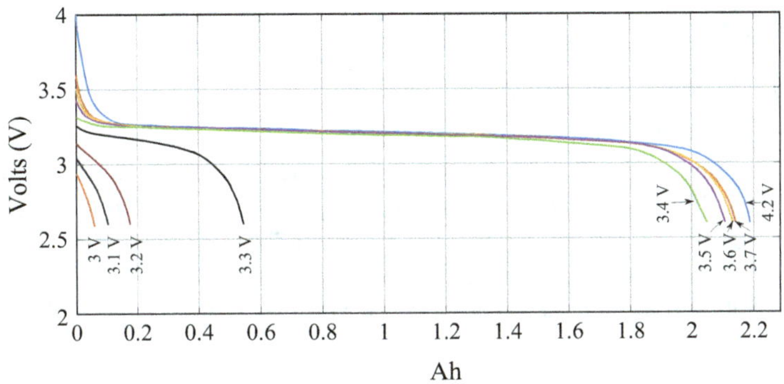


*Figure 3.7 $LiFePO_4$ battery discharge curves*

*Table 3.2 Cycle life as a function of depth of discharge*

| **Depth of discharge (% DoD)** | **Discharge cycles (NMC)** | **Discharge cycles ($LiFePO_4$s)** |
|---|---|---|
| 100 | ~300 | ~600 |
| 80 | ~400 | ~900 |
| 60 | ~600 | ~1500 |
| 40 | ~1000 | ~3000 |
| 20 | ~2000 | ~9000 |
| 10 | ~6000 | ~15 000 |

temperature and high currents also affect cycle life. It can be noted that 100% DoD is a full cycle. Often the batteries have also a 'calendar' warranty of 10 years, at 300 discharges/year of 40%; it corresponds to 3000, the value of the table. $LiFePO_4$ batteries contain neither nickel nor cobalt, both of which are supply-constrained and expensive, and they offer a longer cycle life than other lithium-ion approaches.

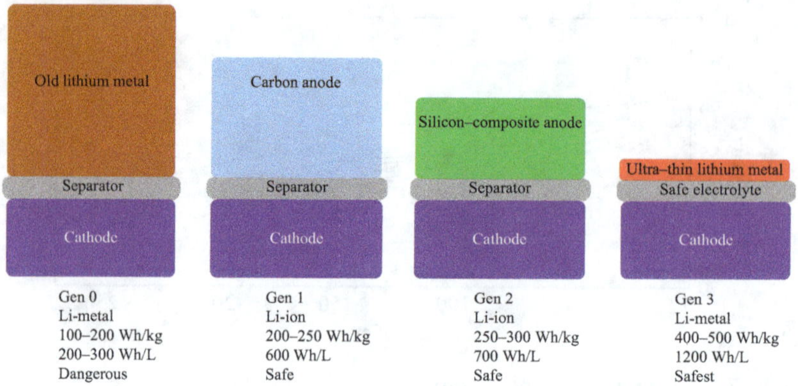


*Figure 3.8 Lithium metal against lithium ion*

### 3.4.4 Lithium metal batteries

Originally, the lithium batteries used lithium metal and this gave a bad name as they could be unstable and catch fire [12]. By reducing the amount of lithium metal to a very thin layer and using a solid electrolyte, the safety has been improved. A lithium metal has a density of only 535 kg/m$^3$ and floats in oil. Some evolution in the technology is observed, where 400 Wh/kg was claimed in [13]. An energy density of 500 Wh/kg is announced at the writing moment of the book (19 March 2021) [14] (Figure 3.8).

### 3.4.5 Sodium-, potassium-, magnesium-, calcium-, zinc-, and aluminium-based batteries

These types of chemistries are promising in cost/kWh stored but do often have less cycles. In sodium ion, chromium and vanadium doping are also considered [15]. Aluminium-ion batteries [16] are conceptually similar to lithium-ion batteries but possess an aluminium anode instead of a lithium anode. For the cathode side, still lithium is used. While the theoretical voltage for aluminium-ion batteries is lower than lithium-ion batteries, 2.65 V and 4 V, respectively, the theoretical energy density potential for aluminium-ion batteries is 1060 Wh/kg in comparison to lithium-ion's 406 Wh/kg limit.

The large difference in energy density potential is due to the fact that aluminium ions have three valence electrons, while lithium ions have only one.

Anode half reaction

$$Al + 7AlCl_4^- \rightleftharpoons 4Al_2Cl_7^- + 3e^-$$

Cathode half reaction

$$2MnO_2 + Li^+ + e^- \rightleftharpoons LiMn_2O_4$$

Combining the two half reactions yields

$$Al + 7AlCl_4^- + 6MnO_2 + 3Li^+ \rightleftharpoons 4Al_2Cl_7^- + 3LiMn_2O_4$$

As usual, in new introduced cells, some contradictory messages:

'Some report up to 10 000 charge–discharge cycles without capacity decay and can be charged within one minute.' 'Aluminium-ion batteries have a relatively short shelf-life. The combination of heat, rate of charge, and cycling can dramatically decrease energy capacity'. This seems to change, but lab experiments have yet not experienced large-quantity production.

According to [17], 'trihigh tricontinuous (3H3C) design' is made to achieve the ideal graphene film (GF-HC) cathode with excellent electrochemical performances. Owing to this targeted 3H3C design, the resulting aluminium–graphene battery (Al-GB) achieves an ultralong cycle life (91.7% retention after 250 000 cycles), unprecedented high-rate capability (111 mAh/g at 400 A/g based on the cathode), a wide operation temperature range (−40 °C–120 °C), unique flexibility, and nonflammability.

We recognize polyimide (a high-temperature polymer) to seal. Note that the energy density is significantly lower if wiring and housing is considered when assembling cells to a 'battery', so improving that aspect may help as well. The battery has a normal cell voltage of 2.2 V/cell. According to [18]: *'Notably, we demonstrated AIBs with high energy and power densities ~200 Wh/kg at 200 W/kg, ~160 Wh/kg at 5000 W/kg that exhibit a slow decrease in energy at higher power densities'.*

In Figure 3.9(a), a schematic representation shows the components of an Al-ion battery (AIB). The upper and lower sheets are aluminized plastic. In Figure 3.9(b),

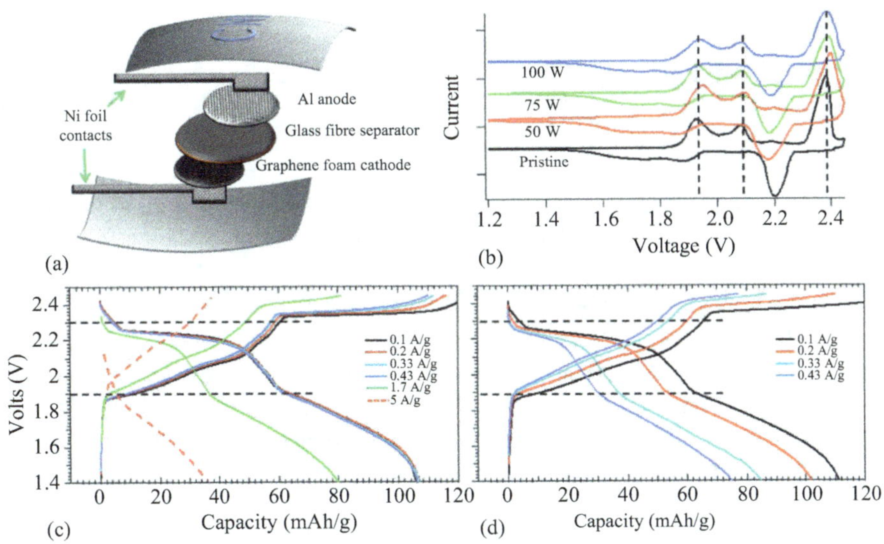


*Figure 3.9 (a) Schematic, (b) voltage vs current, (c) and (d) capacity vs voltage*

cyclic voltammograms of AIBs with non-treated and plasma-treated FLG foam cathodes are presented.

The dashed lines correspond to redox processes in the AIB. In Figure 3.9(c) and (d), charge/discharge characteristics for AIBs with a non-treated and a 100 W plasma-treated cathode are presented, respectively.

## 3.5 Molten salt batteries

These batteries are used at high temperature. It can take several days to freeze and about 12 h to heat from a cold state. Keeping the temperature at its level requires a continuous standby power. The batteries can heat themselves while operating – typically, sodium–sulphur (Na–S) at 300 °C–350 °C, 150 Wh/kg.

$$2Na + 4S \rightarrow Na_2S_4 \quad (E_{cell} \sim 2\ V)$$

Note that sodium metal melts at 97.8 °C. As of 2009, in a lower temperature, solid electrode version was under development in Utah by Ceramatec. They use a NASICON membrane to allow operation at 90 °C with all components remaining solid. Corrosion of the insulators was found to be a problem in the harsh chemical environment as they gradually became conductive and the self-discharge rate increased [19].

Na–$NiCl_2$: Freezes at 157 °C, 90 Wh/kg, 150 W/kg and is often sold under the name 'zebra' from 245 °C to typically 270 °C–350 °C. It is much lower than that of Li types [20]. It consists of Ni–glassy carbon, where nickel is used as cathode and a glassy carbon, anode (Sumitomo) at $>$ 61 °C, 290 Wh/L and 224 Wh/kg, respectively. Often molten salt batteries contain ceramic materials that may break as such or at connections, also a perfect sealing against leaks is not obvious.

## 3.6 Fuel cells

Fuel cells were discovered already in a very early stage of battery development. A lot of fuel cells use hydrogen as a fuel at the cathode (possibly after reforming from another fuel, could be a hydrocarbon, H3B or NH3), and oxygen or air at the anode. A high purity of hydrogen and air is needed, without poisoners such as CO.

### *3.6.1 Low-temperature fuel cells LTFC*

Almost all LTFC need a precious metal such as Pt or Ir. Although much less amount is needed today, it is a limitation for the development of fuel cells and the mines are not very ecologic and need a lot of primary energy. Most of low-temperature fuel cells are not compatible with carbon in the fuel and have a low efficiency: direct methanol fuel cell DMFC [21].

Common operating temperatures are in the range of 50 °C–120 °C, where high temperatures are usually pressurized. Alkali (KOH) fuel cell: efficiency is about 70%, and operating temperature is 150 °C–200 °C, the higher temperature avoids CO poisoning [22]. In Figure 3.10, different types of fuel cells are presented.

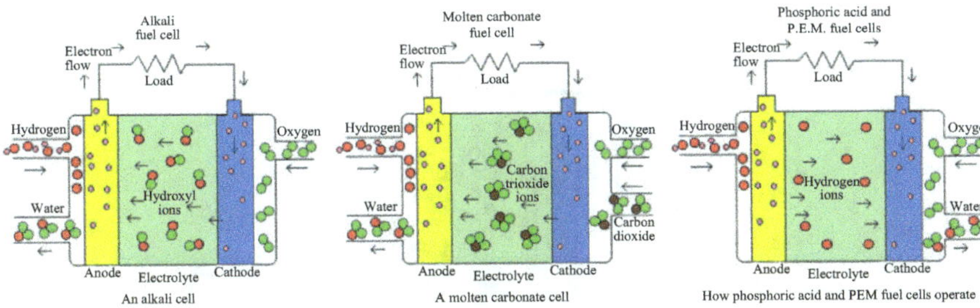


*Figure 3.10 Different types of fuel cells*

### *3.6.2 High-temperature fuel cells HTFC*

Solid oxide fuel cells (SOFC) use a hard, ceramic compound of metal (like calcium or zirconium) oxides (chemically, $O_2$) as electrolyte. The efficiency is about 60%, and operating temperatures are about 1000 °C (about 1800°F). The cells' output is up to 100 kW. At such high temperatures, a reformer is not required to first extract hydrogen from the fuel, and waste heat can be recycled to make additional electricity.

However, the high temperature limits applications of SOFC units and they tend to be rather large. While solid electrolytes cannot leak, they can still crack. This tends to slow temperature gradients. It has the advantage not to need precious metals to operate, and Ni is sufficient as a catalyst. They can operate with $H_2$ as well as with direct hydrocarbon such as $CH_4$ and a mix of them [23]. The thermal energy loss can be used in a bottom cycle, yielding a total efficiency from fuel to electricity of about 70%. Also pure hydrogen may be used in that technology, but then there are alternatives in that case.

## 3.7 Flow battery systems

If the energy is not stored in the electrodes but in a liquid, in principle a higher energy/weight can be obtained. Also refuelling can be faster, or for long-time storage if the catho- and anolyte are cheap products. In Figure 3.11, a flow battery system is presented.

**Vanadium redox flow types:** This type is presented in Figure 3.12. According to [24], 'A few years ago, the cost of a 4 h VRFB system was about $800 per kilowatt-hour (kWhb). These days, it is about half of that, comparable to the cost of a stationary lithium-ion system. But that is not an apples-to-apples comparison. As mentioned earlier, similar to other solid-state batteries, lithium-ion's capacity degrades over time, and its life span becomes shorter. Individually tested VRFBs show more than 14 000 cycles, fully charging and discharging each cycle, and they still perform at 100% capacity. This should translate into a life span of 20 years or more'. At a vanadium price of 20 euro/kg and the lithium-based metal of 10 euro/kg, the advantage is not so clear, but less quantity is needed. Table 3.3 summarizes the permissible charge and discharge temperatures of common rechargeable batteries.

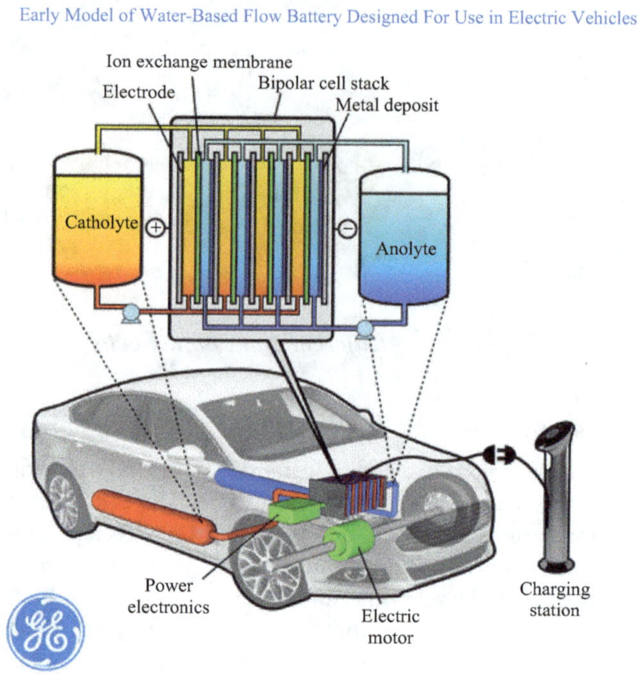


*Figure 3.11 Flow battery system*

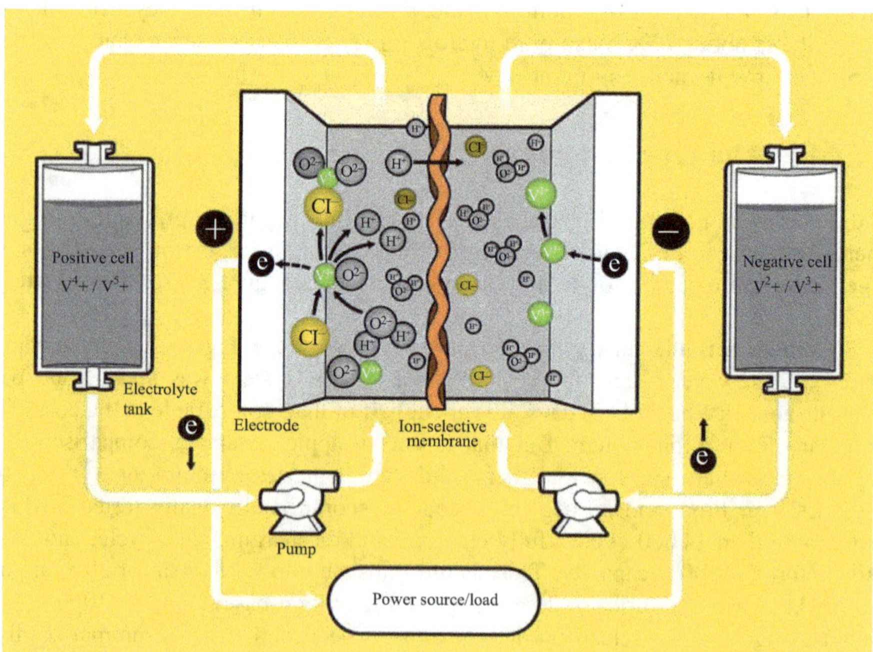


*Figure 3.12 Vanadium redox flow types*

*Table 3.3 Charge and discharge temperatures*

| Battery type | Charge temperature | Discharge temperature | Charge advisory |
|---|---|---|---|
| Lead–acid | −20 °C–50 °C (−4°F–122°F) | −20 °C–50 °C (−4°F–122°F) | Charge at 0.3 C or less below freezing<br>Lower V-threshold by 3 mV/°C when hot |
| NiCd, NiMH | 0 °C–45 °C (32°F–113°F) | −20 °C–65 °C (−4°F–149°F) | Charge at 0.1 C between −18 °C and 0 °C<br>Charge at 0.3 C between 0 °C and 5 °C<br>Charge acceptance at 45 °C is 70%<br>Charge acceptance at 60 °C is 45% |
| Li-ion | 0 °C–45 °C (32°F–113°F) | −20 °C–60 °C (−4°F–140°F) | No charge permitted below freezing<br>Good charge/discharge performance at higher temperature but shorter life |

**An all-iron redox flow battery** based on $Fe_{2+}$ and $Fe_{3+}$ would solve the problem of electrolyte cost and toxicity [25].

Positive electrode: $2FeCl_3 + 2e^- \rightarrow 2FeCl_2 + 2Cl^-$ $E^o = 0.77$ V
Negative electrode: $Fe + 2Cl^- \rightarrow FeCl_2 + 2e^-$ $E^o = -0.44$ V
Overall cell reaction: $Fe + 2FeCl_3 \rightarrow 3FeCl_2$ $E^o_{cell} = 1.21$ V

The cell voltage of 1.21 V is rather low, but the reactants are cheap and available. Today's problem is still the possible formation of hydrogen at the negative pole during charging. The coulombic and voltage efficiency are still not optimal but may be better than hydrogen, or ammonia cycles. About 2 kg of $CO_2$ is emitted for 1 kg of iron production, where the energy is about at 0.17 kWh/kg, but a part is water, and some magnitude of 10 kg $CO_2$/kWh storage could be expected. It is still not perfect for a seasonal storage, except if the materials could be made with excess renewable energy. Also chemical processes might be refined to start directly from iron ore and chloric acid instead of starting from metallic iron [26], which can be directly from ore, or as a by-product of titanium mineral ores [27]. The result is that the all-iron redox flow battery could have one of the lowest carbon footprint. As less materials are involved, recycling could be easier as well [28].

Drawbacks are that energy density is low and that extreme temperatures are not permitted, so it is mainly good for stationary batteries in temperate regions. Antifreeze additives may increase the viscosity of the fluids.

Reactants could also be transported, to get more cycles than once/year, also ships could use it. The development of the more expensive vanadium–iron redox flow battery VRFB for marine applications has started already [29]. It confirms that engineering can reduce the carbon footprint of batteries.

Note that also non-flow iron-air rechargeable batteries exist, the cost of materials is low, the efficiency is rather low and an average cell voltage of about 1.2V, but could be sufficient for weekly storage.

**All-manganese redox flow battery**

In fact, elements with a lot of valence possibilities can deliver potential differences on themselves. This is also true for manganese. It is less expensive than vanadium.

The cell voltage seems to be 2.59 V [30,31]. The metal is not expensive, some 2000–3000 $/tonne. The conclusion about flow batteries is that they are not the lightest solution, but that cost-effective longer term storage solutions might emerge from that corner, where the life is long and recycling easy.

## 3.8 Nernst and electrochemical potential

In chemical cells, the choice of the ions determines the possible potential difference that can be reached. Fundamentally at each electrode redox, reactions occur: so oxidizing on one side and reduction on the other side. To obtain a higher voltage, it helps to choose more extreme elements that are more reducing (alkali metals) and more oxidizing (electronegative gases and rather noble metals). However, they have to be stable as well in contact with the solution, so that zinc or aluminium can be unstable in an aqueous acid electrolyte and gradually dissolve. The case of the primary zinc–carbon cell has been improved using an alkaline pH. Zinc dissolves less between a pH of 7 and 11. Even zinc–air *rechargeable* batteries appeared [32]. Aqueous solutions have the limitation that hydrogen or oxygen gas can be formed (or recombined, giving leakage current).

So progress has been made while using non-aqueous electrolytes (solvents) or solid electrolytes. Also the concentration of the considered solution changes the potential according to the Nernst law [33]. The Nernst equation can be used to calculate the potential of an ion of charge $z$ across a membrane. This potential is determined using the concentration of the ion both inside and outside the cell: when the membrane is in thermodynamic equilibrium (i.e. no net flux of ions), and if the cell is permeable to only one ion, the membrane potential must be equal to the Nernst potential for that ion. Without current through a membrane, and if a permeable membrane only exchanges one type of ions, the potential over that membrane is given is the Nernst potential for that type of ions. It means that different concentrations of the same material, results in a potential over the membrane. Usually the reference is given /mole solution.

$$E = \frac{\mathrm{R}\,T}{z\,\mathrm{F}} \ln \frac{C_{\mathrm{out}}}{C_{\mathrm{in}}} = \frac{T}{z} 86.173 \times 10^{-6} \ln \frac{C_{\mathrm{out}}}{C_{\mathrm{in}}} = \frac{T}{z} 198.421 \times 10^{-6} \log \frac{C_{\mathrm{out}}}{C_{\mathrm{in}}} \tag{3.1}$$

where $E$ is the Nernst voltage drop, [V]; R is the universal gas constant (R = 8.31446 J $\mathrm{K}^{-1}$ $\mathrm{mol}^{-1}$); F is the Faraday constant, the number of coulombs per mole (F = 96485 C $\mathrm{mol}^{-1}$); C is the Coulomb [A s], universal unit; log(10) = 2.3026; $T$ is the absolute temperature in Kelvin; $z$ is the number of electrons transferred in the considered reaction, or half-reaction; $C_{\mathrm{out}}$ is the ion concentration of outer part (far from electrode, or other side of membrane); and $C_{\mathrm{in}}$ is the ion concentration of inner part (at electrode or considered side of membrane).

For a concentration ratio of a factor 10 and one ion/reaction, at 300 K, (26.85 °C) it results in about 59.5 mV drop. So, even the same solution, but in a different concentration, gives a potential across a membrane. It can happen also after a current passed so that the concentration is not the same, which causes a part of the

*Table 3.4 Reaction between elements and the electrochemical potential*

| Element | Reaction between | Electrons | Potential | Ref 1 | Ref 2 | Ref 3 | Ref 4 | Ref 5 | Ref 6 |
|---|---|---|---|---|---|---|---|---|---|
| Lithium | Li $Li^{+}$ | +e | −3.045 | −3.045 | 3.040 | 3.050 | | | 3.02 |
| Potassium | K $K^{+}$ | +e | −3.924 | −2.925 | 2.924 | 2.925 | | | 2.93 |
| Calcium | Ca $Ca^{2+}$ | +2e | −3.866 | −2.866 | 2.870 | 2.870 | | | 2.87 |
| Sodium | Na $Na^{+}$ | +e | −2.714 | −2.714 | 2.710 | 2.714 | 2.712 | | 2.71 |
| Magnesium | Mg Mg | +2e | −2.34 | −2.37 | | 2.370 | 2.34 | | 2.340 |
| Beryllium | Be $Be^{2+}$ | +2e | −1.70 | −1.70 | | | | | |
| Aluminium | Al $Al^{3+}$ | +3e | −1.67 | −1.66 | 1.660 | 1.660 | 1.67 | | 1.67 |
| Manganese | Mn $Mn^{2+}$ | +2e | −1.04 | −1.05 | | | 1.05 | | −1.03 |
| Zinc | Zn $Zn^{2+}$ | +2e | −0.763 | −0.763 | 0.763 | 0.763 | 0.762 | | 0.76 |
| Chromium | Cr $Cr^{3+}$ | +3e | −0.73 | −0.740 | | | 0.710 | | |
| Iron | $Fe^{3+}$ $Fe^{2+}$ | +e | −0.56 | | | | | 0.56 | |
| Sulphur | $S^{2-}$ S | +2e | −0.48 | | | | | 0.480 | |
| Iron | Fe $Fe^{2+}$ | +2e | −0.440 | −0.440 | 0.441 | 0.440 | 0.440 | | |
| Cadmium | Cd $Cd^{2+}$ | +2e | −0.402 | −0.402 | −0.403 | 0.40 | 0.402 | | |
| Cobalt | Co $Co^{2+}$ | +2e | −0.277 | −0.277 | | | 0.277 | | |
| Nickel | Ni $Ni^{2+}$ | +2e | −0.24 | −0.250 | 0.236 | 0.250 | | | 0.23 |
| Tin | Sn $Sn^{2+}$ | +2e | −0.136 | −0.136 | 0.140 | 0.136 | 0.136 | | 0.14 |
| Lead | Pb $Pb^{2+}$ | +2e | −0.126 | −0.126 | 0.126 | 0.126 | 0.126 | | 0.13 |
| Hydrogen | $H^{+}$ H | +e | 0 | 0 | | | | | |
| Copper | Cu $Cu^{2+}$ | +2e | 0.337 | 0.337 | 0.337 | 0.153 | 0.345 | | 0.34 |
| Oxygen | $OH^{-}$ $O_2$ | +e | 0.40 | | | | | 0.40 | |
| Iodine | $I^{-}$ $I_2$ | +e | 0.535 | 0.535 | | | | 0.54 | |
| Mercury | Hg $Hg_2{}^{2+}$ | +2e | 0.790 | 0.790 | | 0.771 | | | |
| Silver | Ag $Ag^{+}$ | +e | 0.80 | 0.799 | | 0.789 | 0.8 | | 0.8 |
| Mercury | Hg $Hg^{2+}$ | +2e | 0.854 | 0.854 | 0.920 | | | | |
| Bromine | Br $Br_2{}^{-}$ \| Pt | +2e | 1.08 | 1.08 | | | | | 1.09 |
| Platinum | Pt $Pt^{2+}$ | +2e | 1.188 | 1.20 | | 1.188 | 1.20 | | |
| Chlorine | $Cl^{-}$ $Cl_2$ \| Pt | +e | 1.36 | 1.36 | | | | 1.36 | |
| Gold | Au $Au^{3+}$ | +3e | 1.50 | 1.50 | | 1.50 | 1.42 | | |
| Manganese | $Mn^{2+}$ $Mn^{3+}$ | +1e | 1.51 | | | | | 1.51 | |
| Cerium | $Ce^{3+}$ $Ce^{4+}$ \| Pt | +e | 1.61 | 1.61 | | | | | |
| Gold | Au Au+ | +e | 1.68 | 1.68 | | | | | |
| Fluorine | $F^{-}$ F2 \| Pt | +e | 2.87 | 2.87 | | | | | |

dynamics of a cell. Note that the ln() takes operates on the reaction quotient, so can contain more ratios if more reactants are involved. A series of electrochemical potentials are given for different metals but are in practice influenced by concentration and drop by current density, see Table 3.4. The most electropositive materials that want to set electrons free are at the top of the list (reducing agents), the bottom of the list are like to capture electrons (oxidizing agents). The reference of the potential is traditionally given by hydrogen, which gets the potential '0 V'. Non-conductive elements need an electrode to measure the potential, usually platinum. The more left in the table of Mendeleev and lighter the more the exchange is at a negative potential (alkali metals), right in the table exchanges at a more positive potential: electronegative gases and more noble metals. Metals give electrons when they dissolve, electronegative gases absorb electrons when they dissolve. Beryllium and cadmium are highly toxic so do not appear in normal designs, also mercury has been almost eliminated. Data are given at 25 °C (77°F).

## Summary

The chapter has been structured depending on the electrolyte more than the active materials: water, organic, molten salt, fuel cells, low and high temperature, and flow batteries. Anyhow it is difficult to be exhaustive, as every week incremental or bigger steps in progress are discovered. The fact that a lot of battery types exist is a hint to say that the ideal battery does not exist (yet). In the water-based types, the lead–acid batteries have still some advantages, having a high cell voltage for water-based batteries, and the best developed recycling. The experience of other batteries improved also the lead–acid. Some niche markets remain for NiMH as that can be fully discharged without damage. There might be a future for the salt battery; at 100 Wh/kg, it is near to the $LiFePO_4$ types. A major breakthrough in energy density has been made with the Li-ion types. Today, the highest energy density combined with a good number of cycles is obtained by the LiPo, but it contains a lot of cobalt, which is detrimental, mainly used in IT. The Li–Mn–Co or LMC is a better compromise, so less Co is mainly used in automotive. A high number of cycles are obtained by the $LiFePO_4$ battery. The energy density is a bit lower but has a high efficiency and a low risk of catching fire. The Li–Ti–O, LTO, is known for a high number of cycles and a good power density. Some alternative ion types that use sodium, potassium, magnesium, or aluminium are emerging as well. Some show a higher energy density, lower cost, or higher number of cycles, but all those good properties should be combined in one battery. Molten salt types have the advantage of using cheaper materials but suffer from a high standby loss due to heat losses. Ceramic materials can be brittle as well. Fuel cells are recharged by new gaseous or liquid fuel. If the fuel and the storage of that fuel has a low cost, a seasonal storage could be imagined, as the storage material has not to be intercalated in the electrodes. More possibilities seem to emerge in flow batteries, which could be used in stationary batteries at the user or at the utility grid level. The future not only seems bright for batteries, but also for the research for improving them.

## References


[1] "Energy Efficiency & Renewable Energy." Vehicle Technologies Office. Washington, DC. Available from: https://www1.eere.energy.gov/vehiclesandfuels/images/facts/fotw607.jpg (accessed August 2020).

[2] dothemath.ucsd.edu. Available from: https://dothemath.ucsd.edu/wp-content/uploads/2012/08/Ragone.png (accessed August 2012).

[3] Sartbaeva A, Kuznetsov V, Wells S, *et al.*, "Hydrogen nexus in a sustainable energy future." *Energy & Environmental Science*. 2008;1(1):79–85.

[4] "Confusion with voltages." Vancouver, Canada: batteryuniversity.com. Available from: https://batteryuniversity.com/learn/article/confusion_with_voltages (accessed May 2013).

[5] "Charging at high and low temperatures." Vancouver, Canada: batteryuniversity.com. Available from: https://batteryuniversity.com/index.php/learn/article/charging_at_high_and_low_temperatures (accessed Sep 2013).

[6] "Rechargeable alkaline battery." USA: wikipedia.org. Available from: https://en.wikipedia.org/wiki/Rechargeable_alkaline_battery. (accessed June 2021)

[7] "Salt water battery." USA: wikipedia.org. Available from: https://en.wikipedia.org/wiki/Salt_water_battery (accessed June 2021).

[8] "Jay Whitacre takes the Lemelson-MIT prize for his saltwater battery." USA: inhabitat.com. Available from: https://inhabitat.com/jay-whitacre-takes-the-lemelson-mit-prize-for-his-saltwater-battery (accessed Sep 2015).

[9] Li Q, Chen J, Fan L, *et al.* "Progress in electrolytes for rechargeable Li-based batteries and beyond." *Green Energy & Environment*. 2016;1(1):18–42.

[10] "How does capacity correlate with charge voltage for lithium iron phosphate batteries?" USA: powerstream.com. Available from: https://www.powerstream.com/lithium-phosphate-charge-voltage.htm (accessed Aug 2021).

[11] "How to prolong lithium-based batteries." Vancouver, Canada: batteryuniversity.com. Available from: https://batteryuniversity.com/index.php/learn/article/how_to_prolong_lithium_based_batteries (accessed Jul 2019).

[12] "Doubling battery power of consumer electronics." USA: news.mit.edu. Available from: https://news.mit.edu/2016/lithium-metal-batteries-double-power-consumer-electronics-0817 (accessed Aug 2016).

[13] "SolidEnergy systems promising to double battery life of lithium-ion batteries with new lithium-metal technology." USA: cleantechnica.com. Available from: https://cleantechnica.com/2016/08/23/solidenergy-systems-promising-double-battery-life-lithium-ion-batteries-new_lithium-metal-technology (accessed Aug 2016).

[14] "Battery revolutions are predicted weekly, but this one might be real." USA: www.designnews.com. Available from: https://www.designnews.com/electronics-test/battery-revolutions-are-predicted-weekly-one-might-be-real (accessed Mar 2018).

[15] "Sodium-ion battery." USA: wikipedia.org. Available from: https://en.wikipedia.org/wiki/Sodium-ion_battery (accessed Jul 2021).

[16] "Aluminium-ion battery." USA: wikipedia.org. Available from: https://en.wikipedia.org/wiki/Aluminium-ion_battery (accessed Jul 2021).
[17] "Zhejiang University. Scientists develop ultrafast battery with quarter-million cycle life." USA: Phys.org. Available from: https://phys.org/news/2018-01-scientists-ultrafast-battery-quarter-million-life.html (accessed Jan 2018).
[18] Childress AS, Parajuli P, Zhu J, *et al.*, "A Raman spectroscopic study of graphene cathodes in high-performance aluminum ion batteries." *Nano Energy*. 2017;39:69–76.
[19] "Sodium–sulphur battery." USA: wikipedia.org. Available from: https://en.wikipedia.org/wiki/Sodium-sulfur_battery (accessed Jul 2021).
[20] "Molten-salt battery." USA: wikipedia.org. Available from: https://en.wikipedia.org/wiki/Molten-salt_battery (accessed Jul 2021).
[21] "Direct methanol fuel cell." USA: wikipedia.org. Available from: https://en.wikipedia.org/wiki/Direct_methanol_fuel_cell (accessed Jul 2021).
[22] Garrison E. "Solid oxide fuel cells." USA: ots.iit.edu. Available from: http://mypages.iit.edu/~smart/garrear/fuelcells.htm (accessed 2015).
[23] Shi N, Xie Y, Yang Y, *et al.*, "Review of anodic reactions in hydrocarbon fueled solid oxide fuel cells and strategies to improve anode performance and stability." *Materials for Renewable and Sustainable Energy*. 2020;9(1):1–18.
[24] Yang ZG. "It's big and long-lived, and it won't catch fire: the vanadium redox-flow battery." USA: spectrum.ieee.org. Available from: https://spectrum.ieee.org/green-tech/fuel-cells/its-big-and-longlived-and-it-wont-catch-fire-the-vanadium-redoxflow-battery (accessed Oct 2017).
[25] "All-iron redox flow battery." USA: www.dornsife.usc.edu. Available from: https://dornsife.usc.edu/labs/narayan/all-iron-redox-flow-battery (accessed Jul 2021).
[26] "Iron(III) chloride." USA: www.wikipedia.org. Available from: https://en.wikipedia.org/wiki/Iron(III)_chloride (accessed Jul 2021).
[27] "Iron(II) chloride." USA: www.wikipedia.org. Available from: https://en.wikipedia.org/wiki/Iron(II)_chloride (accessed Jul 2021).
[28] Manohar AK, Kim KM, Plichta E, *et al.* "A high efficiency iron-chloride redox flow battery for large-scale energy storage." *Journal of The Electrochemical Society*. 2015;163(1):A5118.
[29] "Electrolyte and battery innovation." Canada: www.vanadiumcorp.com. Available from: https://www.vanadiumcorp.com/storage (accessed Jul 2021).
[30] Hutchins M, "Investigating all-manganese flow batteries." Available from: https://www.pv-magazine.com/2021/06/11/investigating-all-manganese-flow-batteries/.
[31] Schmucker M, Gully TA, Schmidt A, *et al.* "Investigations toward a Non-Aqueous Hybrid Redox-Flow Battery with a Manganese-Based Anolyte and Catholyte." *Advanced Energy Materials*. 2021;11:2101261.
[32] "New battery chemistry results in first rechargeable zinc-air battery." USA: www.arstechnica.com. Available from: https://arstechnica.com/science/2020/12/lose-the-alkaline-keep-the-zinc-for-a-new-battery-chemistry (accessed Dec 2020).
[33] "Nernst equation." USA: www.en.wikipedia.org. Available from: https://en.wikipedia.org/wiki/Nernst_equation (accessed Jul 2021).

# Chapter 4
# Purpose of BMS, state of charge, state of health

## 4.1 State of charge, state of health, and state of power

Batteries are widely used as the common electrical energy storage device in vehicles as a replacement of traditional fossil fuels [1]. The capacity of the battery is gradually reduced due to erosion, passivation, outgassing, decomposition of materials, and changes on the electrode surface during its operation [2]. Also, the battery's ageing process increases when the battery is subjected to operate beyond the specified safe operating conditions.

The aim is to guarantee the safe operation and a long life of a battery. It helps to know the battery state, which simply can be described by two parameters: State of Charge (SoC) and State of Health (SoH) [3]. The SoC mostly depends on the current from and towards the battery, and its initial charge condition at a given temperature.

The SoH is a measurement that indicates the ageing of a battery and its ability to deliver the specified performance compared to a fresh battery. Anyhow, any stored energy source has a hidden danger and requires some management to avoid an uncontrolled release of that energy. Therefore, series-connected lithium-ion batteries should be controlled and monitored by an effective Battery Management System (BMS).

Battery monitoring means considering the operational parameters during charging and discharging. The electrical key parameters are voltage, current, battery internal resistance, and temperature during cycling. The monitoring is addressed by an independent system known as the BMS. A BMS can protect the battery from damage, predict battery life, and maintain battery operation for offering high performance.

The main goal of a BMS is to minimize the imbalance within the pack to extend the life of the batteries and to obtain the highest performance. Reducing the wasted energy in a passive balancing method could have a significant effect on the energy throughput of the battery system.

### *4.1.1 State of charge*

As the battery SoC is an important parameter, which reflects the battery performance, so accurate estimation of SoC can not only protect the battery, prevent overcharge or discharge, and improve the battery life but also helps the application to make rationally control strategies to achieve the purpose of saving energy. The SoC estimation is a fundamental challenge for battery use. The SoC of a battery, which is used to describe its remaining capacity, is a very important parameter for a control strategy.

This condition is then used for calculating the terminal voltage of the battery, and the calculated terminal voltage is compared with the actual battery voltage. The error between the calculated and the real battery voltage is then used to calculate the filter gain in the Kalman filter to update the state SoC. Thus, the model is used to estimate the SoC of a battery. SoH is defined in terms of the percentage of nominal capacity. Age and charge–discharge cycles are two major factors that reduce the SoH of a battery. The SoH of an average Li-ion battery is reduced to 80% after 1000 charged–discharge cycles. The commonly used method is given by

$$\text{SoH} = \frac{\text{Nominal capacity} - \text{loss of capacity}}{\text{Nominal capacity}} \tag{4.1}$$

The battery model consists of a controlled voltage source in series with a resistor that represents the internal resistance of the cell. The various mathematical methods of estimation are classified according to methodology. The classification of these SoC estimation methods is different in the various literature [4] as following:

- Direct measurement: this method uses physical battery properties, such as the voltage and impedance of the battery.
- Coulomb counting: this method uses discharging current as of the input and integrates the discharging current over time to calculate the SoC.
- Adaptive systems: the adaptive systems are self-designing and can automatically adjust the SoC for different discharging conditions. Various new adaptive systems for SoC estimation have been developed.
- Hybrid methods: the hybrid models benefit from the advantages of each SoC estimation method and allow a globally optimal estimation performance. The literature shows that hybrid methods generally produce a better estimation of SoC, compared to individual methods. The drawback is that the methods are less transparent to debug and transferred to other battery types.

### *4.1.2 State of health*

The SoH is a 'measurement' that reflects the general condition of a battery and its ability to deliver the specified performance compared with a fresh battery. It takes into account such factors as charge acceptance, internal resistance, voltage, and self-discharge. It is a measure of the long-term capability of the battery and gives an 'indication' not an absolute measurement of how much of the available 'lifetime energy throughput' of the battery has been consumed, and how much is left.

Using the automotive analogy, it can be compared to the 'odometer' display function, which indicates the number of miles travelled since the vehicle was new.

During the lifetime of a battery, its performance or 'health' tends to deteriorate gradually due to irreversible physical and chemical changes, which occur during usage and with age until eventually, the battery is no longer usable or dead. The SoH is an indication of the point that has been reached in the life cycle of the battery and a measure of its condition relative to a fresh battery.

Battery manufacturers do not specify the SoH because they only supply new batteries. The SoH only applies to batteries after they have started their ageing process either on the shelf or once they have entered service. The SoH definitions are therefore specified by test equipment manufacturers or by the user [5].

- SoH for electric vehicle (EV) applications
  - For EV applications, the ability to achieve the range when called upon to do so is most important; hence, the SoH is based on a comparison of the total charge capacity compared to new.
- SoH for hybrid electric vehicle (HEV) applications
  - For HEV applications, the ability to deliver the specified power is important and so SoH can be based on a comparison of the DC resistance (or 1 kHz impedance) with DC resistance (or 1 kHz impedance) when new. But the fact that the SoH is used for estimating power means in fact that the State of Power can be considered a separated parameter as well.

### *4.1.3 State of power*

Depending on the application, the state of charge might be limiting or rather the state of power. The state of charge is rather the available energy at full charge compared to the reference (new) condition: it can be limited by the chemical ability of intercalation or less electrolyte.

In some cases, power is more important, for example, during acceleration or hill climbing, especially in hybrid applications. The SoP State of Power is the ability to deliver power compared to the reference (nominal) condition: the ability to deliver a given peak power during $Tp$ seconds. The state of power can be affected by several things, for example, a very cold battery will have a lower SoP. The SoP can be bad for a single cell. For example, it could have a partial cracked conductor inside and go in undervoltage during peak loads. If this is not recognized, it will chemically degrade as well. The state of power can be a limitation of the available power; users will not be surprised if the car accelerates less at −20 °C. The total mix of SoC, SoH, and SoP will determine the user 'satisfaction' with the performance of the battery.

## 4.2 Sophisticated SoH modelling by BMS

Most of BMS equalize the cell voltage close to full charge or close to empty. At those charge states, the voltage is a good indication of the charge, around 30%–70% charge, the cell voltage hardly changes, so equalizing is not very effective then. By a sophisticated SoH control and individual controllable bidirectional power supplies, more is possible. We take an example of only one cell that has an SoH of 80%, and all the other cells 100%. So that single cell would reduce the whole capacity of the battery by 20%. However, if the intelligent BMS knows it, it can discharge continuously the cell at a rate of 20% of the charging current and charge it at 20% of the average discharge current during discharge. If this is one of ten cells, it reduces the total capacity only by 2% (at 100% efficiency). However, this means that each cell would need a bidirectional converter and that the SoH is perfectly known. The idea seems nice but it relies highly on perfect monitoring and logging. It can be quite obvious in lab circumstances but less in a wide temperature range if the SoH changes due to a change in temperature. It has also an expense of a perfectly controllable bidirectional power supply for each cell. So, it is a matter of cost and reliability of power electronics compared to the cells. Although this type is possible, it will not be the aim of the book.

## 4.3 Battery management system

Batteries should be protected from many conditions, including thermal and mechanical stresses. In addition to that, large battery packs used for electrical vehicles require a complex control circuit to measure battery parameters and conditions. The control circuit ensures that the batteries operate within the desired conditions. According to the IEEE standard 1491, a BMS is a permanently installed system for measuring, storing, and reporting battery operating parameters. In other words, a BMS is an embedded system that monitors the cells in a series-connected string for a number of variables. An implementation of a more complex BMS requires more processing power and costs more than a simpler structure. So there is a cost associated with advanced BMSs; therefore, not all applications implement all features. The cost adds to the price of the battery pack as well.

These variables can include individual battery voltage, SoC, and total pack voltage. Usually the BMS has protection mechanisms built-in for over- and undervoltage situations, but it can also include temperature sensing and failure modes of batteries. The need for a BMS in lithium-based battery systems is very high due to inherent imbalance issues associated with the charging and discharging of the battery pack. Purposes of a BMS can be expressed as described in the following:

- To ensure the safety of the operator of the battery-powered system. The BMS must detect unsafe operating conditions and respond accordingly. This may be just warning the operator by some display or alert, or in more severe cases to disconnecting the battery pack from the load. Note that sometimes the mechanics or hardware do it when servicing as well. Or a disconnection at the rear right for a fireman.
- To protect cells of the battery pack from damage in the case of *failure*. This action may involve active intervention under software control or specialized electronics that can detect failures and isolate the failing components from the rest of the battery pack and from the load it powers.
- To prolong the life cycle of the battery under normal operating conditions. The BMS ensures that the battery pack will not be overcharged or overdischarged.

It also controls the thermal management system, ensuring that the battery pack is kept within its design operational temperature range.

- To maintain the battery pack in a state in which it will not become deeply discharged or highly charged as well. Note that this concerns not the whole pack but in fact each cell. One cell too low or one cell too high is a separate item in this chapter. The pack is as strong as its weakest cell.
- A special function might be needed when servicing: replacing one cell, which is much more or much less charged than the pack. Without taking care, it could damage the new cell by being overcharged or overdischarged even at the first use.

It can be noticed that without BMS, some dangerous conditions may occur. For example, overcharging or internal damage causes a swollen cell. Figure 4.1 shows a swollen cell due to too high temperature, overcharging, or internal damage.

Besides overcharging conditions, overheating is another issue that should be taken into account. Ideally, the battery pack should be monitored by thermal management circuit. Cells may catch fire by overheating or a short circuit. Figure 4.2 shows an exploded cell due to overheating.

To ease the purpose of BMS, the structure of BMS is shown in Figure 4.3. Some features of a BMS can be listed as

*Figure 4.1 A swollen cell*

*Figure 4.2 An exploded cell*

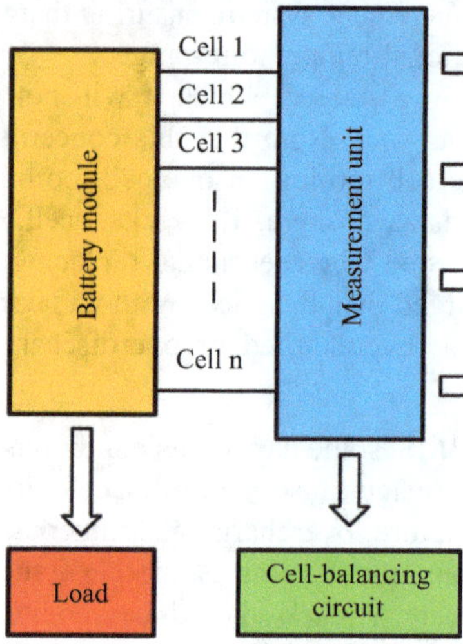


*Figure 4.3 Battery management system*

- to maintain the cell voltage at fully charged voltage,*
- to prolong the life cycle of the battery,
- to equalize battery cells in series in a battery module,
- to detect battery voltage,
- to maintain overvoltage protection,
- to maintain overcurrent protection, and
- to detect battery temperature.

## 4.4 Battery configuration

In a battery pack, the absorbed or delivered electrical power is computed as electric current multiplied by voltage, which can be calculated as $P = I\,V$. Hence, high-power battery packs must deliver either high current, high voltage, or both. To complete a design specification on maximum power, the decisions should be made for the voltage range and peak currents. The chemistry of an individual cell fixes its voltage range.

Therefore, to achieve high voltage, battery packs should be connected in series. So, the pack voltage is then the sum of the individual cell voltages. Therefore, $V_{pack} = NV_{cell}$, where $N$ corresponds to the number of cells in series. The cell structure puts limits on the maximum current that the cell is designed to provide. Hence, for high current packs, the cells should be wired in parallel. The pack

*It is a good idea to charge it fully to be able to equalize easily and then to discharge somewhat to extend the life of the cells.

current is the sum of the currents passing through the individual cells as $I_{pack} = M\ I_{cell}$. The number $M$ is the number of cells in parallel.

The maximum power that a cell can deliver is limited by the cell's internal resistance and it is not easy to quantify. Power is usually determined by enforcing minimum and maximum limits on cell terminal voltage. A lead–acid cell with a voltage of 2 V and 10 Ah has an energy storage capacity of roughly 20 Wh. It is essential to remark that energy and power are also depending on the rate of discharge. Power is the instantaneous rate at which energy is being released.

Power is measured in watts (W) or kilowatts (kW). In Figure 4.4, the battery configuration in parallel and series is shown. The battery with brown colour indicates the parallel connection. This configuration allows a higher current, but no changes in voltage, the individual cell voltage is the same. The battery with blue colour indicates the series connection. This configuration allows a higher voltage, but no changes in current, the individual cell current is the same. When batteries are operated in parallel, some designers parallel the respective cells and use only one BMS, accepting that it will be slower.

It is possible to operate two batteries in series or parallel using a single transistor. Diodes can avoid a high circulating current flow between the batteries. In Figure 4.5, a configuration is shown where two batteries can be but in series or in parallel. One can verify that when the MOSFET is on, both batteries are in series.

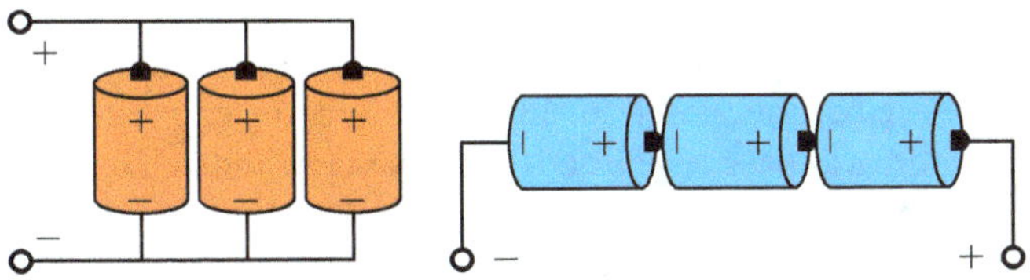

*Figure 4.4 Battery configuration (parallel and series)*

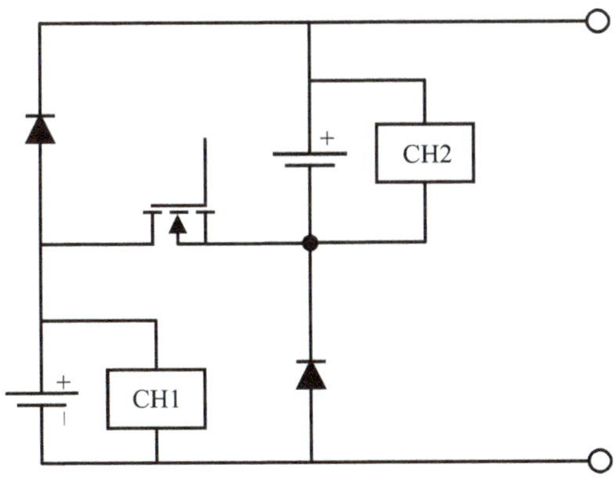


*Figure 4.5 Paralleling using a MOSFET and two diodes*

The system can be used to have lower losses at low speed. It can also be used if one battery is out of order or earlier discharged.

When the MOSFET is off, both batteries deliver a voltage to the output by means of a diode, they are in parallel for positive load currents. When the MOSFET is off, the voltage is still limited for negative load currents to the sum of both batteries, by the body diode of the MOSFET. Most convenient is having two chargers, CH1 and CH2. When the diodes are also replaced by MOSFETs, both current directions are possible also in the lower voltage. But a lot of circulating current can flow when both batteries have not the same emf.

It is advised that the MOSFET control has a desaturation protection set in the range of 1 V–2 V, the battery with the highest voltage will deliver the current as the MOSFETS have still their body diode. The desaturation can catch again when the voltages are equal. The advantage would be two levels of supply with only 3 MOSFETs.

For example, in parallel for driving in a city and in series for highway, having a lower voltage in the city would reduce the PWM losses in motor and transistor, and less ripple in the battery. If the load contains a DC-link capacitor, a means to avoid peak transient currents needs to be provided (to get to a higher voltage, it could use some regenerative braking moment to switch).

## 4.5 Types of battery management systems

We start with the task of balancing cells. There are many types of cell balancing methods, which have been proposed [6–13]. Therefore, the balancing method is divided into two categories as passive and active balancing methods. These two categories also have some subcategories [14,15]. Two types of BMSs are commonly found in commercial and industrial lithium-battery-based systems as shown in Figure 4.6.

Also these two main types are divided into some categories. Each type of system has different benefits that vary among cost, efficiency, and complexity.

1. **Passive balancing**
   i. Shunt resistor controlled balancing
      a. Advantages: easy to implement, low cost.
      b. Disadvantages: low efficiency, possible thermal runaway, might require a heat sink, no energy recovery.
2. **Active balancing**
   i. Capacitor-based balancing
   ii. Inductor-based balancing
   iii. Coupled-inductor-based balancing
   iv. Transformer-based balancing
      a. Advantages: high efficiency, fast balancing time, energy recovery.
      b. Disadvantages: more circuitry, higher cost.

The first type of BMS equalizing is a **passive balancing system**. It discharges too much charged cells by shunting resistors. Battery stacks that are being produced

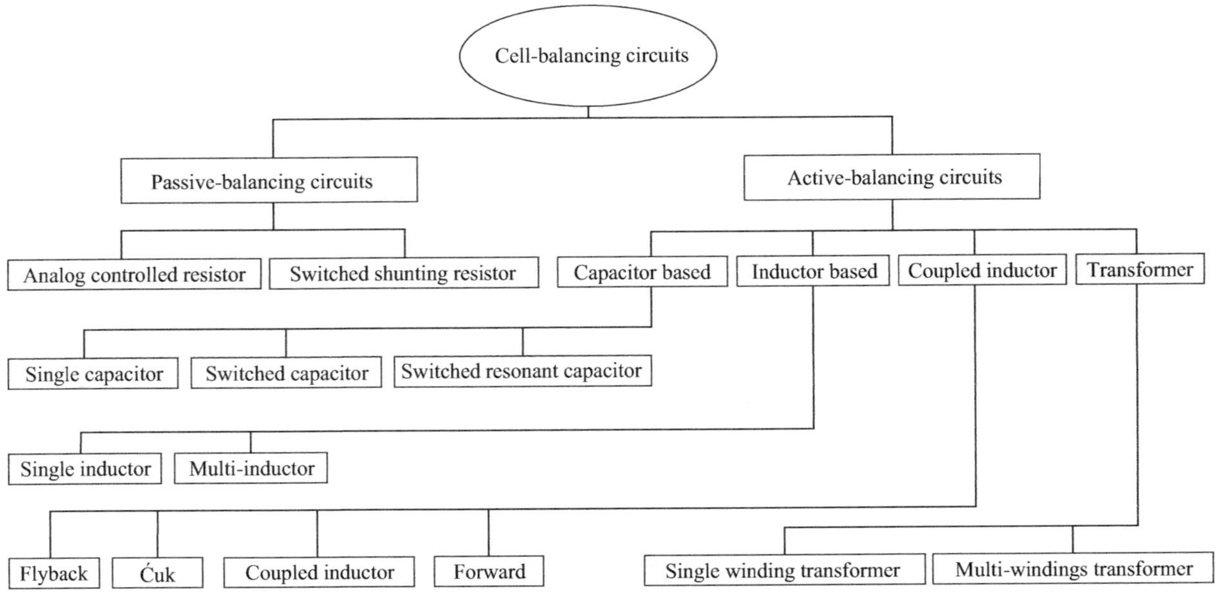


*Figure 4.6 Passive and active cell balancing topologies*

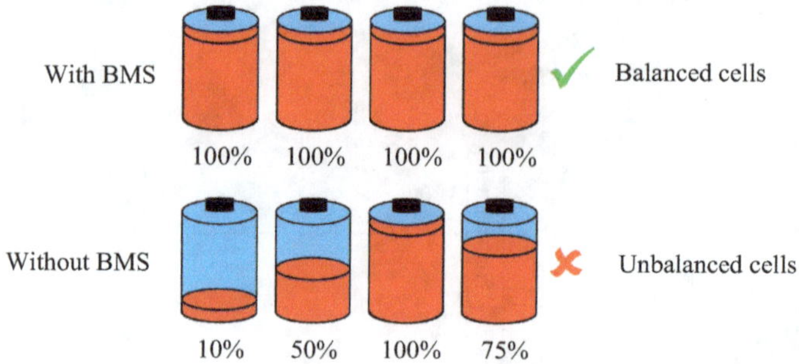


*Figure 4.7 Pack*

today largely use a passive cell balancing method. The passive cell balancing type has simple structure, easy to implement but it reduces the battery pack efficiency. The second type of BMS is the active balancing system. The active management system discharges too highly charged cells to charge less charged ones. It can be far more efficient in charging the cells while taking a shorter amount of time. As the efficiency is higher, it does heat less.

Besides, an **active cell balancing** is proposed to overcome the problem of energy losses in passive balancing, and the aim is to use an external circuit to remove energy among the cells to balance them. The active cell balancing is often used in Li-based batteries systems, as it helps keeping the temperature in a safe operating range.

The BMS has an important role in EV. Without it, the individual cell voltages and SoC will differ over time. Slight differences between the series-connected cell in a lithium-ion (Li-ion) battery pack can produce imbalances in charge, which greatly reduces the charge capacity. These kinds of batteries are not like a lead–acid battery, and it is not advised to trickle charge. This may cause overcharge of some cells and make them ignite. Lead–acid batteries can be brought into overcharge without damage (except loss of water). For Li types, it is important to monitor the battery by a BMS to ensure optimized performance and safety.

The BMS performs various duties such as measuring voltage, current, temperature, SoC, and SoH of the system. Unbalancing cells are important issues in the battery system life. Figure 4.7 shows unbalanced cell against balanced ones, when a BMS is implemented and without.

Without BMS, the voltage cell will drift over time, this results in decreasing the total capacity of the system. Cell unbalance may be caused by components such as manufacturing variance, internal impedance, or differences in self-discharge rate, or external sources such as multirank protection ICs.

### *4.5.1 One cell too low, one cell too high principle*

If one cell is too low, the discharging current should be limited. If one cell is too high, charging should be limited, for example up to the level, the balancing circuit can still manage. If it is not taken care of, that individual cell might be damaged, resulting in a reduction of the life of the whole battery. The number of cycles measured on separate cells is not directly valid for a pack. In practice, when the SoC of the cells are equal, it means that one avoids a cell that is 10% more charged or 10% more discharged and would reduce the charge capability of whole series-connected battery pack by 10%. Often a BMS has a limited current capability, which means that in the condition of 'one cell too high', the charging current should be limited to that level. Too simplified systems hope that once cells are balanced, they will remain balanced. However, if one cell is replaced, it could damage the whole unit, unless each cell is charged individually. If the BMS is consistent with the principle, one cell too high/one cell too low is used, a cell could be replaced and the pack changed without special manual charging precautions.

### *4.5.2 Shunt resistor switching balancing (passive)*

Shunt resistor switching methods are the most straightforward equalization concept. The shunting cell balancing method shunts the charging current around each cell as they become fully charged [6,16–18]. It removes excess energy from the higher voltage cells by bypassing the current of this cell and wait until it is at the same level as low-voltage cells.

The resistor is controlled by a transistor or switch, in a controlled manner as shown in Figure 4.8. When an overvoltage is detected by a sensor by measuring the voltage across cells, it decides which resistor should be shunted. The disadvantages of this method are the requirement for a large power dissipating resistor, high current switches, and thermal management requirements.

### *4.5.3 Capacitive shunting balancing (active)*

This method is called active cell balancing as it employs an active charge shunting element to remove energy from one cell to give it to another. It can use

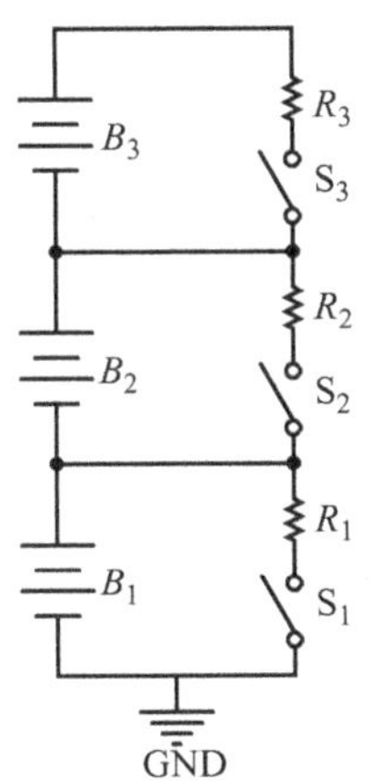


*Figure 4.8 Shunting switched resistor balancing circuit*

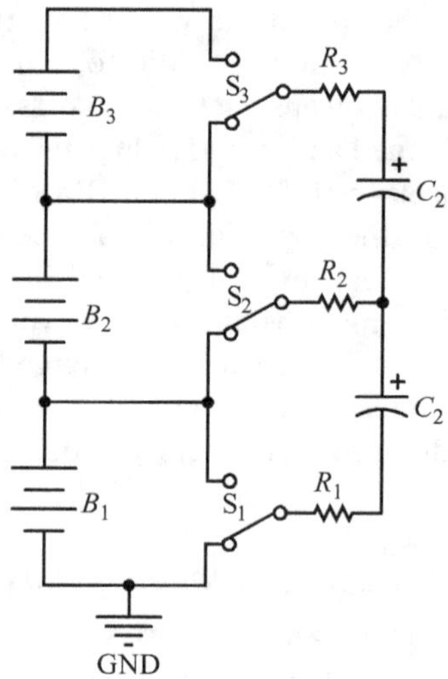


*Figure 4.9 Switched capacitor balancing circuit*

an analogue- or digital-controlled system. The capacitor balancing has a simpler structure but it has slower balancing time as compared to the inductor and transformer balancing types. The capacitor shunting can be brought into three categories such as switched capacitor, single switched capacitor, double-tiered switched capacitor [7,8,19–30].

The switched capacitor as shown in Figure 4.9 has two conditions: charging and discharging operation. A group of capacitors is used to shift charge along cells. A single switched capacitor needs only one capacitor to balance cells. The double-tiered switched capacitor balancing method is a derivation of the switched capacitor but it uses two capacitors for energy shuttling. The advantage is that the second capacitor reduces the balancing time. In general, two transistor voltage drops are used for charging and two for transfer, so, finally it has four drops, even if the voltage change of the capacitor is neglected. But depending on the topology, much more drops can occur. The switched resonant converter balancing [31] will improve balancing, as it reduces the effect of ripple voltage on the capacitor.

### *4.5.4 Inductor balancing (active)*

In this kind of method, an inductor is used to remove energy from one cell to another. Because of relatively high balancing current, they offer a smaller balancing executing time. It also has a disadvantage such as high cost and magnetic losses. Due to the high switching frequency, a filter capacitor is suited to filter the high-frequency components. To transfer energy, two elements are needed: one to charge the inductor and the other to discharge, so minimal two drops.

There are some types of inductor balancing system that are single- or multi-switched inductor balancing [9,32]. It measures the voltage of the cells and selects two cells that be used for energy transferring and it has fast equalization time. The multi-switched inductor can sense voltage difference of two adjacent cells, then with applying PWM in such a way that the higher cell must be switched at first. The disadvantage of this type is that it takes a long time to transfer energy and has long equalization time. The multi-switched inductor balancing circuit is shown Figure 4.10. However, the coupled inductor balancing improves balancing time and reduces the cost [33].

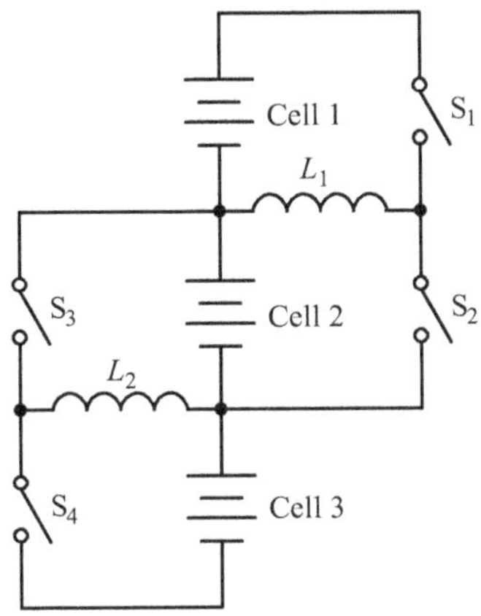


*Figure 4.10 Inductor balancing circuit*

## 4.5.5 Transformer balancing (active)

Another type is transformer balancing, which can be divided into two categories such as single winding transformer balancing and multi-secondary windings transformer balancing [10,34–47]. The first topology is based on carrying the energy from the whole battery through the switching transformer and transferring it to the weak cell shown in Figure 4.11.

The transformer output is rectified through a diode and delivered into a cell, which is determined by the setting of switches. The control system is required to select the cell and set switches. It allows a rapid balancing; however, it has some disadvantages which are high complexity, high parts count in terms of control and switches, and low efficiency due to magnetic losses and switching losses. The multi winding transformer cell balancing has a single magnetic core with one primary winding and multi secondary windings for each cell [48,49].

The principle is that the current from cell stack is switched into the transformer primary and induces a current in each of the secondary transformers. The secondary with the lowest emf and/or the least leakage inductance will have the most induced current. Figure 4.12 shows the structure of multi-winding secondary transformer cell balancing. The shared transformer can rapidly balance a multicell pack with minimal losses. The disadvantages are complex magnetics and a high parts count due to each secondary's rectifier and the control of it. This method could be suitable for EVs.

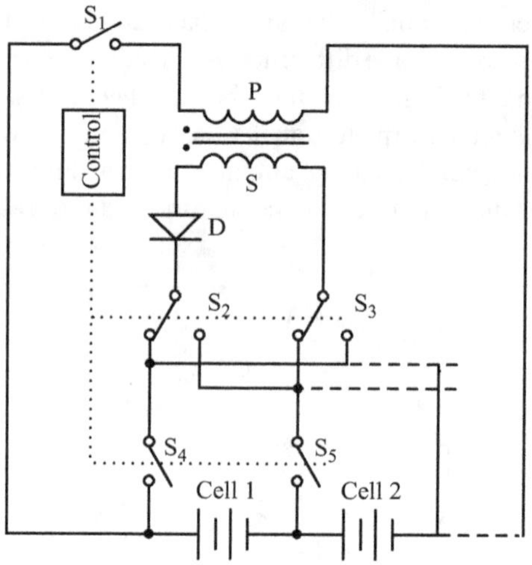


*Figure 4.11 Single winding transformer balancing circuit*

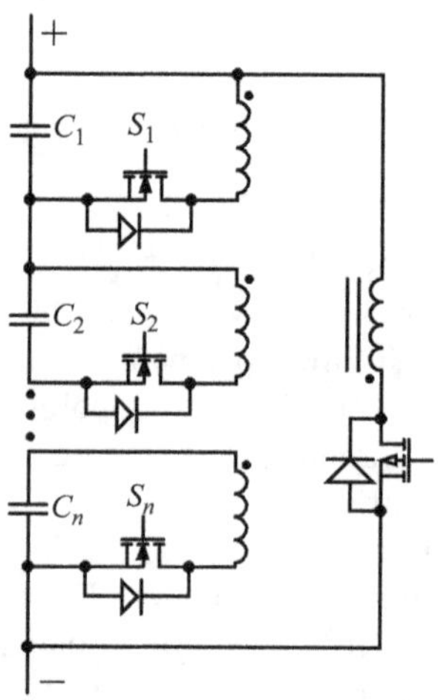


*Figure 4.12 Transformer-based balancing circuit (flyback structure)*

### 4.5.6 Ćuk converter balancing (active)

The Ćuk converter transfers energy between two adjacent cells; therefore, it takes a long equalization time, especially for long series batteries [11,50,51]. Buck and boost converters are more used in cell balancing. It can be divided into many categories such as buck dc converter, a boost converter, or a buck–boost converter. For the buck type, it transfers energy from battery to the cell with low energy and in boost type, it

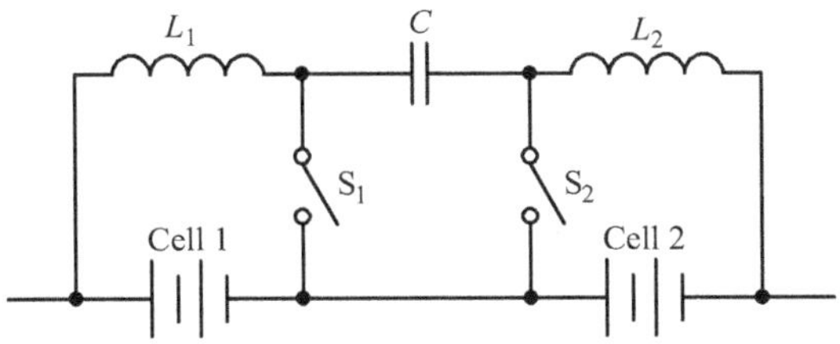


*Figure 4.13 Ćuk converter balancing circuit*

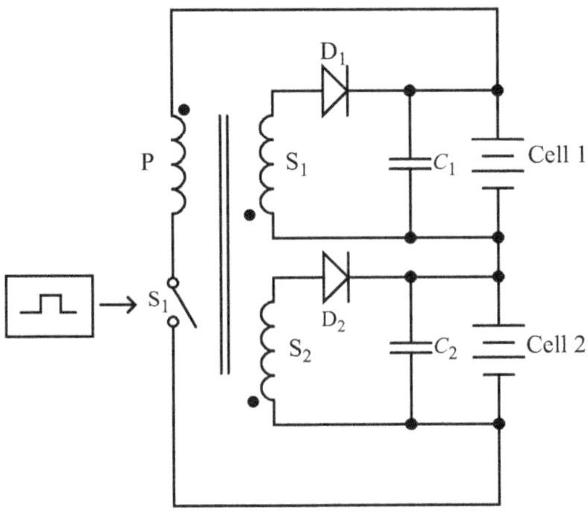


*Figure 4.14 Flyback balancing circuit*

removes energy from a single cell to the battery. The Ćuk converter has a bidirectional energy transferring capability. The traditional Ćuk converter balancing method using separate inductors requires $2N-2$ switches for $N$ cells. The coupled inductor type uses coupled inductors instead of un-coupled ones. The circuit in [52] has fewer components and losses. In [53], a Ćuk converter with coupled inductors is used, which improves balancing time. In Figure 4.13, a Ćuk converter balancing circuit is shown.

### 4.5.7 Flyback converter balancing (active)

The flyback converter balancing method has a fast balancing time compared to the capacitor-based balancing circuit. It has $N$ secondaries for $N$ cells. It only requires one switch and $N$ diodes at the output for each cell [12,54–56]. Therefore, the number of active switches is reduced. Flyback converters are used in an isolated structure and can be bidirectional. The flyback stores energy during the on-time of the transistor and transfers it in the remainder of the period to the cells that have a lower emf and a low inductive part. The core of the flyback has an air gap to store energy. A flyback converter balancing circuit is presented in Figure 4.14. The

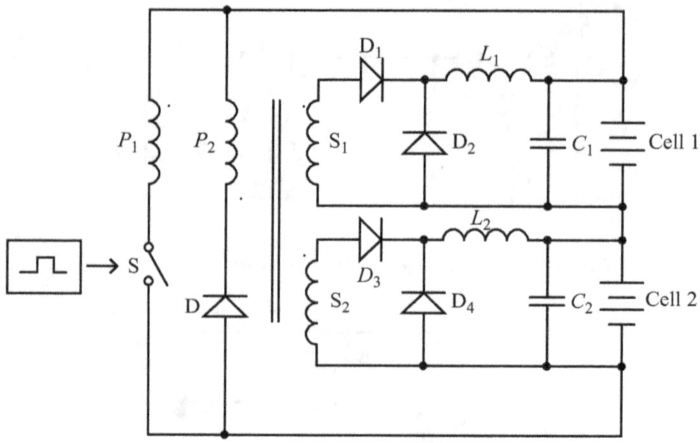


*Figure 4.15 Forward balancing circuit*

bidirectional flyback converter is more flexible in energy transmission since the energy can also be transferred from the pack to the cells. As a flyback transformer contains an air gap, the field is not homogeneous and an uniform voltage is not guaranteed, also magnetic losses and eddy currents in conductors close to the air gap are other drawbacks. There are losses associated with leakage since leakage energy is not recovered.

### *4.5.8 Forward converter balancing (active)*

The forward balancing circuit has a fast balancing time and some simpler transformer as no air gap is needed [13,63–67]. As an air gap might result in a different emf, there is less risk to have a different emf at different cells compared to a flyback converter. It has one magnetic core with one primary winding and multiple secondaries. In Figure 4.15, a forward converter balancing circuit is shown.

They are mainly used for low-voltage and high-current applications, even though the forward converter has some disadvantages. The main disadvantage is that a transistor-voltage-limiting supplementary circuit is needed. If that circuit is badly designed, the core of the transformer could be saturated generating peak currents in the transistor with poor functioning. The energy stored in the magnetizing inductor should be reset to avoid the transformer saturation when the switch is off. That supplementary circuit can be implemented by either an extra reset winding (lossless), RCD snubber circuit (loss-reset), or a two-switch forward type, or a quasi-resonant type. In [68], the forward converter has a single transistor used in zero-voltage switch; therefore, it does not require a supplementary circuit. It is also capable of equalizing multiple battery modules in series.

*Table 4.1 List of features of some active equalization topologies*

| Topologies | Capacitor | Diode | Switch | Inductor | Transformer | Balancing time |
|---|---|---|---|---|---|---|
| Switched capacitor | $N-1$ | × | $2N$ | × | × | • |
| Double-tiered | $2N-3$ | × | $2N$ | × | × | •• |
| Resonant | $N-1$ | × | $2N$ | $N-1$ | × | ••• |
| Ćuk coupled | $N/2$ | × | $N$ | $N-3$ | × | •••• |
| Inductor | × | × | $2N-2$ | $N-1$ | × | ••••• |
| Coupled inductor | × | × | $N$ | $N/2$ | × | ••••• |
| Multi-winding transformer | × | $N$ | 4 | × | 1:$N$ | •••••• |
| Transformer double-forward | × | $N$ | 4 | × | 1:$N/2$ | •••••• |
| Flyback | × | $N$ | 1 | × | 1:$N$ | •••••• |
| Forward | × | $N$ | 1 | × | 1:$N$ | •••••• |

Labels: very poor (•), poor (••), fair (•••), good (••••), very good (•••••), excellent (••••••).

## Summary

The state of charge, state of health and state of power are introduced. An overview of cell balancing methods is given and discussed. The battery management system is described and the needs for BMS are clarified. Some special attention is given to the 'one cell too low one cell too high' principle. At the end, all types of cell-balancing circuits are compared to distinguish between them due to balancing time, the number of components. In Table 4.1, a list of features of active equalization topologies is presented. The following chapters will discuss specific implementations of balancing circuits.

## References

[1] Sen C and Kar NC. Battery pack modeling for the analysis of battery management system of a hybrid electric vehicle. In: 2009 IEEE Vehicle Power and Propulsion Conference. IEEE; 2009. p. 207–212.

[2] Hussein AA. Capacity fade estimation in electric vehicle Li-ion batteries using artificial neural networks. *IEEE Transactions on Industry Applications*. 2014;51(3):2321–2330.

[3] Chaoui H and Ibe-Ekeocha CC. State of charge and state of health estimation for lithium batteries using recurrent neural networks. *IEEE Transactions on Vehicular Technology*. 2017;66(10): 8773–8783.

[4] Chang WY. The state of charge estimating methods for battery: a review. *ISRN Applied Mathematics*. 2013;2013.

[5] Lawson B. *State of health (SOH) determination*; 2005. Available from: https://www.mpoweruk.com/soh.htm#history.

[6] Gallardo-Lozano J, Romero-Cadaval E, Milanés-Montero MI, *et al.* Battery equalization control based on the shunt transistor method. *Electrical, Control and Communication Engineering.* 2014;7(1): 20–27.

[7] Baughman AC and Ferdowsi M. Double-tiered switched-capacitor battery charge equalization technique. *IEEE Transactions on Industrial Electronics.* 2008;55(6): 2277–2285.

[8] Kobzev G. Switched-capacitor systems for battery equalization. In: Proceedings of the 6th International Scientific and Practical Conference of Students, Post-graduates and Young Scientists. Modern Techniques and Technology. MTT'2000 (Cat. No. 00EX369). IEEE; 2000. p. 57–59.

[9] Cui X, Shen W, Zhang Y, *et al.* A fast multi-switched inductor balancing system based on a fuzzy logic controller for lithium-ion battery packs in electric vehicles. *Energies*. 2017;10(7):1034.

[10] Li S, Mi CC, and Zhang M. A high-efficiency active battery-balancing circuit using multiwinding transformer. *IEEE Transactions on Industry Applications.* 2012;49(1): 198–207.

[11] Lu X, Qian W, and Peng FZ. Modularized buck-boost+ Ćuk converter for high voltage series connected battery cells. In: 2012 Twenty-Seventh Annual IEEE Applied Power Electronics Conference and Exposition (APEC). IEEE; 2012. p. 2272–2278.

[12] XueZhe W, Xiaopeng Z, and Dai H. The application of flyback DC/DC converter in Li-ion batteries active balancing. In: 2009 IEEE Vehicle Power and Propulsion Conference. IEEE; 2009. p. 1654–1656.

[13] Gerislioglu B, Ozturk F, Sanli AE, *et al.* The multi-windings forward structure battery balancing. In: 2014 11th International Conference on Electronics, Computer and Computation (ICECCO). IEEE; 2014. p. 1–4.

[14] Daowd M, Omar N, Van den Bossche A, *et al.* Passive and active battery balancing comparison based on MATLAB simulation. In: 2011 IEEE Vehicle Power and Propulsion Conference. IEEE; 2011. p. 1–7.

[15] Cao J, Schofield N, and Emadi A. Battery balancing methods: a comprehensive review. In: 2008 IEEE Vehicle Power and Propulsion Conference. IEEE; 2008. p. 1–6.

[16] Ismail K, Nugroho A, Kaleg S, *et al.* Passive balancing battery management system using MOSFET internal resistance as balancing resistor. In: 2017 International Conference on Sustainable Energy Engineering and Application (ICSEEA). IEEE; 2017. p. 151–155.

[17] Aizpuru I, Iraola U, Canales J, *et al.* Passive balancing design for Li-ion battery packs based on single cell experimental tests for a CCCV charging mode. *In: 2013 International Conference on Clean Electrical Power (ICCEP).* IEEE; 2013. p. 93–98.

[18] Vitols K. Redesign of passive balancing battery management system to active balancing with integrated charger converter. In: 2014 14th Biennial Baltic Electronic Conference (BEC). IEEE; 2014. p. 241–244.

[19] Friansa K, Haq IN, Leksono E, *et al.* Battery module performance improvement using active cell balancing system based on switched-capacitor boost converter (S-CBC). *In: 2017 4th International Conference on Electric Vehicular Technology (ICEVT)*. IEEE; 2017. p. 93–99.
[20] Pascual C and Krein PT. Switched capacitor system for automatic series battery equalization. In: Proceedings of APEC 97-Applied Power Electronics Conference. vol. 2. IEEE; 1997. p. 848–854.
[21] Kim MY, Kim CH, Kim JH, *et al.* A chain structure of switched capacitor for improved cell balancing speed of lithium-ion batteries. *IEEE Transactions on Industrial Electronics*. 2013;61(8): 3989–3999.
[22] Ye Y, Cheng KWE, Fong YC, *et al.* Topology, modeling, and design of switched-capacitor-based cell balancing systems and their balancing exploration. *IEEE Transactions on Power Electronics*. 2016;32(6): 4444–4454.
[23] Ye Y and Cheng K. An automatic switched-capacitor cell balancing circuit for series-connected battery strings. *Energies*. 2016;9(3):138.
[24] Yuanmao Y, Cheng K, and Yeung Y. Zero-current switching switched-capacitor zero-voltage-gap automatic equalization system for series battery string. *IEEE Transactions on Power Electronics*. 2011;27(7): 3234–3242.
[25] Daowd M, Antoine M, Omar N, *et al.* Single switched capacitor battery balancing system enhancements. *Energies*. 2013;6(4): 2149–2174.
[26] Daowd M, Antoine M, Omar N, *et al.* Battery management system—balancing modularization based on a single switched capacitor and bi-directional DC/DC converter with the auxiliary battery. *Energies*. 2014;7(5): 2897–2937.
[27] Kim MY, Kim CH, Kim JH, *et al.* Switched capacitor with chain structure for cell-balancing of lithium-ion batteries. In: IECON 2012—38th Annual Conference on IEEE Industrial Electronics Society. IEEE; 2012. p. 2994–2999.
[28] Qi J and Lu DDC. Review of battery cell balancing techniques. In: 2014 Australasian Universities Power Engineering Conference (AUPEC). IEEE; 2014. p. 1–6.
[29] Goodarzi S, Beiranvand R, Mousavi SM, *et al.* A new algorithm for increasing balancing speed of switched-capacitor lithium-ion battery cell equalizers. In: The 6th Power Electronics, Drive Systems & Technologies Conference (PEDSTC2015). IEEE; 2015. p. 292–297.
[30] Wang Z, Chen C, Zhang Y, *et al.* A resonant switched-capacitor balancing strategy for series battery pack. In: 2014 IEEE Conference and Expo Transportation Electrification Asia-Pacific (ITEC Asia-Pacific). IEEE; 2014. p. 1–5.
[31] Moghaddam AF and Van den Bossche A. A cell equalization method based on resonant switched capacitor balancing for lithium ion batteries. In: 2018 9th International Conference on Mechanical and Aerospace Engineering (ICMAE). IEEE; 2018. p. 337–341.
[32] Moghaddam AF and Van den Bossche A. An active cell equalization technique for lithium ion batteries based on inductor balancing. *In: 2018 9th International Conference on Mechanical and Aerospace Engineering (ICMAE)*. IEEE; 2018. p. 274–278.

[33] Farzan Moghaddam A and Van den Bossche A. An efficient equalizing method for lithium-ion batteries based on coupled inductor balancing. *Electronics*. 2019;8(2):136.

[34] Kim MY, Kim CH, Cho SY, *et al.* A cell selective charge equalizer using multi-output converter with auxiliary transformer. In: 8th International Conference on Power Electronics-ECCE Asia. IEEE; 2011. p. 310–317.

[35] Chen Y, Liu X, Cui Y, *et al.* A multiwinding transformer cell-to-cell active equalization method for lithium-ion batteries with reduced number of driving circuits. *IEEE Transactions on Power Electronics.* 2015;31(7): 4916–4929.

[36] Li S, Mi C, and Zhang M. A high efficiency low cost direct battery balancing circuit using a multi-winding transformer with reduced switch count. In: 2012 Twenty-Seventh Annual IEEE Applied Power Electronics Conference and Exposition (APEC). IEEE; 2012. p. 2128–2133.

[37] Lee KM, Lee SW, Choi YG, *et al.* Active balancing of Li-ion battery cells using transformer as energy carrier. *IEEE Transactions on Industrial Electronics.* 2016;64(2): 1251–1257.

[38] Ji X, Cui N, Shang Y, *et al.* Modularized charge equalizer using multi-winding transformers for Lithium-ion battery system. In: 2014 IEEE Conference and Expo Transportation Electrification Asia-Pacific (ITEC Asia-Pacific). IEEE; 2014. p. 1–5.

[39] Shang Y, Xia B, Zhang C, *et al.* A modularization method for battery equalizers using multiwinding transformers. *IEEE Transactions on Vehicular Technology.* 2017;66(10): 8710–8722.

[40] Abeywardana DB, Manaz M, Mediwaththe M, *et al.* Improved shared transformer cell balancing of Li-ion batteries. In: 2012 IEEE 7th International Conference on Industrial and Information Systems (ICIIS). IEEE; 2012. p. 1–6.

[41] Lim CS, Kim RY, and Hyun DS. Battery voltage sensorless charge equalizer using the multi-winding transformer. In: 2012 IEEE Vehicle Power and Propulsion Conference. IEEE; 2012. p. 789–793.

[42] Bui T, Kim CH, Kim KH, *et al.* A modular cell balancer based on multi-winding transformer and switched-capacitor circuits for a series-connected battery string in electric vehicles. *Applied Sciences*. 2018;8(8):1278.

[43] Xue T, Cheng X-m, Fang Y-q, *et al.* Advanced multi-winding transformer equalizer for electric vehicle battery system. In: Proceedings of the 32nd Chinese Control Conference. IEEE; 2013. p. 7662–7667.

[44] Shen H, Zhu W, and Chen W. Charge equalization using multiple winding magnetic model for lithium-ion battery string. In: 2010 Asia-Pacific Power and Energy Engineering Conference. IEEE; 2010. p. 1–4.

[45] Park SH, Park KB, Kim HS, *et al.* Single-magnetic cell-to-cell charge equalization converter with reduced number of transformer windings. *IEEE Transactions on Power Electronics.* 2011;27(6): 2900–2911.

[46] Jeon S, Yun JJ, and Bae S. Active cell balancing circuit for series-connected battery cells. In: 2015 9th International Conference on Power Electronics and ECCE Asia (ICPE-ECCE Asia). IEEE; 2015. p. 1182–1187.

[47] Yun J, Yeo T, and Park J. High efficiency active cell balancing circuit with soft-switching technique for series-connected battery string. In: 2013 Twenty-Eighth Annual IEEE Applied Power Electronics Conference and Exposition (APEC). IEEE; 2013. p. 3301–3304.

[48] Moghaddam AF and Van den Bossche A. Multi-winding equalization technique for lithium ion batteries for electrical vehicles. In: 2018 7th International Conference on Renewable Energy Research and Applications (ICRERA). IEEE; 2018. p. 139–143.

[49] Farzan Moghaddam A, Van den Bossche A. A single transformer for active cell equalization method of lithium-ion batteries with two times fewer secondaries than cells.*Electronics*. 2019;8(9):951.

[50] Lee YS, Duh CY, Chen GT, *et al.* Battery equalization using bi-directional Ćuk converter in DCVM operation. In: 2005 IEEE 36th Power Electronics Specialists Conference. IEEE; 2005. p. 765–771.

[51] Qi J and Lu DDC. A preventive approach for solving battery imbalance issue by using a bidirectional multiple-input Ćuk converter working in DCVM. *IEEE Transactions on Industrial Electronics.* 2017;64(10): 7780–7789.

[52] Moghaddam AF and Van den Bossche A. A battery equalization technique based on Ćuk converter balancing for lithium ion batteries. In: 2019 8th International Conference on Modern Circuits and Systems Technologies (MOCAST). IEEE; 2019. p. 1–4.

[53] Farzan Moghaddam A and Van den Bossche A. A Ćuk converter cell balancing technique by using coupled inductors for lithium-based batteries. *Energies*. 2019;12(15):2881.

[54] Moghaddam AF and Van den Bossche A. Flyback converter balancing technique for lithium based batteries. In: 2019 8th International Conference on Modern Circuits and Systems Technologies (MOCAST). IEEE; 2019. p. 1–4.

[55] Imtiaz AM and Khan FH. "Time shared flyback converter" based regenerative cell balancing technique for series connected Li-ion battery strings. *IEEE Transactions on power Electronics.* 2013;28(12): 5960–5975.

[56] Shin JW, Seo GS, Chun CY, *et al.* Selective flyback balancing circuit with improved balancing speed for series connected lithium-ion batteries. In: The 2010 International Power Electronics Conference-ECCE ASIA. IEEE; 2010. p. 1180–1184.

[57] Yang D, Li S, and Qi G. A bidirectional flyback cell equalizer for series-connected lithium iron phosphate batteries. In: 2015 6th International Conference on Power Electronics Systems and Applications (PESA). IEEE; 2015. p. 1–5.

[58] Kim JW, Shin JW, and Ha JI. Cell balancing control using adjusted filters in flyback converter with single switch. In: 2013 IEEE Energy Conversion Congress and Exposition. IEEE; 2013. p. 287–291.

[59] Kim CH, Kim MY, and Moon GW. Individual cell equalizer using active-clamp flyback converter for Li-ion battery strings in an electric vehicle. In: 2012 IEEE Vehicle Power and Propulsion Conference. IEEE; 2012. p. 327–332.

[60] McCurlie L, Preindl M, Malysz P, *et al.* Simplified control for redistributive balancing systems using bidirectional flyback converters.In: 2015 IEEE Transportation Electrification Conference and Expo (ITEC). IEEE; 2015. p. 1–6.

[61] Shang Y, Xia B, Zhang C, *et al.* An automatic battery equalizer based on forward and flyback conversion for series-connected battery strings.In: 2017 IEEE Applied Power Electronics Conference and Exposition (APEC). IEEE; 2017. p. 3218–3222.

[62] Liu X, Sun Y, He Y, *et al.* Battery equalization by fly-back transformers with inductance, capacitance and diode absorbing circuits.*Energies*. 2017;10 (10):1482.

[63] Shang Y, Xia B, Zhang C, *et al.* An automatic equalizer based on forward–flyback converter for series-connected battery strings. *IEEE Transactions on Industrial Electronics*. 2017;64(7): 5380–5391.

[64] Hussain A, Lee H, and Sul SK. Forward fly-back voltage balancing circuit for series connected super capacitors using digital control.In: 2013 International Conference on Renewable Energy Research and Applications (ICRERA). IEEE; 2013. p. 377–382.

[65] Lee Y, Yun JJ, and Bae S. Active cell balancing circuit based on the forward converter with an RCD circuit.In: 2015 9th International Conference on Power Electronics and ECCE Asia (ICPE-ECCE Asia). IEEE; 2015. p. 2887–2892.

[66] Remes CL, Rosa MB, and Oliveira SVG. A two-switch forward converter application for battery charging. In: 2015 IEEE 13th Brazilian Power Electronics Conference and 1st Southern Power Electronics Conference (COBEP/SPEC). IEEE; 2015. p. 1–6.

[67] Sun J, Lu R, Wei G, *et al.* A high efficiency equalizer based on forward converter for series connected battery string. In: 2012 IEEE Vehicle Power and Propulsion Conference. IEEE; 2012. p. 376–379.

[68] Farzan Moghaddam A and Van den Bossche A. Forward converter current fed equalizer for lithium based batteries in ultralight electrical vehicles. *Electronics*. 2019;8(4):408.

*Part II*

# Passive cell balancing circuits

# Chapter 5

# A smart high-voltage cell detecting and equalizing

## 5.1 Introduction

Nowadays, batteries are the most common electrical energy storage device in vehicles as a replacement for traditional fuels (Gasoline) [1–4]. The capacity of a battery is gradually reduced due to erosion, passivation, outgassing, temperature, decomposition of materials, and changes on the electrode surface during its operation [5,6]. Also, the battery's ageing process can be accelerated when the battery is subjected to operate beyond its specified safe operating conditions. The battery state is also used to estimate the expected lifetime of the battery and can simply be described by two parameters: state of charge (SOC) and state of health (SOH). In smaller vehicles, monitoring the voltage and current of each cell is a compromise between reliability and the cost of the increased amount of circuitry. Also, these circuits require some power to operate; therefore, they induce a self-discharge. The battery monitoring may have inputs as the total voltage, current, and temperature but requires a reference point to know that all cells are full. Achieving a long life of the battery requires an appropriate battery management system (BMS) to equalize the cells in a battery string. A BMS plays an important role in electrical vehicles to achieve a reasonable lifetime. Beyond a number of years even the reliability of the BMS should be considered as well, how servicing of the battery pack and the BMS can be achieved, but it starts with a good and transparent method.

A monitoring system is required to check if any cell voltage is high or low. In this chapter, an equalizing and monitoring system for an ultra-light electric vehicle is investigated. It permits to pin important information if one cell is high, charging should be limited to some trickle charge, and if all cells are high to stop charging. The proposed monitoring system detects if one cell is fully charged or all cells are fully charged and the equalizing system tops each cell at the desired voltage. To this issue, a light-emitting diode (LED) bandgap is used, as a voltage reference, another LED is used to inform the user if any cell is at its high voltage. A smart monitoring displays a text message on the liquid crystal display (LCD) screen, if one cell is high or all cells are high. This detection also provides a signal to the microcontroller to turn on/off the charger if all cells are high. Also, a Bluetooth module is designed to command the microcontroller to turn on/off the charger via voice/text message by using a smartphone. A major feature of the system is to draw a very low standby current, so that the system does not contribute significantly to the self-discharge of the battery.

## 5.2 Possible technologies

Also, other technologies were tried out, such as supply level detectors, but for an even higher cost, the leakage current was higher and the voltage accuracy was lower. Typical Zener diodes do not have a very sharp knee as the threshold voltage for equalization (discrete or integrated circuit (IC), e.g. LM329). The art is to obtain a low leakage current at low voltage (below threshold voltage) and high equalizing current at a constant voltage, or simply put, a sharp knee in the *I-V* characteristic. The most often used IC in commercially available BMS systems is a three-terminal TL431, in which the third terminal is used to set the voltage. However, a minimum regulating current of 400 μA is required, with a voltage divider needed to get an adjustable output voltage. With a 2.495 V reference (1%) and the assumption of a 10 kΩ reference resistor in the voltage divider, this requires 0.2495 mA (together 0.6495 mA). This causes a 5.689 Ah capacity decrease in 1 year, and usually, there are also other leakage currents in a system. The simplified schematic of TL431 and cathode current vs cathode voltage is presented in Figure 5.1.

For the newer technologies such as the JFET pinch-off references, the same problem arises with a 560–800 μA quiescent current for the ADR 430 series. While obtaining a high accuracy, low-temperature coefficients, and noise, JFET pinch-off references cost more. One of the most recent entries in the voltage reference sweepstakes is a floating gate voltage reference. The idea here is to introduce some charge into the buried and well-insulated gate of a MOSFET during manufacturing, thereby putting it at a certain voltage. The ISL21009 series exhibit a typical regulating current of 100 μA, which is acceptable.

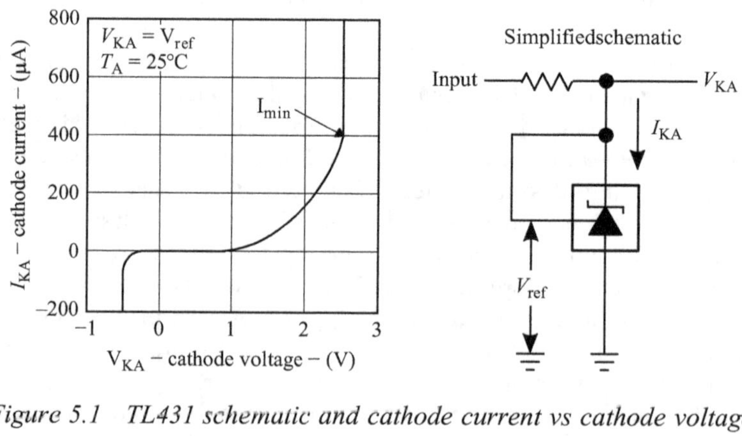


*Figure 5.1 TL431 schematic and cathode current vs cathode voltage*

*Section 1*

# High-voltage cell detecting and equalizing circuit

## 5.3 Proposed smart high-voltage cell equalizing detection

In this section, the principle of high-voltage equalizing detection (HVED) via an LED bandgap, based on the voltage reference, will be considered. This proposes both an equalizing and monitoring system. The monitoring system detects if one cell is fully charged or all cells are fully charged and the equalizing system tops each cell at the desired voltage. To achieve this, an LED bandgap is used, as a voltage reference to detect the cell voltages. It will be used to inform the user if any cell is at its high voltage. Also, a smart monitoring displays on LCD, if one cell is at high voltage level or all cells are at high voltage level by displaying the text messages, 'one cell is high' or 'all cells high' on the LCD. This detection also gives a signal to the microcontroller to turn on or to turn off the charger, if all cells are at high voltage level. In addition, a Bluetooth module is designed to command the microcontroller to turn on or to turn off the charger via voice or text message by using a smartphone. The operational principle for one cell is illustrated in Figure 5.2. Assume that the voltage over the considered cell rises above a predefined maximum value. At this moment, the two LEDs in series ($D_1$, $D_2$) start conducting, enabling the transistor Q1, BJT (PNP) to be turned on. The gate of the N-channel MOSFET Q2 gets pulled up and current flows through the 4 W resistor. This is a power resistor since it needs to dissipate the surplus battery power for some time. An extra LED is foreseen to inform users that the system is in high-voltage equalization.

The LED driver/constant current regulator limits the current to 10 mA and has a negative *T*-coefficient to protect the LED ($D_3$) from thermal runaway at high currents. Note that it tops all cells at a predefined voltage. It should be noted that diodes $D_1$ and $D_2$ are hardly visible as they are used just above their leakage current. A detection is added to enable actions such as balancing and charger cut-off when the cells go beyond the threshold voltage. This detection is located on the right-hand side of the schematic. The over-voltage detection is implemented through an optoisolator (optocoupler) shown in Figure 5.3. For servicing, it is better

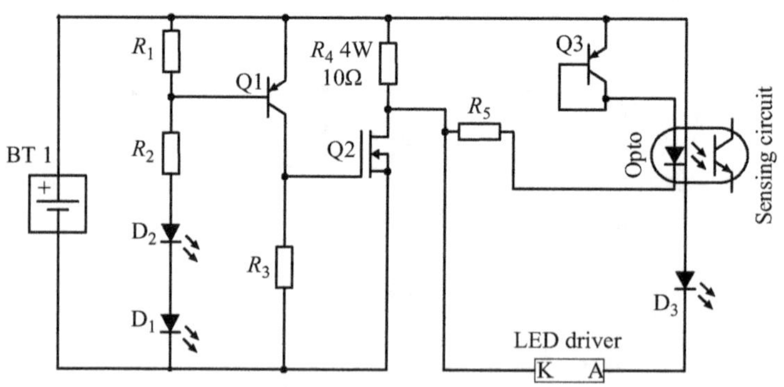


*Figure 5.2 High-voltage equalizing detection circuit of one cell*

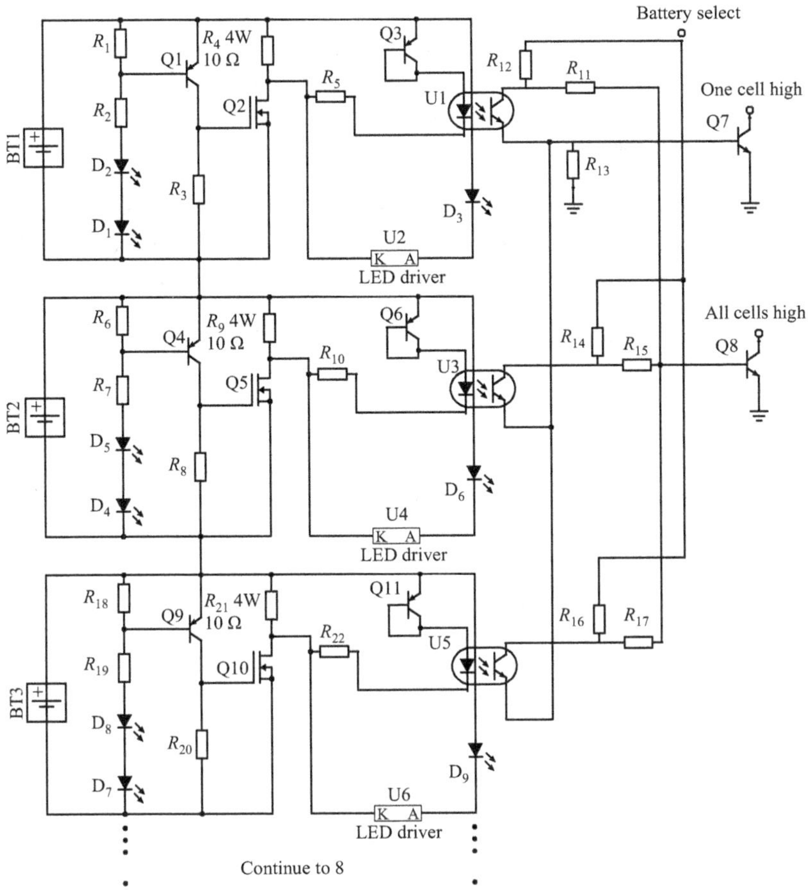


*Figure 5.3 Voltage sensing circuit of HVED for eight cells in series*

to know which cell did reach the full charge; thus, a visible LED has been implemented. As a cell gets a high voltage, the phototransistor conducts. Note that the voltage drop of an infrared LED in the optocoupler is lower than a red LED, so some transistor voltage drop has been added to light the red LED before the optocoupler, to avoid the hidden ones that stop the fast charging.

A wired 'OR' circuit using transistor Q7 allows detecting if one cell is high and a wired 'AND' circuit using Q8 detects if all cells are high. This current originates from a bank- or battery select, which is a microcontroller signal (5 V) enabling monitoring of the different battery packs at a certain frequency. In this way, the amount of signals for four packs is reduced from four to one as the packs are not simultaneously monitored but sequentially. The frequency should not be too low to make sure that during non-monitored dissipation, the cell will not be damaged. Also, a Bluetooth module is predicted to command the microcontroller to turn on or off the charger when all cells are high via a voice or text message by using a smartphone. Depending on the fastest charger type (within the current limits of the batteries), the resistor might vary. Assume only one HVED circuit (assume again cell 1) in the pack has reached equalizing mode, or simply put one cell is high, then the collector of the Q7 BJT is pulled down, enabling a microcontroller to interpret this pull-down as a warning that one cell is high. This warning can be seen on an LCD screen.

Collecting the signals reduces a lot of wiring and software to detect which cell is high. An additional LED '$D_3$' has been provided for maintenance and fault detection. This information can then be used to disable the charger temporarily and balance the cells amongst each other through a parallel balancing method. During regenerative braking, this method is also used. When all cells are high, the collector of the Q8 BJT is pulled down and the microcontroller is warned that all cells are high. This case will probably only happen when regenerative braking. In a charging situation, most chargers (without individual cell monitoring) will stop when the average peak voltage is high enough, and some cells might be rather low and at the same time other cells too high. By doing this, an important reduction of wires is obtained towards the processor. It can even be further reduced by multiplexing data from series-connected battery sets.

## 5.4 Practical implementation

In this section, the practical implementation of HVED circuit will be discussed. $LiFePO_4$ cells were chosen as an example for the practical implementation. The manufactures specifies 3.65 V$\pm$0.05 V as a maximum voltage for long life. The LEDs were chosen in such a way that the threshold voltage for equalization was very close to 3.65 V. This was done by selecting gallium–arsenide–phosphide LEDs. However, in $LiFePO_4$ technology batteries, this limit is less strict than in Li-ion batteries since the electrolyte does not decompose quickly and there is much less risk of thermal runaway. This makes a trade-off between a slight overcharge and a resulting decrease in cycle-life possible. There are some reasons to select a higher threshold voltage than 3.65 V. First of all, assume that the batteries are to be charged to their full charge. This means all cells should reach 3.65 V. Equalization at this threshold is rather hard to obtain. Anyhow, there is a variety of LED types to be able to get the right effect. Assume that the normal charging stops at one cell high. A trickle charging can give a reduced current that HVED circuits can manage. A way is to switch to a trickle charge mode, where a limited current for example 1% of nominal Ah is given. In this way, the voltage limiting can fully act without the risk of overcharging. At the end of trickle charging, all HVED units are warm, so rather the higher temperature threshold counts.

The LEDs ($\approx -1.5$ mV/K for Al-GaInP) and BJT base-emitter junction ($\approx -2$ mV/K) both have a negative temperature coefficient causing the threshold voltage to decrease when the temperature rises. The HVED will equalize earlier than required at higher temperatures. It can be seen as a positive effect since a higher temperature puts stress on the cell. As explained before, it makes it harder to reach full charge and limits the obtainable range of the vehicle. One could argue that the open-circuit voltage vs SOC characteristic is also temperature dependent, but not in the same temperature range. $LiFePO_4$ batteries have a higher tolerance to overcharge.

The type of LED is a choice if some designers want to use a lower threshold LED, such types also exist. A rather high threshold voltage is chosen by using aluminium–gallium–indium–phosphide LEDs. Since the purpose of the LEDs is not to emit light, it is not the forward voltage drop mentioned in the data sheet (1.95 V–2.5 V at 20 mA and 25 °C) that has to be considered, since the HVED will start working at lower currents and thus lower forward voltages. One could determine the forward voltage drop making a use of the Schottky diode equation and Ebers–Moll model, but this would unnecessarily complicate matters. The estimate for the LED drop is made from the following characteristic as supplied by the manufacturer. At low current, the voltage drop depends on the semiconductor type and the current density and is almost independent of the metallization resistivity. The semiconductor type is recognized by the nm wavelength in the data sheet.

The reverse leakage current in the data sheet is also an indication of the forward leakage current. Next to the LED drop, the threshold voltage is also determined by the base-emitter voltage drop of the PNP BJT (Q1). The LED supplier specified a reverse LED leakage of 10 μA. In the assumption of a forward leakage 10 μA, the BJT has to be biased sufficiently above that value so the leakage does not influence the equalization. A 10 kΩ base-emitter resistor is chosen to be safe in that regard. The current will stay well below 1 mA and the LED drop can thus safely be assumed around 1.65 V according to Figure 5.4. The transistor works at a limited collector current of <0.1 mA, the threshold $V_{BE}$ can be estimated at 0.55 V according to Figure 5.5.

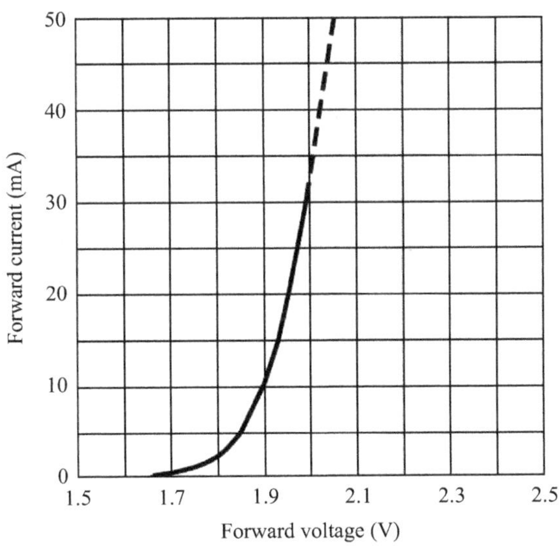


*Figure 5.4 I–V characteristic of the LED*

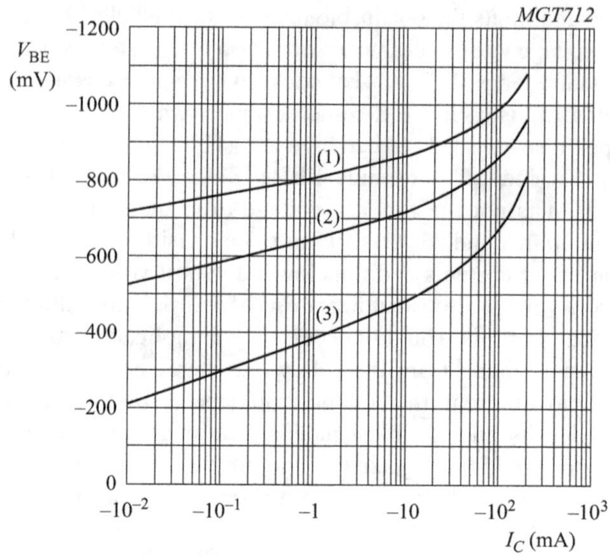


**BC857A;** $V_{CE} = -5$ V.
(1) $T_{amb} = -55$ °C.
(2) $T_{amb} = 25$ °C.
(3) $T_{amb} = 150$ °C.

*Figure 5.5 Base-emitter voltage as a function of collector current*

The threshold voltage obtained by this configuration is estimated as

$$V_{threshold} = 2 \times V_{LED} + V_{BE} = 2 \times 1.65\text{ V} + 0.55\text{ V} = 3.85\text{ V} \tag{5.1}$$

Due to the use of the 10 Ω dissipating resistor, and $R_{DS\ (on)}$ of the MOSFET, the equalizing current is limited below 370 mA. One could consider using a lower resistance to speed up the dissipation of energy from the high cell; however, there is no need to quickly dissipate and a larger dissipation current would require more cooling size of the printed circuit board (PCB) heat sink. A 370 mA equalizing current means a 1.3% dissipative equalization per hour. The HVED will also be active during regenerative braking. A 370 mA equalization current gives a 45.58 W power dissipation. This is definitely not sufficient for going downhill with a full battery pack. So the user should be informed that he better uses the mechanic brake.

## 5.5 Experimental results

In this section, the functionality of the previous design is tested and conclusions concerning the results are made. In all of the following tests, capacitors were connected in parallel to each HVED circuit to simulate the series impedance of a battery cell. Since one of the design goals was to obtain a voltage reference that distinguishes itself compared to other voltage references in its quiescent behaviour, the HVED is first tested at sub-threshold voltages. A 1 MΩ resistor in series with a 56 kΩ resistor was used in parallel with the 20 MΩ internal resistance of the voltmeter to obtain a 1.003 MΩ current measuring resistance. The cell voltage was measured on the PCB. The following result of a single HVED cell was obtained.

From Figure 5.6, it is clear that the quiescent current at the nominal voltage is below 10 μA (9.87 μA to be precise). On a yearly basis, this means a capacity decrease of 0.3%, which is considered to be an outstanding result. From there on, the knee starts to form with a 41 μA current at 3.65 V, sufficiently small to assume a sharp knee. To observe the characteristic at higher currents, initially, a single HVED cell was tested to above 400 mA. A large number of data points were selected over a 10 min measurement to indicate the temperature influence on the HEVD cell voltage, and the result is shown in Figure 5.7.

The cold threshold voltage amounts to 3.86 V, which is almost equal to the 3.85 V, estimated in the previous section. Due to the temperature coefficient of the LED and

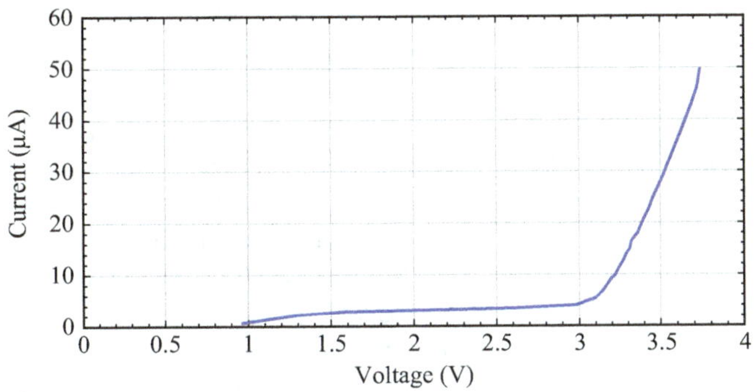


*Figure 5.6 Cell voltage vs balancing current (non-equalizing)*

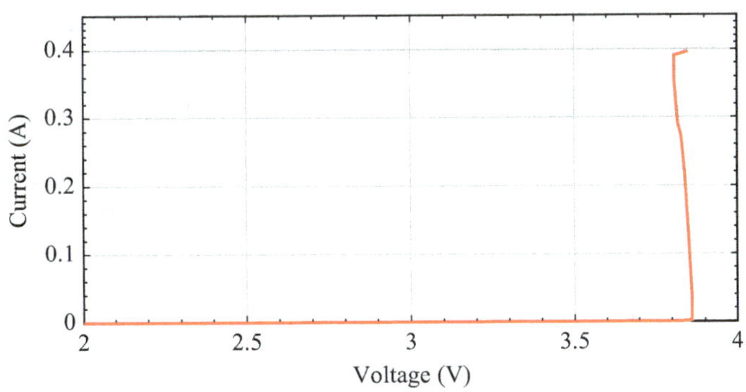


*Figure 5.7 Characteristic of equalizing mode*

BE-junction of the BJT, the HVED cell voltage at 390 mA is reduced to 3.81 V. This relates to a 10 °C temperature increase. Above 380 mA, the voltage rises again as is foreseen due to the presence of the 10 Ω dissipation resistor. A 10 °C temperature increase is something very common in any vehicle if parked in the sun for example. Temperature differences of up to 30 °C are likely. This would lead to a 0.15 V reduction in the threshold voltage. For the design with a 3.65 V threshold, the warm threshold would thus be 3.5 V, which is still well above the nominal voltage.

Depending on the influence the overcharge has on the batteries, there can thus still be opted to decrease the threshold voltage. This influence is to be evaluated in a later phase. To conclude an entire pack of HVED cells, all eight cells are tested simultaneously to observe if the threshold voltage and *I–V* characteristic show mutual differences. If there is a faulty cell between the normal cells, it will be detected by LED at the end of the circuit for each cell, Figure 5.2, as well as the voltage-sensing circuit as shown in Figure 5.3. A large voltage range is represented in order to examine possible differences between cells in quiescent and equalizing mode. The obtained result is represented in Figure 5.8, and in Figure 5.9, the *I–V* characteristic of one cell is shown.

In this case, the lowest HVED cell voltage at a supply current of 294 mA is 3.65 V in cell 4 (C4), with the highest being 3.72 V in cell 8 (C8). There is a noticeable increase in mutual differences between cells at higher currents. When assessing the differences, it is clear that cells, located in the middle of the PCB, are more affected by the heat production of the surrounding cells and show a bigger temperature effect than the border cells. One would expect that the voltage of cell 8 and cell 1 show little difference, which does not correspond to the measurement. This is due to the testing set-up. The resistor used to obtain the supply current also generates heat with increasing current. In the test, this resistor was located nearest to cell 1, thereby explaining its increased voltage drop.

To verify the feasibility of the system, the equalization results of eight battery cells are presented. In Figure 5.10(a), cell voltage waveforms vs time are shown. The cell voltages vary from 2.67 V to 3.67 V. In this case, all cells are above 3.25 V, only one cell is 2.67 V. The cells were fully charged to 3.45 V. As it can be

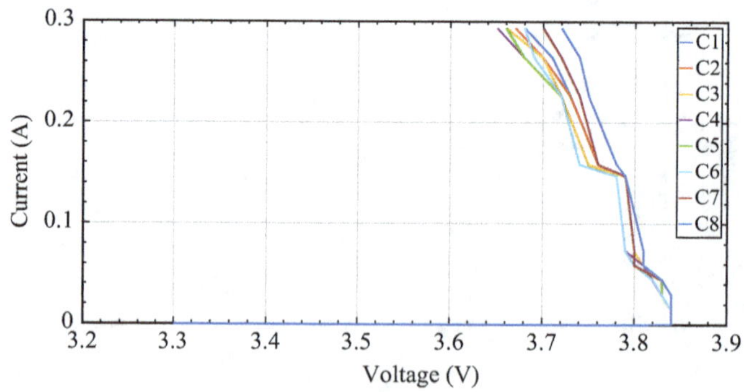


*Figure 5.8 Cell voltage vs balancing current*

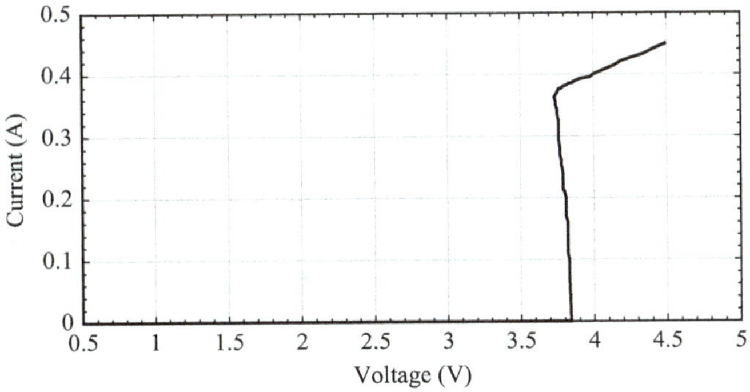


*Figure 5.9 Cell voltage vs balancing current of one cell*

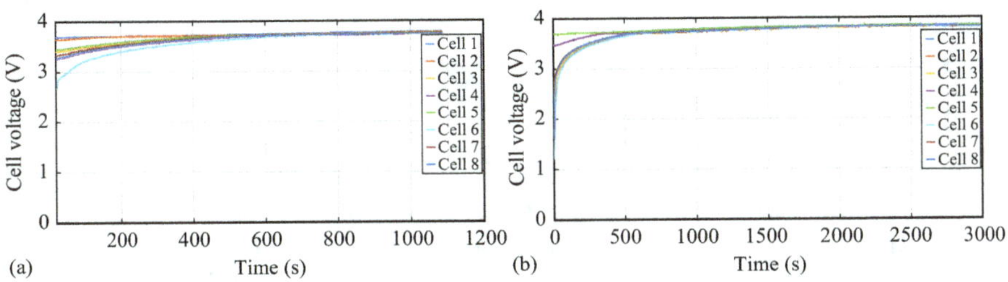


*Figure 5.10 Experimental results of the HVED circuit: (a) cell voltages vs time (fully charged) and (b) cell voltages vs time (threshold)*

seen in the figure, all cells are fully charged and they are converged at the end of balancing process. To test the limit voltage level of the proposed circuit and have more cell imbalances, the circuit was tested with more voltage imbalances. The cell voltages vary from 1.21 V to 3.7 V. The results are shown in Figure 5.10(b). It can be noticed that all cells are charged to the threshold voltage of 3.85 V according to (5.1).

A picture of the testing layout is represented in Figure 5.12(a). As is shown in the picture, an LCD is predicted to warn the user if one cell is higher than the desired voltage or if all cells are higher than this voltage. To avoid too much self-discharge by this feature, it can turn off itself after 10 days. If it draws <1 W for 10 days, it is 0.24 kWh, which is reasonable. It also avoids that the battery stays at its maximum charge level when for some reason it is not used during the month. In this book, we are aware of it and take it in account, try to make the best of it. In general the user should be informed about the self-discharge of battery and BMS combination. It would avoid loosing valuable batteries when stored for a longer time. This would be by preference on a visible label y the user and not just at some 'page 51' of a user manual. Some people may not use or recharge an appliance during 6 months or do store it for more than one year.

## 5.6 Communication between battery and charger via Bluetooth

Also BMSs can benefit from technology related to the Internet of Things (IoT). A contribution to it is done by a Bluetooth module HC-05 module. The Bluetooth module HC-05 is added to the circuit to turn off the charger with voice message from smartphone which is programmed by Arduino Uno if all cells are fully charged and all cells are at high level of charge and also turn on the charger if the cells are at low voltage level. In order to control the charger via Bluetooth module, a mobile app 'BT Voice Control for Arduino' is used. This app is available in the Google play store. If a cell is high, it will be detected by circuit at the end of each battery section transistor Q7 as shown in Figure 5.3, and if all cells are high it will be detected by transistor Q8. These transistors provide a signal to the analogue pin of microcontroller (Arduino Uno) to turn off the charging process.

These detentions can be seen on LCD by the text 'all cells high' or 'one cell is high'. It also can be observed by LED for each cell, when LED is turned on, it means this cell is at its high level. If one LED is not turned on, it means that the cell is not high or it is damaged. The algorithm flowchart of smart monitoring is presented in Figure 5.11.

In Figure 5.12(b), a picture of the testing layout with the Bluetooth module is shown. The SOC of the cells is also detected by the microcontroller, so if all cells are at a high level of charge, it automatically turns off the charger by sending a signal to the charger. It can be done both automatically by the microcontroller and also by voice message from the smartphone. The voice message can be generated by the smartphone with a mobile app 'AMR-Voice', which is available in the Play Store, then this message will be received by Bluetooth module HC-05 which is connected to the Arduino; thus, Arduino will disconnect the charger. In this chapter, eight cells are considered, but the number of cells can be extended by adding more parts to the whole circuit, which are all the same and connected in series. A picture of smart monitoring system of the proposed circuit is shown in Figure 5.13. As can be seen in the figure, first the signals from transistor Q7 or Q8,

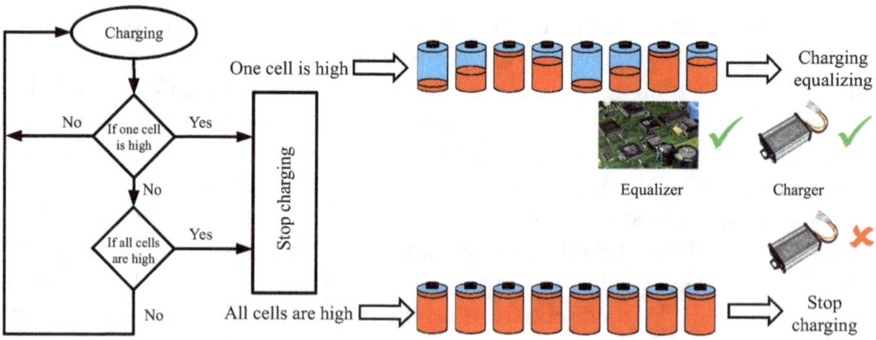


*Figure 5.11 Algorithm flowchart of the smart monitoring*

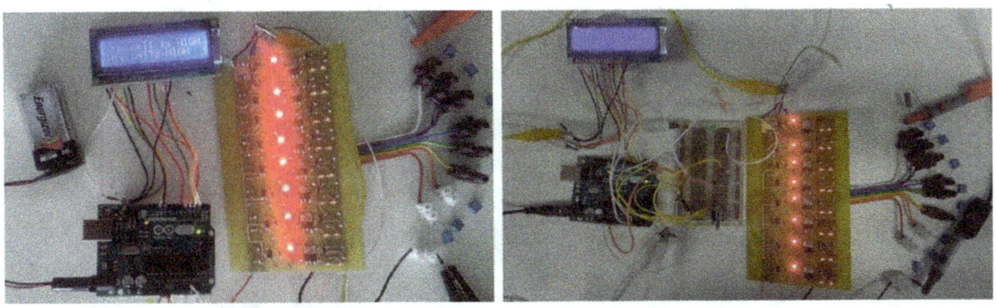

(a) (b)

*Figure 5.12 The PCB prototype of HVED circuit with/without Bluetooth module: (a) PCB prototype and (b) PCB prototype with Bluetooth module*

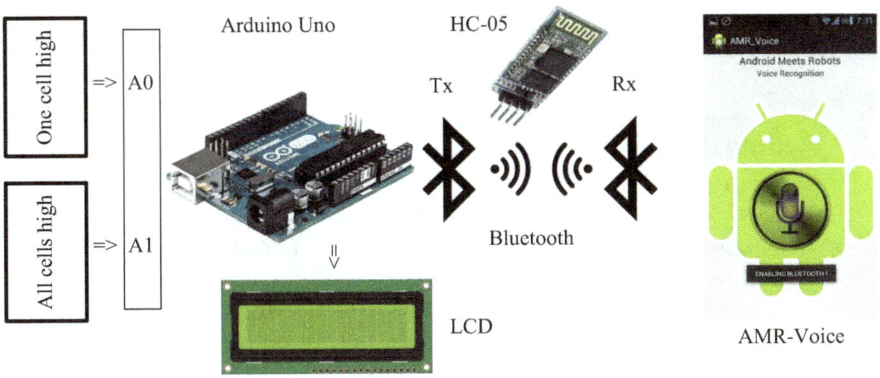


*Figure 5.13 Smart monitoring system*

which are collected by the sensing circuit at the end of each battery section, will be received by analogue pin of the microcontroller (A0 and A1). Then the collected information will be displayed on LED as following. If for instance, one cell is high, it will be displayed as 'One cell high', and if all cells are high, it will be displayed as 'All cells high' on the screen. The LCD is an alphanumeric 2×16 (2 rows by 16 characters) screen. The smartphone can also transmit the data in order to command microcontroller to turn on or turn off the charger by voice message via Bluetooth module if it is desired.

In Figure 5.14, a picture of the practical implementation of the proposed circuit is presented. As can be seen in the figure, eight batteries are connected in series. The cell voltage varies between 3 V and 3.85 V. When a cell reaches the maximum allowed voltage, which is designed to be 3.85 V, the LED will be forward biased and turned on; therefore, the circuit gives a signal to the analogue pin of the controller to cut off the charger. In the figure, for example, cells 1 and 2 are not fully charged; therefore, the proper LED is turned off, while the other cells are fully charged and the proper LED is turned on.

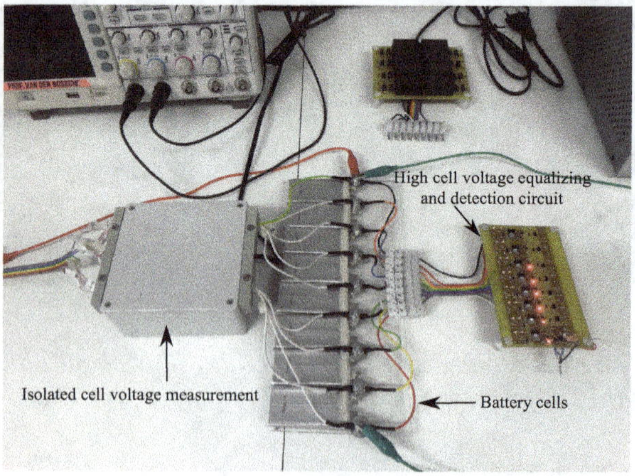


*Figure 5.14 The practical implementation of the proposed circuit (HVED)*

## 5.7 Reduced temperature coefficient set-up

If we consider stationary application, the temperature change is not that high and some temperature effect can be allowed. If wide temperature changes occur, it might be good to compensate for temperature effects. The question is also what the temperature coefficient is of the cells. Internet is not giving a clear view, so some little tests have been carried out with some smaller Li-ion batteries. A careful way in development is good to check the temperature coefficient of the battery as well. It was convenient to do so with smaller cells – a fully charged voltage of 4.2 V at 4115 mAh. Between −20 °C and 65 °C, less than 10 mV change was observed, so at fully charged batteries. Around 3.75 V at 20 °C, 1000 mAh Li-ion cells, we observed a very small change some 30 mV less at 60 °C compared to 20 mV more at −20 °C, so slightly negative; this was changing the temperature without charging. If we look at typical charge curves in literature, at 50% charge a voltage of about 3.8 V and almost not temperature coefficient is very little between 10 °C and 70 °C. At −20 °C: 3.4 V and at 0 °C: 3.7 V and at lower temperatures, it is a slope of 0.3 V over 20 °C, so about +15 mV/°C. So it means that it is not good to use the voltage around 50% charge to equalize, as it may reduce the charge on the higher temperature cells. From this point, we again look at the $LiFePO_4$ cell case. If nothing is done, a circuit using two red LEDs and a transistor has about three times, and this is at about −2 mV/K, which results in −6 mV/K. If a compensating circuit is designed using an NTC resistor, the voltage drop of two red LEDs is too much and infrared LEDs have too little drop. The solution was to use one blue LED. An adapted design has been made to see if the temperature drift of the simple circuit could be lowered. First, a measurement has been done of a blue LED at about 6 μA. At that current, it gives already enough light to be visible, except in the full sun. It is well above the leakage current of about 0.1 μA. The LED seems to have

−1.4 mV/K slope. A transistor BC557B is measured at about 0.2 mA collector current, where the MOSFET threshold is expected. The transistor has a measured −2.08 mV/K.

Figure 5.15 shows the scheme with compensating network connected to HVED circuit. Both transistor and LED are used in the logarithmic part of their characteristic, so that the voltage drop depends on the type of semiconductor and the current density, and almost not on resistive effects, which makes the voltage drop quite reliable. Components with a quite high-temperature coefficient are NTC resistors or NTC thermistors. Their 'nominal' value is given at 25 °C. Around that temperature the temperature coefficient is about −4.5%/K, which is rather high. The temperature dependence of the NTC can be reduced by putting a resistor in series and then resistor in parallel. Two other resistors are needed to act as a voltage divider. The PNP transistor needs about 1 μA at the base. The used NTC 'ERTJ0EP473J' has a $B = 4100$ K, and a nominal value of 47 K at 25 °C. A more precise approximation that describes the resistance as a function of the temperature of an NTC is

$$R = R_0 e^{B((1/T)-(1/T_0))} \tag{5.2}$$

A choice of resistors has been made using only the E6 series. The resistance circuit at the base NTC, $R_1$, $R_2$, $R_3$ has the following resistance:

NTC = 47 K, $R_1 = 22$ K, $R_2 = 100$ K, $R_3 = 22$ K, $R_4 = 68$ K, $R_5 = 10$ K
LED: 150060BS75000 Blue, SMD, 0603, 30 mA, 3.2 V, 470 nm InGaN
NTC: ERTJ0EP473J NTC THERMISTOR, 47K, 0402 $B$-value = 4100 K

$$R_{\text{base}} = R_3 + \frac{1}{\dfrac{1}{R_1} + \dfrac{1}{R_2 + R_{\text{NTC}}}} \tag{5.3}$$

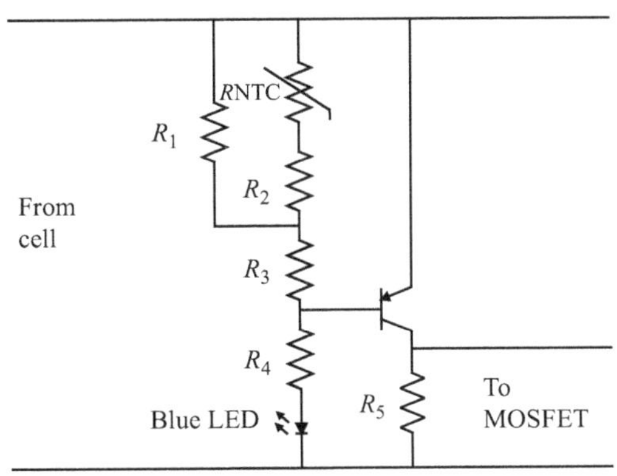


*Figure 5.15 Scheme with compensating network*

The total threshold drop is (at an estimated base current of 1 μA)

$$V_{\text{drop}} = V_{\text{BE}}\left(1 + \frac{R_4}{R_{\text{base}}}\right) + R_4 \times 1\,\mu\text{A} + V_{\text{LED}} \tag{5.4}$$

The measured result is in yellow in Figure 5.16. The measured characteristic corresponds with the calculated one, this means that the simplified equations are enough to find good candidate values for resistors. For measuring the drop, the LED and transistor were the same type, but not the same specimen, and measured in a different set of measurements. An SMD NTC was used as they have an automotive temperature range: −40 °C to 125 °C. The LED has a range of −40 °C up to 85 °C, and the upper limit is generally given by an increased opacity of the lens, which is not a problem here.

If the value of NTC is unknown, we can use a voltage divider to measure the NTC resistor $R_T$. The principle is shown in Figure 5.17. The voltage $V_T$ by voltage

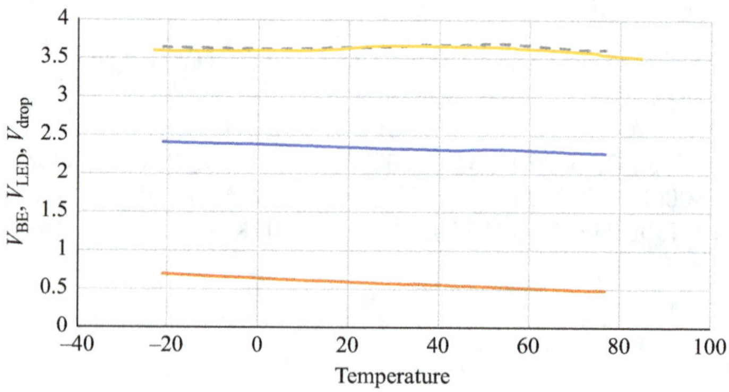


*Figure 5.16 Voltage drops $V_{BE}$ (red), $V_{LED}$ (blue), calculated drop (dashed black), measured (yellow)*

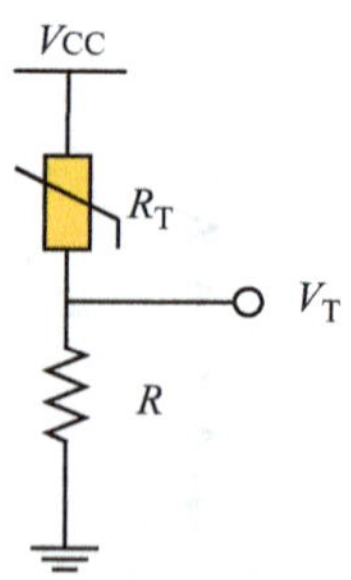


*Figure 5.17 Voltage divider with NTC*

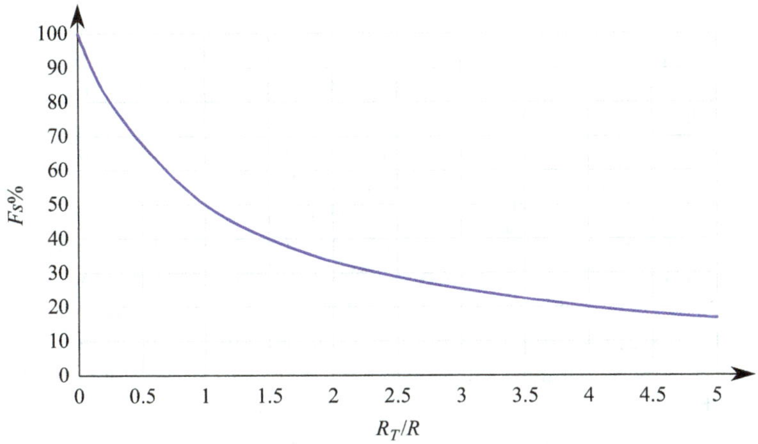


*Figure 5.18 Sensitivity Fs vs $R_T/R$*

divider is calculated as shown later. To solve for NTC resistor $R_T$, we get the following equation:

$$V_T = \frac{R}{R + R_T} V_{CC} \quad \Rightarrow \quad R_T = R\frac{V_{CC} - V_T}{V_T} \tag{5.5}$$

To select the value of $R$ to maximize measurement sensitivity, we can rewrite (5.5).

$$\frac{V_T}{V_{CC}} = \frac{R}{R + R_T} = \frac{R}{R + \left(\frac{R_T}{R}\right)R} = \frac{1}{1 + \frac{R_T}{R}} \tag{5.6}$$

Therefore, the sensitivity of the $R$ can be expressed in (5.6), where $Fs$ is full-scale reading in percentage. Figure 5.18 shows the measurement sensitivity for different values of $R$.

$$\frac{V_T}{V_{CC}} = \frac{100}{1 + \frac{R_T}{R}} \% Fs \tag{5.7}$$

**Summary**

A high cell voltage detection and equalizing circuit for an $LiFePO_4$ battery for an ultra-light electric vehicle is proposed. The proposed system has a simple circuit and very low standby leakage current. The battery pack consists of eight numbers of cells in series with a nominal voltage of 3.3 V. By applying a two LED bandgap, a sufficiently correct voltage is obtained, 1.65 V for each LED (connected in series). It shows a sharp knee and very low leakage current compared to other known techniques so that a self-discharge due to the BMS becomes negligible. On one hand, if the voltage of an individual cell is higher than expected value, the excess power is delivered to the 4 W resistor (operated at 2 W). It also gives a signal that we have reached to the limited voltage. On the other hand, if all cells are high, it also generates a signal to the optocoupler. These detections are displayed by an LCD and also with LEDs mounted on each cell to inform the user that in which condition is the battery charge and giving a signal to the charger to revert to a trickle charge of 0.3 A if one cell is high or a switch off charging completely if all cells are high. The detection will appear on LED with the text message, 'One cell high' is one of the cells are at a high level or 'all cells high', also the LEDs mounted at the end of each circuit will be turned on if the proper cell is high. To make the application smart and compatible with IoT, a Bluetooth module is added to command the microcontroller to turn on or off the charger via a voice or text message by using a smartphone when it is desired to be turned off or turned on. This voice message is provided by phone and it will be received by Bluetooth module which is connected to the microcontroller (Arduino). If a cell is replaced in a battery that has not the same charge as the other cell, it will not make a problem, rather it just takes more time to equalize, but no cell will be damaged. A significant feature of the circuit is to draw a very low standby current as compared with the existing commercially BMS systems for an instant, TL431 which requires a minimum regulating the current of 400 μA. Two resistive equalizers have been proposed, one using two red LEDs as a reference voltage drop. It is simple an sufficient if no extreme temperatures are expected. A second solution was added using a blue LED as a voltage reference and a temperature compensation. It has also a very low current consumption in standby at low cell voltage. In general the user should be informed about the self-discharge of battery and BMS combination, by preference on a visible label, e.g. of the drawn current of a BMS or the maximal time between two full charges. The smart application can also be used in the next chapters. Differential amplifiers described in later chapters can be used here to monitor and display each cell voltage or to detect as well a too low or too high cell voltage.

# *Section 2*
# Individual cell voltage measurement circuit using differential amplifiers and a smartphone

## 5.8 Introduction

To monitor the individual cell voltage, a new, low standby consumption, smart monitoring system based on differential amplifier and Bluetooth has been made. The system provides users to monitor the cell voltages of the battery pack from a safe distance. The smart monitoring system played an important role during the last decades [7–9]. The system has been designed using Arduino Nano as a microcontroller, a Bluetooth module (HC-05), and a smartphone. This microcontroller is inexpensive and capable of measuring eight-cell voltages by eight analogue pins (A0–A7), which is suitable for our special purpose. The Bluetooth module is a serial port protocol (SPP) module, designed for transparent wireless serial connection setup. The Bluetooth module adds a two-way (full-duplex) wireless functionality. This module can be used to communicate between two microcontrollers like Arduino or communicate with any device with Bluetooth functionality like a smartphone or laptop. There are many android applications that are already available, which makes this manner a lot more comfortable. It is pretty straightforward to pair the HC-05 module with a microcontrollers as it operates using the SPP. Simply power the module with +5 V and connect the Rx pin of the module to the Tx of a microcontroller unit (MCU) and Tx pin of the module to Rx of MCU. Some other applications of the Bluetooth module are

- wireless communication between two microcontrollers,
- communicate with laptop desktops and mobile phones,
- data logging application,
- consumer applications,
- wireless robots,
- home automation, and
- wireless mouse, wireless keyboard.

This Bluetooth module can also be used for other applications like wireless headset, game controllers, and many more consumer applications. The operating range is up to <100 m, which depends upon transmitter and receiver, atmosphere, geographic conditions. The Bluetooth serial modules allow all serial-enabled devices to communicate with each other using Bluetooth. The HC-05 pins is described as follows:

- **Key/EN:** it is used to bring Bluetooth module in AT commands mode. If the pin is set to high, then this module will work in command mode. Otherwise by default, it is in data mode. The default baud rate of HC-05 in command mode is 38 400 bps and 9600 in data mode.
- $V_{cc}$**:** Connect 5 V or 3.3 V to this pin.
- **GND:** Ground pin of module.
- **TxD:** Transmit serial data.
- **RxD:** Receive data serially.
- **State:** It tells whether module is connected or not.

The Bluetooth module-05 has a red LED which indicates connection status, whether the Bluetooth is connected or not. Before connecting to HC-05 module, this red

LED blinks continuously in a periodic manner. When it gets connected to any other Bluetooth device, its blinking slows down to 2 s.

To monitor the cell voltages, first, the individual cell voltages are measured by analogue pins of Arduino Nano via isolated voltage probes, and then the data will be transmitted wirelessly by Bluetooth module to a new android application. The smartphone will receive the data from the Bluetooth module, and the measured cell voltages can be displayed on the smartphone screen. The schematic of the isolated cell voltage measurement is shown in Figure 5.19.

In order to measure each individual cell voltage, a voltage measuring circuit is built up based on an operational amplifier which is used as a differential amplifier. In this way, the electronic circuit for each cell remains the same. The resistors have the same value because the voltage of each cell is less than 5 V, so the amplification factor can be just 1. The cell voltage can be measured by connecting to the inverting and non-inverting input of the op-amp. However, care has to be taken to remain within the common mode voltage range of the operational amplifier. The output of the op-amp will be measured by analogue pins of the microcontroller (A0–A7), in order to calculate each individual cell voltage. A Bluetooth module "HC-05" will be used for measurements. The properties are presented in Table 5.1. The measured voltage is transmitted via this Bluetooth to the smartphone and it will be displayed on the screen every second. To have more resolution, the measured

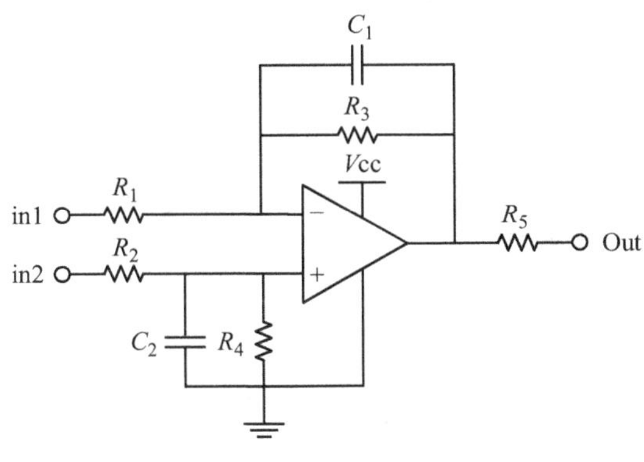


*Figure 5.19 Differential cell voltage measurement circuit*
$R_1$, $R_2$, $R_3$, $R_4$: 1 M; 0.1% $R_5$: 10 K

*Table 5.1 Properties of the Bluetooth module HC-05*

| Parameter | Values |
|---|---|
| Frequency | ISM band, 2.4 GHz |
| Synchronous | 1 Mbps/1 Mbps |
| Power supply | +3.3 VDC, 50 mA |
| Working temperature | (−25 to 75)°C |

voltage will be shown in five digits. The properties of Arduino Nano are presented in Table 5.2.

In Figure 5.20, the schematic of Arduino Nano is presented. The microcontroller has eight analogue pin 'inputs' which is enough to measure eight cell voltages.

The proposed monitoring system is shown in Figure 5.21. This recent android smartphone application is developed by MIT App Inventor 2. This application is an open-source platform available on Google. During the past few years, mobile devices have developed very quickly and have become more than just a means of communication between individuals due to the different technologies that these devices now

*Table 5.2 Arduino nano properties*

| Parameter | Values |
|---|---|
| Microcontroller | ATmega328 |
| Operating voltage | 5 V |
| Clock speed | 16 MHz |
| Analogue I/O pins | 8 |
| DC current per I/O pins | 40 mA (I/O pins) |
| Input voltage | 7 V–12 V |
| Digital I/O pins | 22 |
| PWM output | 6 |

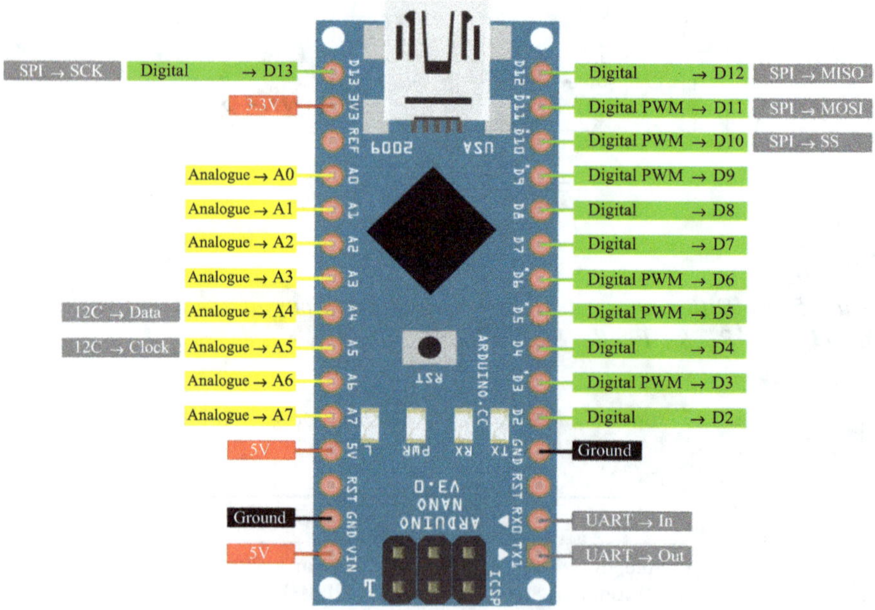


*Figure 5.20 Arduino Nano*

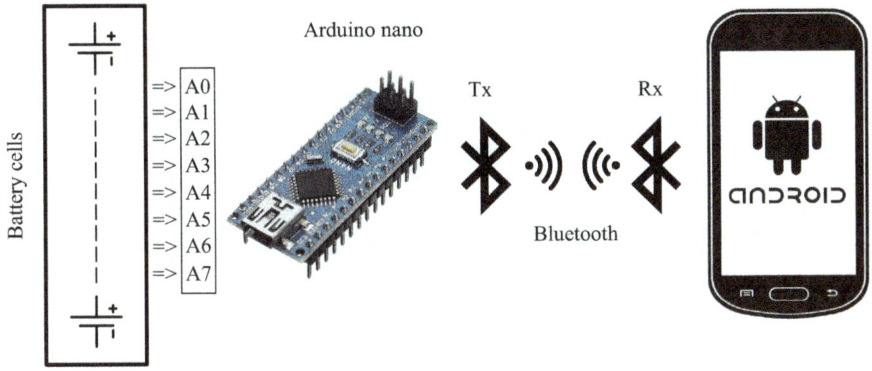


*Figure 5.21 The monitoring system of the battery cells*

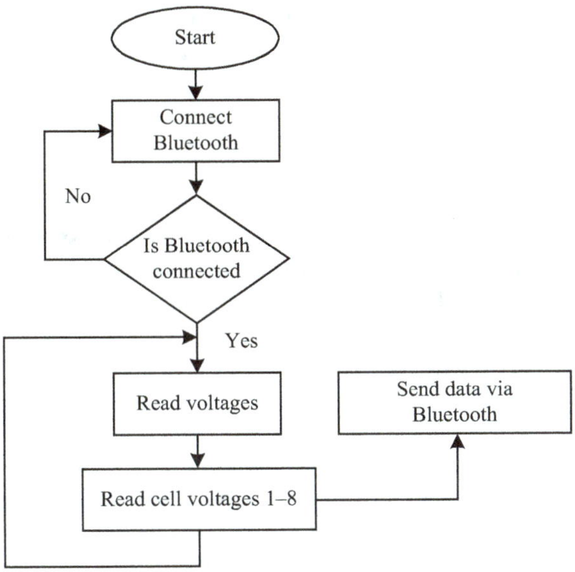


*Figure 5.22 Flowchart of the monitoring system*

have that enable them to communicate with surrounding devices. Some of these technologies are present in mobile devices such as Wi-Fi, Bluetooth, and Infrared.

The flowchart of the proposed monitoring system is demonstrated in Figure 5.22. First, the measured voltage values by the differential amplifier will be sensed by Arduino input pins. Second, the Bluetooth bottom of the mobile app screen should be pressed to connect the Bluetooth to the mobile. Next, if it is connected, it will read the voltages and the data will be displayed on the screen.

Different types of communication technologies were integrated with the smart monitoring of measurements and control devices [10–13]. Besides, many types of control systems based on Arduino are presented in the literature [14–16]. Measuring

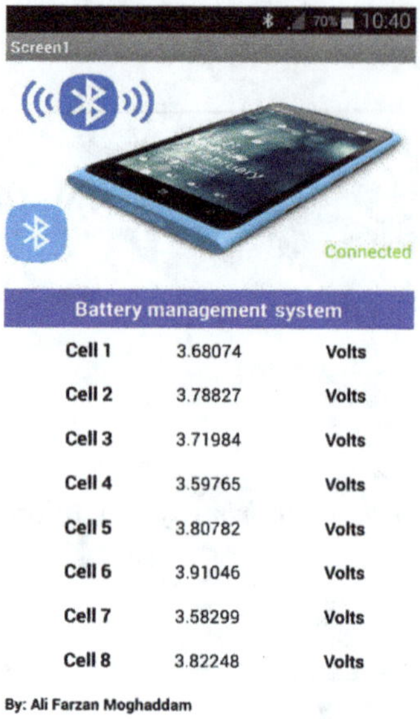


*Figure 5.23 Cell voltage values received by smartphone via Bluetooth*

and protection circuits are one of the primary aspects of all BMSs. The cell voltage monitoring results are shown in Figure 5.23. The monitoring system will update the measured cell voltages each second and display them on the screen. If different packs are enabled in a sequence, the value of a higher voltage battery can be read out.

## 5.9 Estimation of the standby current

For the accuracy, resistors of 1 MΩ are chosen, as even 0.1% is affordable. Errors in resistance are amplified by the fact that a high common mode voltage is present. The supply of the Arduino can be turned off after for example 3 days to reduce the standby consumption, also the supply of the op-amp, but the resistors remain. If the clamping diodes of the op-amp clamp the resistors to 0 V, the sum is $8+(7+6+5+4+3+2+1)\times 2\times V\text{cell}/1\ \text{M}\Omega$. At $V\text{cell} = 3.6$ V, it is $64\times 3.6 = 209\ \mu\text{A}$ for the lowest cell, so the discharge would be 1.767 Ah after 1 year. It could be lowered by using higher resistance values or eliminated while using low current SSR solid state relays (also bipolar transistors could do that job) that disconnect the measuring points.

## 5.10 Variant for 16 channels

Using eight Li-ion cells of 3.6 V is 28.8 V nominal. If 16 cells are used, it corresponds to 57.6 V. Apparatus of 48 V, with enough voltage tolerance, can be designed. Also for that voltage, it opens the way to telecom, mild hybrid in cars, or to recycle such modules for other applications. Using 16 channels can be supported by an Arduino Mega. The Arduino can have its reference in the middle. The upper eight cells can have a similar circuit as mentioned earlier, the op-amps supplied by +20 V and 0 V. The lower eight cells need operational amplifiers that are supplied by +5 V and –16 V. It can still fulfil the specs of rail-to-rail input and output op-amps. The same principle of putting the processor at middle potential could be used as well for eight cells only to lower the standby current by a factor 4. It is good to provide a (PTC) fuse to the middle connection, as some people may think that processors are always connected to the lower voltage pin of a battery.

## Summary

This section aimed at designing a low-cost individual cell voltage measurement circuit and new Android smartphone application to monitor the measuring voltage from a safe distance. The smart measurement system was designed based on an Arduino Nano board used as a microcontroller to read the cell voltage. The system can be divided into three parts: measurement circuit, wireless communication system, and Android smartphone monitoring application. First, the individual cell voltage is measured by an operational amplifier which is used as a differential amplifier. Next, the measured data are transferred via Bluetooth to the smartphone. In the end, the data which are the cell voltages will be displayed on the smartphone screen. Bluetooth was used as a wireless communication instrument between Arduino and smartphone. The smartphone monitoring application was a new application designed by the open-source developed program (MIT App Inventor 2). Moreover, the standby current can be kept low, below 0.5 mA, or even below 0.1 mA. The main advantages of the system are low cost, low standby, and safety.

## References


[1] Saxena N, Hussain I, Singh B, *et al.* Implementation of a grid-integrated PV-battery system for residential and electrical vehicle applications. *IEEE Transactions on Industrial Electronics.* 2017;65(8):6592–6601.

[2] Gong C, Liu J, Han Z, *et al.* Analysis on uncontrolled generation in electrical vehicles and a battery protection method. In: 2017 IEEE 3rd International Future Energy Electronics Conference and ECCE Asia (IFEEC 2017-ECCE Asia). IEEE; 2017. p. 1606–1610.

[3] Camacho OMF, Nørgård PB, Rao N, *et al.* Electrical vehicle batteries testing in a distribution network using sustainable energy. *IEEE Transactions on Smart Grid.* 2014;5(2):1033–1042.

[4] Miniguano H, Raga C, Barrado A, *et al.* A comparative study and parameterization of electrical battery models applied to hybrid electric vehicles. In: 2016 International Conference on Electrical Systems for Aircraft, Railway, Ship Propulsion and Road Vehicles & International Transportation Electrification Conference (ESARS-ITEC). IEEE; 2016. p. 1–6.

[5] Hussein AA. Capacity fade estimation in electric vehicle li-ion batteries using artificial neural networks. *IEEE Transactions on Industry Applications.* 2014; 51(3):2321–2330.

[6] Xintian L, Guojian Z, Yao H, *et al.* Research on the capacity fading characteristics of a Li-ion battery based on an equivalent thermal model. In: Proceedings of 2015 International Conference on Intelligent Computing and Internet of Things. IEEE; 2015. p. 145–150.

[7] Salamone F, Belussi L, Danza L, *et al.* An open source "smart lamp" for the optimization of plant systems and thermal comfort of offices. *Sensors.* 2016;16(3):338.

[8] Alsibai MH and Siang HM. A smart driver monitoring system using android application and embedded system. In: 2015 IEEE International Conference on Control System, Computing and Engineering (ICCSCE). IEEE; 2015. p. 242–247.

[9] González F, Villegas O, RamÍrez D, *et al.* Smart multi-level tool for remote patient monitoring based on a wireless sensor network and mobile augmented reality. *Sensors.* 2014;14(9):17212–17234.

[10] Joaquinito R and Sarmento H. A wireless biosignal measurement system using a SoC FPGA and Bluetooth low energy. In: 2016 IEEE 6th International Conference on Consumer Electronics-Berlin (ICCE-Berlin). IEEE; 2016. p. 36–40.

[11] Varshney V, Goel RK, and Qadeer MA. Indoor positioning system using Wi-Fi & Bluetooth low energy technology. In: 2016 Thirteenth International Conference on Wireless and Optical Communications Networks (WOCN). IEEE; 2016. p. 1–6.

[12] Kumar S and Solanki S. Voice and touch control home automation. In: 2016 3rd International Conference on Recent Advances in Information Technology (RAIT). IEEE; 2016. p. 495–498.

[13] Sabani AH and Jailani R. Android based control and monitoring system for leg orthosis. In: 2015 IEEE 11th International Colloquium on Signal Processing & Its Applications (CSPA). IEEE; 2015. p. 40–45.

[14] Müller L, Mohammed M, and Kimball JW. Using the Arduino Uno to teach digital control of power electronics. In: 2015 IEEE 16th Workshop on Control and Modeling for Power Electronics (COMPEL). IEEE; 2015. p. 1–8.

[15] Wang Y and Chi Z. System of wireless temperature and humidity monitoring based on Arduino Uno Platform. In: 2016 Sixth International Conference on Instrumentation & Measurement, Computer, Communication and Control (IMCCC). IEEE; 2016. p. 770–773.

[16] Kovačević H and Stojanović Ž. Buck converter controlled by Arduino Uno. In: 2016 39th International Convention on Information and Communication Technology, Electronics and Microelectronics (MIPRO). IEEE; 2016. p. 1638–1642.

*Part III*

# Active cell balancing circuits

# Chapter 6
# Resonant switched-capacitor balancing

## 6.1 Introduction

This chapter presents the concept of a cell balancing method for lithium-ion batteries for electric vehicles, based on a resonant switched-capacitor converter (RSCC) balancing technique. It equalizes eight numbers of cells in a series. In Li-ion batteries, voltage differences always exist between cells due to the charging and discharging process; therefore, a battery management system is required to ensure that all cells are equally charged or discharged and it also increases the life time of the battery. An equalizing method is essential to achieve the best performance.

A number of cell balancing methods have been presented. Among the active balancing topologies, the switched-capacitor converter (SCC) is popular as it can be partly integrated in IC technology. The conventional capacitor-based balancing circuit is shown in Figure. 6.1. For simplicity, only two cells are presented. They use only capacitors in the power stage for power storage and transferring energy. Its features are low volume, low cost. However, a low balancing speed is an important problem of the SCC, especially when the number of cells is increased. One of the charge transport devices is a row of capacitors which are alternately connected to neighbouring batteries.

In this method, the switches are controlled in such a way that switches operate backwards and forwards between batteries at a fixed frequency without regard to the state of charge. The advantages of this system are that no sensing or closed-loop control is needed and that the process is self-limiting. When voltage equalization is complete, the switching of the capacitors consumes minimal energy. The switched-capacitor balancing method requires $2N$ switches and $N-1$ capacitors per $N$ cells.

To improve the cell balancing speed, several switched-capacitor methods were investigated in the literature [1,2]. In order to raise the balancing speed of capacitor-based balancing, a double-tiered switched-capacitor topology has been considered in [3]. The technique can decrease balancing time by using supplementary bridging. The probability of cell imbalance among outer cells is high because of the different gradient of ambient temperature. In this chapter, a resonant switching capacitor is proposed to reduce the balancing time and reduce energy loss. The simulation results of both a conventional switched-capacitor and an RSCC are performed to verify the feasibility of the system.

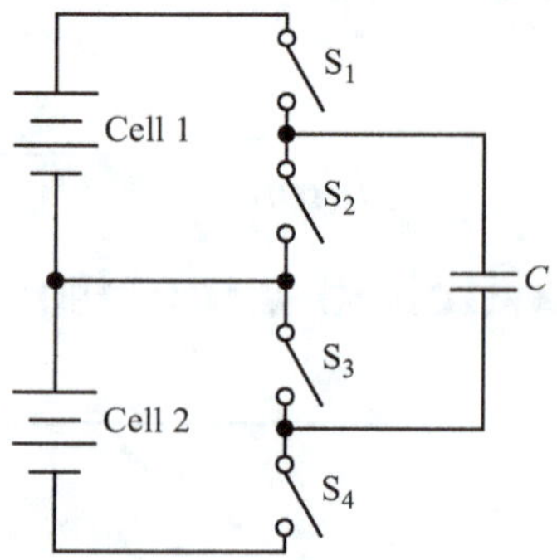


*Figure 6.1 The conventional capacitor-based balancing circuit*

## 6.2 The conventional switched-capacitor balancing

The conventional SCC circuit with eight batteries is shown in Figure 6.2. It consists of 16 switches $S_1$–$S_{16}$. All switches are controlled to be turned on and turned off simultaneously with the desired switching frequency with a duty cycle of 50%. According to the switching connection, this circuit can be divided into two modes. To analyse the circuit, assume that the voltages of the adjacent battery cells are different; thus, the charge will be transferred between adjacent cells through the common capacitor.

So the common capacitor connected between the two cells is connected to the upper cell and the lower cell with the same duty ratio of 50%. The voltage of the common capacitor is equal to the average voltage of the two adjacent cells. In the conventional SCCs circuits, the equalizing speed is low, especially when the number of cells increases. Figure 6.2 shows the equivalent circuit of the conventional switched-capacitor balancing circuit.

In stage one, the odd switches are turned on while the others are turned off. The $R_{DS(on)}$ stands for 'drain-source on resistance', or the total resistance between the drain-source of the MOSFETs during turn-on, $R_{ESR}$ is the equivalent series resistance of the capacitors. If the MOSFETs are similar, the $R_{DS(on)}$ can be equal. For charging, typically two $R_{DS(on)}$ are in series.

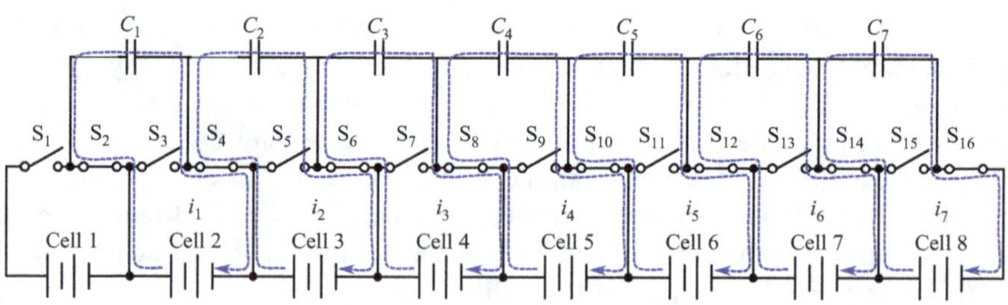


*Figure 6.2 The conventional capacitor balancing circuit*

Typically, multilayer ceramic capacitors are used, as electrolytic capacitors have too much series resistance and hence slow down the equalizing action. Also electrolytic capacitors might reduce the lifetime of the BMS and hence maybe the battery also. If 10 μF capacitance is used, and 0.05 Ω MOSFET, two are in fact in series, it results in 1 μs time constant. If 50 kHz is used, it corresponds with a half period of 10 μs. So the capacitor has about 10 time constants to charge, and can be considered as fully charged. If much higher frequencies are used, more EMC (radio interference) is generated and more gate charge losses occur. The charging process involves two MOSFET voltage drops, the discharging as well, so 4 in total, but the current is limited by the charge of the capacitor. In Figure 6.2, the conventional capacitor balancing, the transfer is very poor from C1 to C7 as it involves virtually only one-seventh of the capacitance. It is better in Figure 6.3 with the double tired switched capacitor scheme. The designer should take in account that multilayer capacitors can be very voltage dependent, and that the second and third capacitor row may need other voltage types.

In stage two, the even switches are turned on while the others are turned off. Each time, between adjacent cells, a charge is transmitted of C ΔV in the case the time constant is much smaller than a half period. The capacitor value acts as a "resistance" of:

$$Z_{eqi} = f\, C_i \tag{6.1}$$

If $T_{on}$ is not very large, the first half period will charge a part of the voltage difference fraction:

$$1 - e^{\frac{T_{on}}{\tau}} \tag{6.2}$$

The time constant might be (two MOSFETS in series to charge the voltage difference)

$$\tau = 2\, R_{dson}\, C \tag{6.3}$$

In steady state the voltage is again the same after one period, if that is expressed, we get for steady state:

$$Z_{eqi} = \frac{2}{f\, C_i \dfrac{1}{1 + \dfrac{1}{1 - e^{\frac{T_{on}}{\tau}}}}} \tag{6.4}$$

Note that $\tau$ contains two MOSFET resistances. The $Z_{eqi}$, for very big capacitors, gets its lowest value, 4 times the MOSFET resistance. It is correct as two MOSFETs in series and two other discharge the capacitor. A more mathematical approach can be found in [4].

## 6.3 Double-tiered switched-capacitor balancing

It this section, the operational principle of the double-tiered switched-capacitor balancing circuit will be investigated. As it has been discussed before, in switched-capacitor balancing, one capacitor is connected between every two neighbouring cells through switches. Therefore, the charge will be transferred between higher cells to lower cells. The switched-capacitor balancing circuit requires 2$N$ switches, and $N-1$ capacitors to equalize $N$ cells. This method is also known as the adjacent cell-to-cell method because the energy transferring is from one cell to the neighbouring cell. The circuit does not require an intelligent control circuit for the switches. The circuit is easy to implement, but it has a long equalization time. To overcome this difficulty, a double-tiered switched-capacitor balancing circuit is studied. The circuit is shown in Figure 6.3, respectively. The capacitors, $C_1$–$C_7$, are adjacent cell capacitors, and capacitors $C_8$–$C_{13}$ are nonadjacent capacitors. The circuit operation and control are the same as the switched-capacitor method, but it has extra capacitors in parallel with the conventional switched-capacitors method. By using the extra capacitors, the charge transfer between nonadjacent cells can be led in one switching cycle. For instance, the capacitor $C_8$ exchanges the charge between cell 1 and cell 3. Therefore, the charge is transferred to distant cells in a small number of switching cycles.

In the first stage, the odd switches $\mathbf{S}_1-\mathbf{S}_{15}$ will be turned on while the other switches are kept turned off as shown in Figure 6.3. The fact that more distant loops are present, improves a lot the 'far-away' equivalent capacitance, resulting in a 'backbone'.

In the second stage, when odd switches are off, the even switches will be turned on as shown in Figure 6.3. As a result, the balancing time of the double-tiered switched-capacitor balancing circuit can be reduced as compared to the switched-capacitor balancing. On the other hand, it requires more components (capacitors) to transfer the energy for equalizing.

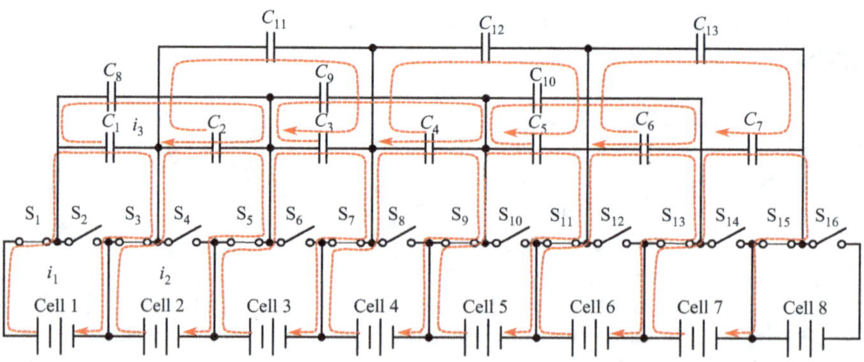


*Figure 6.3 Double-tiered switched-capacitor balancing*

## 6.4 The proposed resonant switched-capacitor balancing

In Figure 6.4, the block diagram of the proposed circuit is presented. All switches are N-channel MOSFETs with body diodes and controlled by a pair of complementary signals in synchronous trigger patterns with the duty cycle of 50%. The operational principle of the circuit can be divided into two stages. For half of the switching period, the capacitors are connected through the switches to the upper cells, which is called stage one, and for the second half of the switching period, they are connected to the lower cells as defined stage two.

In stage one, the odd switches $S_1$–$S_{15}$ are turned on to store energy to the resonant tanks while the even switches $S_2$–$S_{16}$ are kept turned off. When the switches $S_1$–$S_{15}$ are turned on, the energy of the cell 1, cell 3, cell 5, and cell 7 will be transferred to the resonant tanks; thus, the resonant tank will be connected in parallel with odd cells 1–7. Thus, the capacitors $C_1$–$C_7$ will be charged. The voltage of the capacitors will be increased and resonant current increases. The mesh equations can be expressed in the following equation:

In stage two, the even switches $S_2$–$S_{16}$ are turned on, while the odd switches are turned off. To investigate the system, it is assumed that the system is in steady-state and the initial capacitor voltage dissipated or charged.

All switches resistances are modelled as $R_{DS(on)}$, and capacitors equivalent series resistances as ESR. To reduce the energy loss produced during the balancing process, the resonant switched-capacitor balancing topology is introduced. Thus, by adding the resonant inductor, the switched-capacitor balancing circuit is improved. The resonant current for the first resonant tank $L_1$ $C_1$ in parallel with cell 1 and equivalent voltage equation can be expressed as

$$i_{(\text{resonant})} = C_1 \frac{dV_{C_1}}{dt} \quad \text{and} \quad L_1 \frac{di_{(\text{resonant})}}{dt} + V_{C_1} = V_{\text{cell}_1} \tag{6.5}$$

Therefore, the solution for (6.5) can be expressed as

$$V_{C_1} = V_{\text{cell}_1} + V \cos(\omega t + \phi) \tag{6.6}$$

$$i_{(\text{resonant})} = I_{\max} \sin(\omega t + \varphi) \tag{6.7}$$

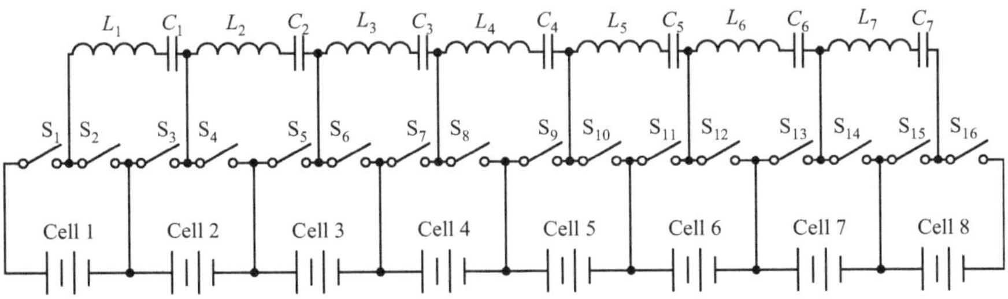


*Figure 6.4 The proposed resonant switched-capacitor balancing circuit*

where $I_{max}$ is the amplitude of the resonant current, $V$ is the amplitude of the $V_{C1}$, $\omega$ is the resonant angular and $\varphi$ is initial angle of the resonant state. The resonant angular frequency can be driven as

$$\omega = \frac{1}{\sqrt{LC}} \tag{6.8}$$

In stage two, the even switches $S_2$–$S_{16}$ will be turned on while the odd switches $S_1$–$S_{15}$ are kept turned off. Therefore, the resonant tanks will be connected in parallel with the even cells 2–16. Thus, the even cells will be charged and resonant current will increase, but this time in the opposite direction.

A more deep mathematical approach can be found in [4], but some simplified approach is given here. During the *first half period*, the capacitor voltage is $V_{cell1}$, so the capacitor will be charged and the amplitude of the resonant current can be expressed as, the typical transient of an LC circuit:

$$I_{max} = (V_{cell1} - V_{cell2})\sqrt{\frac{C_1}{L}} \tag{6.9}$$

However the capacitor has a bigger charge after that and will generate larger resonant voltages in the next half periods, a typical behaviour of series resonant converters. In practice the transfer will be limited by the voltage drop across switched- as well as internal capacitor resistor (ESR). Note that for using the resonance well, the transistors should be 'ON' during half a resonant period, and not just 50% duty ratio. The total equivalent series of the circuit can be expressed as $R_{eq} = R_{DS(on)} + R_{ESR}$.

In practice it means that the equivalent impedance of the capacitance does not appear any more in the transfer at resonance, but some equivalent series impedance of the L-C connection, as the impedance of C vanishes against L. A much faster equalizing may be obtained at the same frequency, compared to capacitors alone, but at the expense of the inductors.

Note that the resonance half period can be set shorter than the half switching period, and if the duty ratio is also lowered, still half wave sine periods can be obtained. The impedance rises then with 0.5 $T/T_{on}$.

For details, a simulation is still worthwhile. But it is better to do it, while using a lower equivalent capacitance than the real battery to have a limited simulation time and still detailed waveforms.

## 6.5 Simulation results

To verify the feasibility of the proposed circuit, the simulation results are expressed. In order to model the proposed circuit in MATLAB®, the cells are modelled with the capacitors with the value of 50 mF and the resonant tanks are modelled with the 10 μH inductors and 10 μF capacitors. All the switches are MOSFETs with the body diodes, and they are triggered with the synchronous pulse, generated by a pulse generator with the duty ratio of 50% and 45% to investigate different conditions. The battery voltages are chosen to be from 3 V to 3.6 V unequal to present the significant voltage difference of the cells. The resonant frequency is adjusted to 15.915 kHz, which originated from the pulse generator. The simulation parameters are shown in Table 6.1. The resonant current changes from zero at the beginning of each state. Therefore, to have zero current, the resonant frequency should be multiple of switching frequency. The resonant frequency with the inductor value of $L = 10\ \mu\text{H}$ and capacitor values of $C = 10\ \mu\text{F}$ can be calculated as

$$f = \frac{1}{2\pi\sqrt{LC}} \tag{6.10}$$

In order to control the MOSFETs, a synchronous pulse with the duty ratio of 50% is applied to the gate of the MOSFETs with the switching frequency of $f$s = 15.9 kHz ($f$s = $f$resonant). To compare between the proposed RSCC circuit, the conventional SCC, and also double-tiered switched-capacitor balancing, the simulation result of the all conventional circuits are presented. In Figure 6.5(a) and (b), cell voltages of the proposed RSCC circuit are presented.

To analyse the conventional switched-capacitor balancing circuit, the SCC is modelled with eight number of cells in series. The cell voltage is chosen to be from

*Table 6.1 Simulation parameters*

| **Parameter** | **Values** |
|---|---|
| Equivalent capacitors | 50 mF |
| Initial capacitor voltages | 3 V–3.6 V |
| Switching frequency | 50 kHz |
| Resonant capacitors | 10 μF |
| Resonant inductors | 10 μH |
| Resonant frequency | 15.9 kHz |

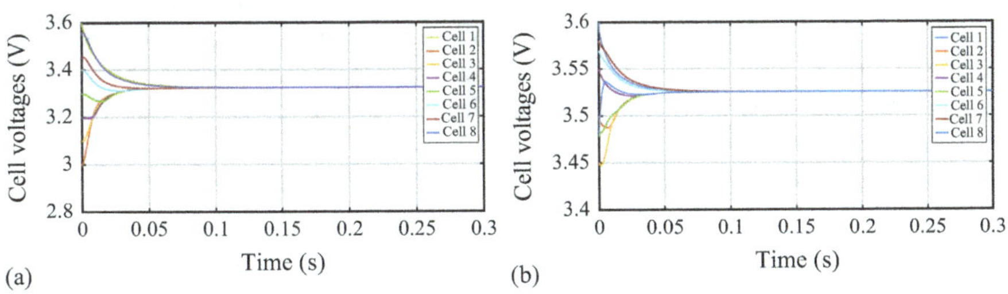


*Figure 6.5 Cell voltages of proposed resonant switched-capacitor converter circuit: (a) proposed (RSCC) balancing and (b) proposed (RSCC) balancing (more cell differences, to compare with Figure 6.6)*

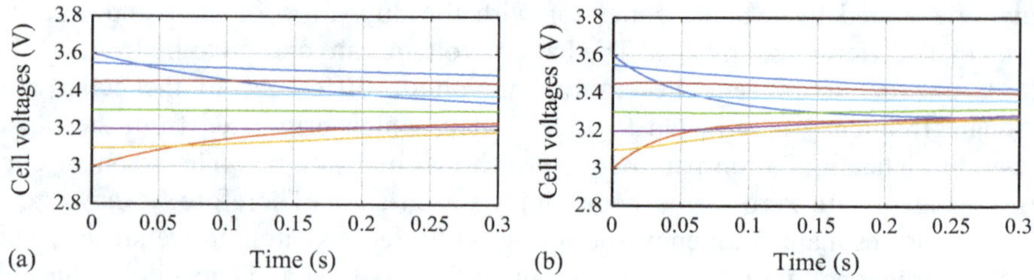


*Figure 6.6 Cell voltages of conventional switched-capacitor balancing: (a) conventional (SCC) balancing f = 15.9 kHz and (b) conventional (SCC) balancing f = 50 kHz*

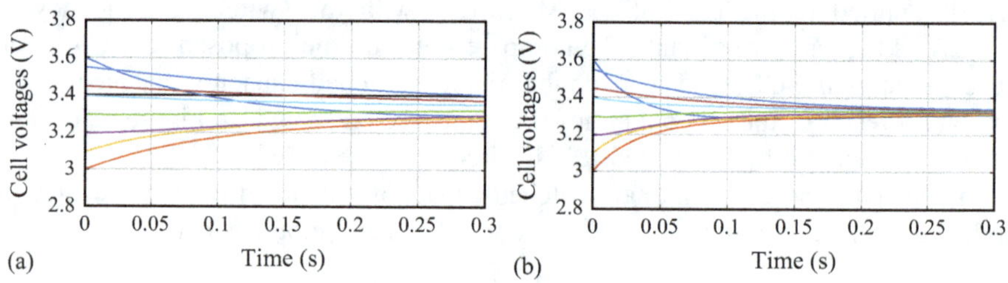


*Figure 6.7 Cell voltages of double-tiered switched-capacitor balancing: (a) doubletiered (SCC) balancing f = 15.9 kHz and (b) double-tiered (SCC) balancing f = 50 kHz*

3 V to 3.6 V to have a significant voltage difference between the cells. The capacitor values are 10 μF, the switching frequency is 15.9 kHz with 50% duty ration. The switches are MOSFETs with body diodes and are controlled with the synchronous pulse. The simulation results of the cell voltages are shown in Figure 6.6(a).

In Figure 6.6(b) the simulation results of the cell voltages with increased switching frequency from 15.9 kHz to the frequency of (50 kHz) is presented. It can be noticed that the balancing time is decreased due to increasing of the switching frequency. As it can be observed, the equalization time is slow compared to the proposed (RSCC) circuit as shown in Figure 6.5. All the parameters for the components are the same as the RSCC circuit in order to consider the difference between the two methods.

Finally, to analyse the double-tiered switched-capacitor circuit, the simulation results of the cell voltages are shown in Figure 6.7(a) and (b), respectively. As before, first the result was driven with the frequency of 15.9 kHz then with the frequency of 50 kHz. All the parameters for the components are the same as the RSCC circuit and conventional switched-capacitor circuit. It can be observed that the equalization time is slow compared to the proposed (RSCC) circuit, but it is faster than conventional circuit. Also like before, the balancing time increases at increasing frequency.

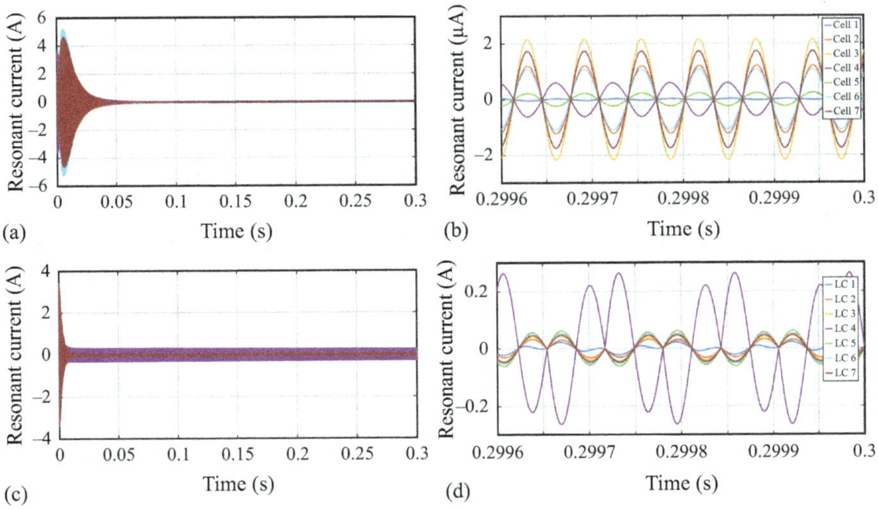


*Figure 6.8 The resonant currents of the resonant tanks 1–7: (a) resonant currents ($fs = fresonant$), (b) resonant currents in detail ($fs = fresonant$), (c) resonant currents ($fs = 1/2\,fresonant$), (d) resonant currents in detail ($fs = 1/2\,fresonant$)*

The resonant currents of the resonant tanks $L_1C_1$ to $L_1C_7$ are presented in Figure 6.8(a). It can be seen that in the transient period when the cells are not equalized the current is oscillating, then after the cells are balanced equally, the resonant current becomes zero. Figure 6.8(b) shows the resonant currents of the resonant tanks in detail. In the proposed circuit as explained before, the switching frequency and the resonant frequency are chosen to be the same in order to obtain zero current, when the cells are equalized. To consider the difference response, the switching frequency $fs$ is chosen to be half of the resonant frequency $fresonant$ ($fs = 1/2\,fresonant$). Figure 6.8(c) shows the resonant currents when the switching frequency is half of the resonant frequency. It can be observed that the current is not equal to zero in the steady state because of the resistor in the resonant circuit which causes oscillation. In Figure 6.8(d) the resonant currents in detail are expressed.

## Summary

In this chapter, the concept of cell equalization technique based on resonant switched-capacitor converter (RSCC) is investigated. The switched-capacitor-based equalizer has many advantages, such as easy implementation, low volume, low cost, and high power density compared to other equalization methods. The conventional switched-capacitor balancing method has a disadvantage of long equalization time. The double-tiered switched-capacitor balancing circuit has faster balancing time compared with the switched-capacitor balancing, but it requires extra capacitors; therefore, resulting in more components. In order to overcome this problem, the proposed topology improves the balancing time compared to the conventional method. This method uses the resonant tank, including the resonant inductor and capacitor to equalize the cells. It equalizes eight number of cells in a one battery pack, faster compared to the conventional topology due to zero current turn-on and off, thus less switching loss is achieved. To verify the feasibility of the system the simulation results are performed.

## References

[1] Uno M and Tanaka K. Single-switch multioutput charger using voltage multiplier for series-connected lithium-ion battery/supercapacitor equalization. *IEEE Transactions on Industrial Electronics*. 2012;60(8):3227–3239.

[2] Emadi A, Lee YJ, and Rajashekara K. Power electronics and motor drives in electric, hybrid electric, and plug-in hybrid electric vehicles. *IEEE Transactions on Industrial Electronics*. 2008;55(6):2237–2245.

[3] Baughman AC and Ferdowsi M. Double-tiered switched-capacitor battery charge equalization technique. *IEEE Transactions on Industrial Electronics*. 2008;55(6):2277–2285.

[4] Moghaddam AF and Van den Bossche A. A cell equalization method based on resonant switched capacitor balancing for lithium ion batteries. In: 9th International Conference on Mechanical and Aerospace Engineering (ICMAE). IEEE; 2018. p. 337–341.

# Chapter 7
# Active equalizer using separate inductors

## 7.1 Introduction

Since the Li and other ion battery types get a better energy/weight at a more affordable level, many applications did emerge. Lithium-ion batteries are widely used in many applications such as electric vehicles, and motorbikes [1,2], due to their advantages such as high power density, low self-discharge rate, lower weight, higher safety, no memory effect [3–5]. The open-circuit voltage of the battery is low; thus in many applications, Li-ion batteries are connected in series [6,7]. Due to the charging and discharging cycles, the charge of the cells may differ.These differences are due to the chemical and electrical characteristics, such an asymmetrical degradation with ageing, production tolerance, and internal impedance [8]. These imbalances reduce the life cycle and utilization rate of the battery [9,10]. The lead-acid battery and NiMH show an increased leakage at high cell voltage. Li-ion-types do not show an increased leakage current but degrade or may become unstable. A battery management system can avoid too much or too little cell charge, and a number of other tasks as monitoring temperature or limiting peak currents. Many battery equalization methods have been proposed in the literature [11–15]. To overcome the problem of the energy loss, an active cell balancing technique is proposed. These are divided into two categories: charge pass-through type and multi-winding transformer type. In the first type, it transfers excessive energy from overcharged cells to the uncharged cells by using switched-capacitor [16–20], inductor [21–23], and multi-winding transformers [24–29]. The most important disadvantage of the switched-capacitor converter is a long balancing time and low speed compared to the inductor-based balancing method. The transformer balancing may have smaller balancing time but may have disadvantages such as higher cost and magnetic losses. The switched inductor balancing method utilizes one inductor for transferring energy between the whole pack. They are featuring a fast equalization time.

*Section 1*

# Inductor balancing circuit 'uncoupled inductors'

## 7.2 Conventional inductor balancing circuit (non-coupled)

The conventional inductor based balancing is shown in Figure 7.1. The principle operation is divided into two states. In the first states, the switches $S_1$, $S_2$, and $S_5$ are turned on in order to store the energy in the inductors from cells while the switches $S_3$, $S_4$, and $S_6$ are kept turned off. In the second stage, switches $S_1$, $S_2$, and $S_5$ are turned off while the switches $S_3$, $S_4$, and $S_6$ are turned on in order to transfer the energy which is stored in the inductors to the neighbouring cells.

## 7.3 A faster inductor balancing circuit (non-coupled)

The block diagram of the proposed circuit is shown in Figure 7.2. This topology consists of $N-1$ inductors per $N$ cells. All switches are MOSFETs with body diodes and controlled by a pair of complementary signals in synchronous trigger patterns with the duty cycle of 50%. The operational principle of the circuit can be divided into two stages.

1. In stage one, the switches $S_1-S_7$ are turned on to store energy to the inductors while the switches $S_8-S_{14}$ are turned off. When the switches $S_1-S_7$ are turned on the energy of cells 1, 3, 5, and 7 will be transferred to the inductors $L_1-L_7$. The inductor voltage of $L_1$ can be driven as

$$VL_1 = L_1 \frac{\mathrm{d}i_{L_1}}{\mathrm{d}t} = V_{\mathrm{cell}_1} \quad V_{L_K} = L_K \frac{\mathrm{d}iL_k}{\mathrm{d}t} = V_{\mathrm{cell}_n} \tag{7.1}$$

and similarly inductor voltages of $V_{L_2}$, $V_{L_3}$, and $V_{L_4}$ can be expressed the same, where $k = 2, 3, 4$, and $n = 3, 5, 7$. The inductor voltages of $V_{L5}$ and $V_{L6}$ can be

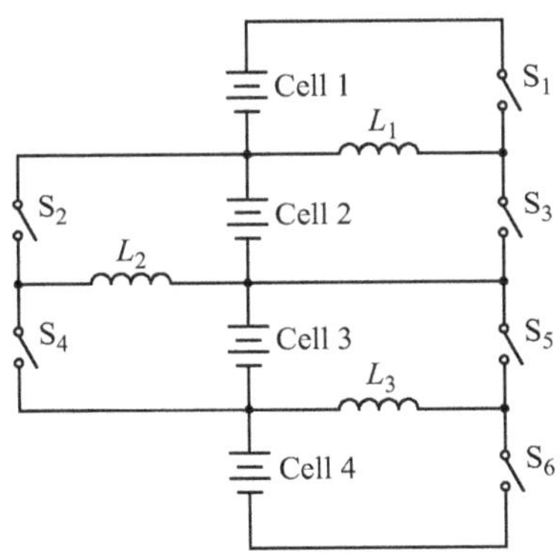


*Figure 7.1 The conventional inductor based balancing circuit*

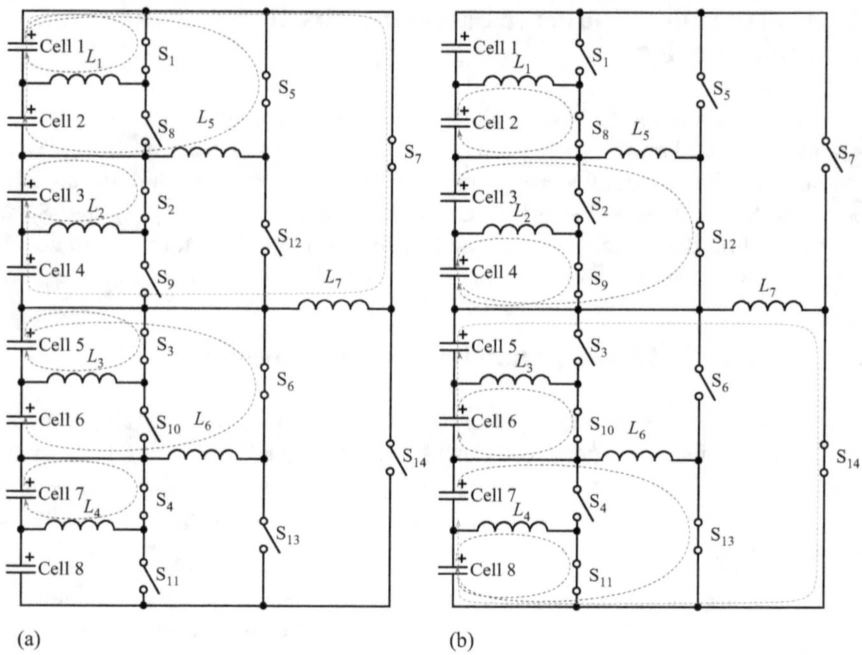


*Figure 7.2 The proposed circuit based on inductor balancing: (a) first stage and (b) second stage*

calculated as

$$VL_5 = L_5\frac{di_{L_5}}{dt} = V_{\text{cell}_1} + V_{\text{cell}_2} \quad V_{L_6} = L_6\frac{diL_6}{dt} = V_{\text{cell}_5} + V_{\text{cell}_6} \tag{7.2}$$

Finally the inductor voltage of $V_{L_7}$ can be driven as

$$VL_7 = L_7\frac{di_{L_7}}{dt} = V_{\text{cell}_1} + V_{\text{cell}_2} + V_{\text{cell}_3} + V_{\text{cell}_4} \tag{7.3}$$

Equation (7.1) can be rewritten as

$$\frac{V_{\text{cell}_1}}{L_1} = \frac{di_{L_1}}{dt} = \frac{\Delta i_{L_1}}{DT} \tag{7.4}$$

where $D$ is the duty cycle. The inductor current when $S_1$ is ON can be written as

$$\Delta i_{L_{1(\text{on})}} = \frac{V_{\text{cell}_1}}{L_1}DT \quad \Delta i_{L_{k(\text{on})}} = \frac{V_{\text{cell}_n}}{L_k}DT \tag{7.5}$$

Similarly the inductor current of $i_{L_2}$, $i_{L_3}$, and $i_{L_4}$ can be expressed the same, where $k = 2, 3, 4$, and $n = 3, 5, 7$.

The inductor current of $i_{L_5}$ and $i_{L_6}$ can be expressed as

$$\Delta i_{L_{5(\text{on})}} = \frac{V_{\text{cell}_1} + V_{\text{cell}_2}}{L_5} DT \quad \Delta i_{L_{6(\text{on})}} = \frac{V_{\text{cell}_5} + V_{\text{cell}_6}}{L_6} DT \tag{7.6}$$

And finally the inductor current of $i_{L_5}$ can be expressed as

$$\Delta i_{L_{7(\text{on})}} = \frac{V_{\text{cell}_1} + V_{\text{cell}_2} + V_{\text{cell}_3} + V_{\text{cell}_4}}{L_7} DT \tag{7.7}$$

When the switch $S_1$ is turned off, then the switch $S_8$ continues to carry the current. Hence, the energy stored in $L_1$ will be transferred to cell 2.

$$VL_1 = L_1 \frac{di_{L_1}}{dt} = -V_{\text{cell}_2} \quad V_{L_K} = L_K \frac{diL_k}{dt} = -V_{\text{cell}_n} \tag{7.8}$$

Similarly for the inductor voltages of $V_{L_2}$, $V_{L_3}$, and $V_{L_4}$, where $k = 2, 3, 4$, and $n = 4, 6, 8$. The inductor voltages of $V_{L_5}$ and $V_{L_6}$ can be calculated as

$$VL_5 = L_5 \frac{di_{L_5}}{dt} = -V_{\text{cell}_3} - V_{\text{cell}_4} \quad V_{L_6} = L_6 \frac{diL_6}{dt} = -V_{\text{cell}_7} - V_{\text{cell}_8} \tag{7.9}$$

Also the inductor voltage of $V_{L_7}$ can be driven as

$$VL_7 = L_7 \frac{di_{L_7}}{dt} = -V_{\text{cell}_5} - V_{\text{cell}_6} - V_{\text{cell}_7} - V_{\text{cell}_8} \tag{7.10}$$

The inductor current when $S_1$ is turned off and $S_2$ turns on can be written as

$$\Delta i_{L_{1(\text{off})}} = \frac{-V_{\text{cell}_2}}{L_1}(1-D)T \quad \Delta i_{L_{k(\text{off})}} = \frac{-V_{\text{cell}_n}}{L_k}(1-D)T \tag{7.11}$$

Similarly the inductor current of $i_{L_2}$, $i_{L_3}$, and $i_{L_4}$ can be written the same. Where $k = 2, 3, 4$, and $n = 4, 6, 8$. The inductor current of $i_{L_5}$ and $i_{L_6}$ can be expressed as

$$\Delta i_{L_{1(\text{off})}} = \frac{-V_{\text{cell}_3} - V_{\text{cell}_4}}{L_5}(1-D)T \quad \Delta i_{L_{6(\text{off})}} = \frac{-V_{\text{cell}_7} - V_{\text{cell}_8}}{L_6}(1-D)T \tag{7.12}$$

And finally the inductor current of $i_{L_7}$ can be expressed as

$$\Delta i_{L_{7(\text{off})}} = \frac{-V_{\text{cell}_5} - V_{\text{cell}_6} - V_{\text{cell}_7} - V_{\text{cell}_8}}{L_7}(1-D)T \tag{7.13}$$

As is known, in steady state the inductor current changes in a period is zero

$$\Delta i_{L_{(\text{on})}} + \Delta i_{L_{(\text{off})}} = 0 \tag{7.14}$$

so substituting (7.5) and (7.11) into (7.14) can be expressed as

$$\frac{V_{\text{cell}_1}}{L_1}DT + \frac{-V_{\text{cell}_2}}{L_1}(1-D)T = 0 \tag{7.15}$$

The transfer function $H_{21}$ can be defined as

$$H_{V_{L_1}} = \frac{V_{\text{cell}_2}}{V_{\text{cell}_1}} = \frac{D}{1-D} \tag{7.16}$$

correspondingly, the transfer function $H_{V_{L2}}$, $H_{V_{L3}}$ and $H_{V_{L4}}$ can be defined in a similar way. The transfer function $H_{V_{L5}}$ can be expressed as

$$H_{V_{L_5}} = \frac{V_{\text{cell}_3} + V_{\text{cell}_4}}{V_{\text{cell}_1} + V_{\text{cell}_2}} = \frac{D}{1-D} \quad H_{V_{L_6}} = \frac{V_{\text{cell}_7} + V_{\text{cell}_8}}{V_{\text{cell}_5} + V_{\text{cell}_6}} = \frac{D}{1-D} \tag{7.17}$$

The transfer function $H_{V_{L7}}$ can be expressed as

$$H_{V_{L_7}} = \frac{V_{\text{cell}_5} + V_{\text{cell}_6} + V_{\text{cell}_7} + V_{\text{cell}_8}}{V_{\text{cell}_5} + V_{\text{cell}_6} + V_{\text{cell}_7} + V_{\text{cell}_8}} = \frac{D}{1-D} \tag{7.18}$$

2. In stage two, the switches $S_1$–$S_7$ are turned off while the switches $S_8$–$S_{14}$ will be turned on to transfer the energy stored in the inductors to the adjacent cells. As explained in the previous mode, the principle is similar except that in this mode, cells 2, 4, 6, and 8 act as an input and the energy will be transferred in the opposite direction.

## 7.4 Simulation results

To verify the feasibility of the system, the simulation results are expressed. The MOSFETs are triggered with a synchronous pulse with a duty ratio of 50%. The battery voltages are chosen from 3 V to 3.6 V to present a significant voltage difference of the cells. At the real capacitances of batteries, the simulation would take very long, to be able to evaluate faster equivalent capacitors that are used, but the timescale is just proportional to the battery capacitance, so it can be evaluated for other cases as well. The switching frequency is 100 kHz, the equivalent capacitor values are 50 mF, and the inductor values are 10 μH. This proposed circuit has advantages that it equalizes the middle cells and top and bottom cells. Therefore, it does not equalize cell by cell, whereas equalization speed is faster compared to convention-based inductor balancing topology which equalizes cell by cell. The simulation parameters are presented in Table 7.1.

As shown in Figure 7.2, during first cycle the inductors $L_1$–$L_4$ will store energy through cell 1, 3, 5, and 7, respectively, and also $L_5$ will store energy through cells 1 and 2, so the stored energy is twice compared to the energy stored in $L_1$–$L_4$. The inductor $L_7$ will store energy through cells 1–4; therefore the amount of energy is four times more than $L_1$–$L_4$. This means in fact that the inductor is not the same, but larger. Still, if two inductors are put in series and then in parallel, the same inductor type could be used, but is a larger number.

*Table 7.1 Simulation parameters*

| Parameter | Values |
|---|---|
| Equivalent capacitors | 50 mF |
| Initial capacitor voltages | 3 V–3.6 V |
| Switching frequency | 100 kHz |
| Inductors | 10 μH |

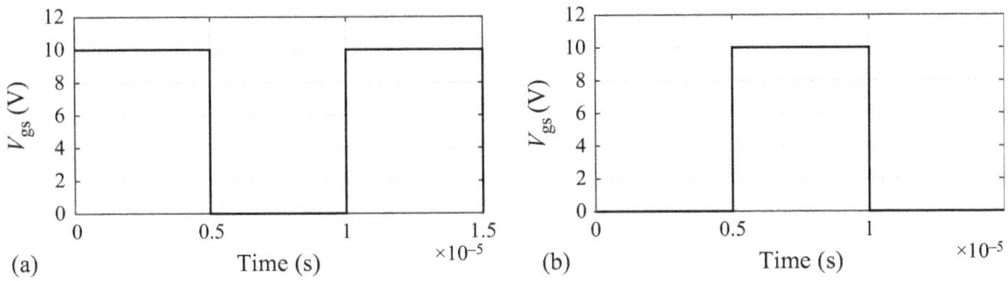


*Figure 7.3 Gate-to-source voltage to the switches 1–7 and switches 8–14: (a) first cycle and (b) second cycle*

Figure7.3(a) shows the gate-to-source voltage to MOSFETs 1–7. All the odd MOSFETs are turned on and off together with a duty ratio of 50%. In Figure 7.3(b) the gate-to-source voltages (*V*gs) to the MOSFETs 8–14 are shown, and similarly the even MOSFETs will be turned on and off with a 50% duty ratio.The inductor voltage waveforms are shown in Figure 7.4(a). The purple one is the inductor voltage waveform of $L_7$, the red one is the inductor voltage waveforms of $L_5$ and $L_6$, and the green one is the inductor voltage waveforms of $L_1$–$L_4$. Figure 7.4(b) shows the inductor currents with a switching frequency of (100 kHz). The red one is the inductor current waveform of $L_7$, the blue one is the inductor current waveforms of $L_5$ and $L_6$, and the purple one is the inductor current waveforms of $L_1$–$L_4$.

The conventional inductor balancing circuit shares the same switching topology as switched-capacitor balancing. The conventional inductor balancing is known as cell-to-cell contiguous method. It is easy to control, but it has long balancing time for balancing end-to-end cell in high string pack. The cell voltage waveforms of the conventional inductor balancing are shown in Figure 7.4(c). As can be seen, the equalization time is slower compared to the proposed topology. To overcome the problem of conventional inductor balancing time, the proposed inductor balancing is introduced. The cell voltages of the proposed inductor balancing circuit (cells 1–8) with switching frequency (100 kHz) are shown in Figure 7.4(d). It can be seen that the balancing time is decreased compared to the conventional inductor balancing.

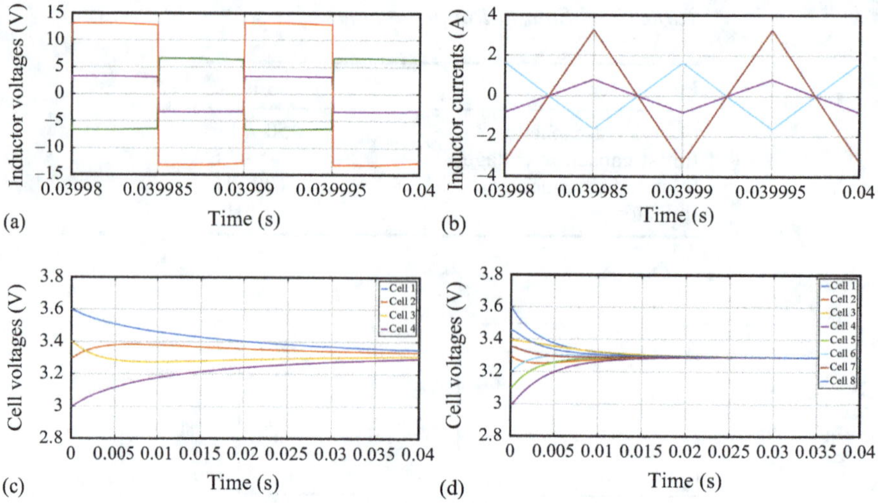


*Figure 7.4 Simulation results of inductor balancing circuit: (a) inductors voltage waveforms, (b) inductors current waveforms, (c) cell voltage vs time (conventional), and (d) cell voltage vs time (proposed, 100 kHz)*

## Summary

In this section, the concept of cell equalization technique based on inductor balancing is investigated. The inductor-based equalizer has an advantage of fast equalization time compared to other equalization methods such as active capacitor balancing. The conventional inductor balancing method has a problem that the energy is transferred cell by cell, which results in a long equalization time for balancing end-to-end cell in a high string pack. In order to overcome this problem, the proposed topology of Figure 7.2 improves the balancing time compared to Figure 7.1. This method will equalize the middle cells as fast as the top and bottom ones. It also equalizes eight cells faster compared to the conventional topology which equalizes four cells. To verify the feasibility of the system, the simulation results are performed.

*Section 2*

# Ćuk balancing circuit 'uncoupled inductors'

## 7.5 Ćuk converter balancing

In this section, the proposed Ćuk converter balancing circuit for lithium-ion batteries will be investigated. The Ćuk converter balancing circuit has good efficiency with non-dissipative currents and a bidirectional energy transferring capability [30,32]. It accepts some inductance at the wiring to the cells, without resulting in large losses or ringing, as in the topology, inductors are in series with the wiring. The bidirectional modified Ćuk converter is easy to be integrated and extended and compared with other kinds of active equalizing circuits, and the modified Ćuk converter has a relatively lower cost. As compared with the conventional Ćuk balancing method, which requires 2*N*–2 switches for *N* cells, the proposed method uses *N* switches and $N + 1$ inductors for *N* number of cells, which results in fewer components and losses.

## 7.6 The proposed Ćuk converter balancing

The schematic of the proposed Ćuk circuit is shown in Figure 7.5. This circuit equalizes eight cells in a battery string connected in series. It has a buck-boost characteristic using a simple pulse with a modulation of a 50% duty cycle. The operational principle has two main stages.

In the first stage, switches Q1, Q3, Q5, and Q7 will be turned on, while the other switches are off. In the next stage, switches Q2, Q4, Q6, and Q8 will be turned on while the others will be turned off. To demonstrate the operational principle, a circuit with two cells is investigated instead of eight cells, as shown in Figure 7.6. In the first stage, Q1 is turned on while Q2 is turned off, by applying KVL around loop ($L_1$, BT1, and Q1) and loop (BT2, $C_1$, and $L_2$), which are indicated as red line, so the equivalent voltage equation can be written as

$$V_{\text{cell}_1} = L_1 \frac{\mathrm{d}i_{L_1}}{\mathrm{d}t} \tag{7.19}$$

$$V_{\text{cell}_2} = -L_2 \frac{\mathrm{d}i_{L_2}}{\mathrm{d}t} + V_{C_1} \tag{7.20}$$

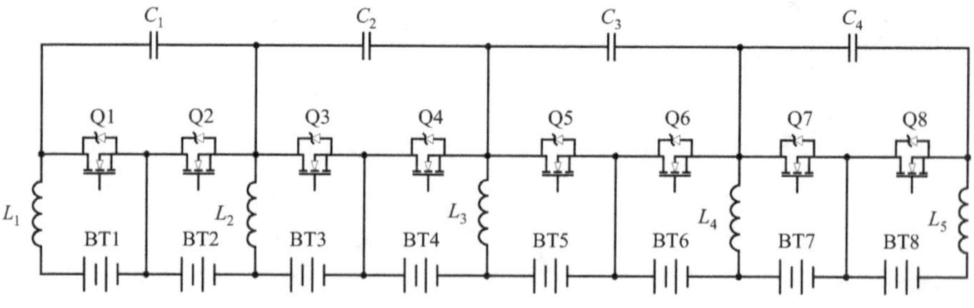


*Figure 7.5 Schematic of the proposed Ćuk converter balancing circuit*

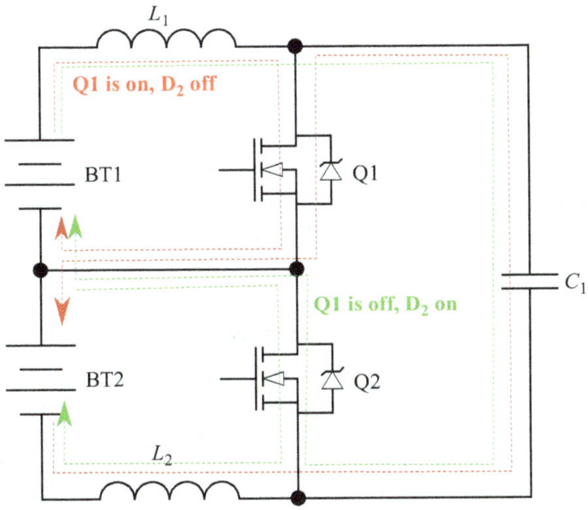


*Figure 7.6 Simplified Ćuk converter circuit with two cells*

In next mode, when Q1 is turned off and the body diode of Q2 is turned on, the equivalent equation can be expressed as

$$V_{\text{cell}_1} = L_1 \frac{\mathrm{d}i_{L_1}}{\mathrm{d}t} + V_{C_1} \tag{7.21}$$

$$V_{\text{cell}_2} = -L_2 \frac{\mathrm{d}i_{L_2}}{\mathrm{d}t} \tag{7.22}$$

Applying volt-second balance across $L_1$

$$V_{\text{cell}_1} DT_s + (V_{\text{cell}_1} - V_{C_1})(1 - D)T_s = 0 \tag{7.23}$$

Similarly, volt-second balance across $L_2$

$$(-V_{\text{cell}_2} + V_{C_1})DT_s + (-V_{\text{cell}_2})(1 - D)T_s = 0 \tag{7.24}$$

Expression for average inductor current can be obtained from the charge balance of $C_1$ as

$$i_{L_1}(1 - D)T_s - i_{L_2} DT_s = 0 \tag{7.25}$$

where $D$ is duty cycle and $T_s$ is switching frequency. The average inductor current for $L_1$ and $L_2$ can be expressed as

$$i_{L_1} = \left[\frac{1}{2}\left(\frac{V_{\text{cell}_1}}{L_1} D^2 + \frac{-V_{C_1} + V_{\text{cell}_1}}{L_1}(1 - D)^2\right)\right] T_s \tag{7.26}$$

$$i_{L_2} = \left[\frac{1}{2}\left(\frac{V_{C_1} - V_{\text{cell}_2}}{L_2} D^2 + \frac{-V_{\text{cell}_2}}{L_2}(1 - D)^2\right)\right] T_s \tag{7.27}$$

Similarly, for the next stage, when Q1 is turned off and Q2 is turned on, the equivalent equation can be expressed as before.

## 7.7 Simulation results

In this section, the simulation results are expressed in order to testify the proposed topology. All the switches are N-channel MOSFETs with body diodes and they are triggered with a synchronous pulse with the duty ratio of 50%, as shown in Figure 7.7. The gate-to-source voltage (*V*gs) of the odd MOSFETs is a blue as shown in Figure 7.7(a), while the gate-to-source voltage for even MOSFETs is red as shown in Figure 7.7(b). The switching frequency is set to 100 kHz, the capacitor values are 100 μF, and the inductor values are 10 μH. The battery voltages are chosen from 3.1 V to 3.6 V to present significant voltage difference of the cells. The batteries are modelled with capacitors, C = 0.5 F. The simulation parameters are given in Table 7.2.

The cell voltages of the proposed circuit are shown in Figure 7.8(a). As it can be seen in the figure, the balancing time is fast with the significant voltage differences. At the time 0.2 s, all cells are equalized. In Figure 7.8(b), the inductor voltage waveforms $L_1$–$L_5$ in steady state are shown. It has been known that the average inductor voltage in a period is zero, which can be observed in the figure. The inductor current waveforms are presented in Figure 7.8(c). The figure indicates that the inductors with negative current belong to the cells with

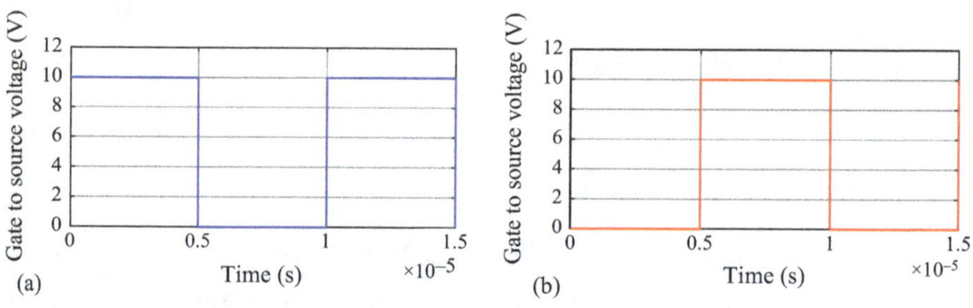


*Figure 7.7 Gate-to-source voltage (Vgs) of (a) the odd and (b) even MOSFETs*

*Table 7.2 Simulation parameters*

| Parameter | Values |
|---|---|
| Equivalent capacitors | 50 mF |
| Initial capacitor voltages | 3.1 V–3.6 V |
| Ćuk capacitors | 100 μF |
| Switching frequency | 100 kHz |
| Inductors | 10 μH |

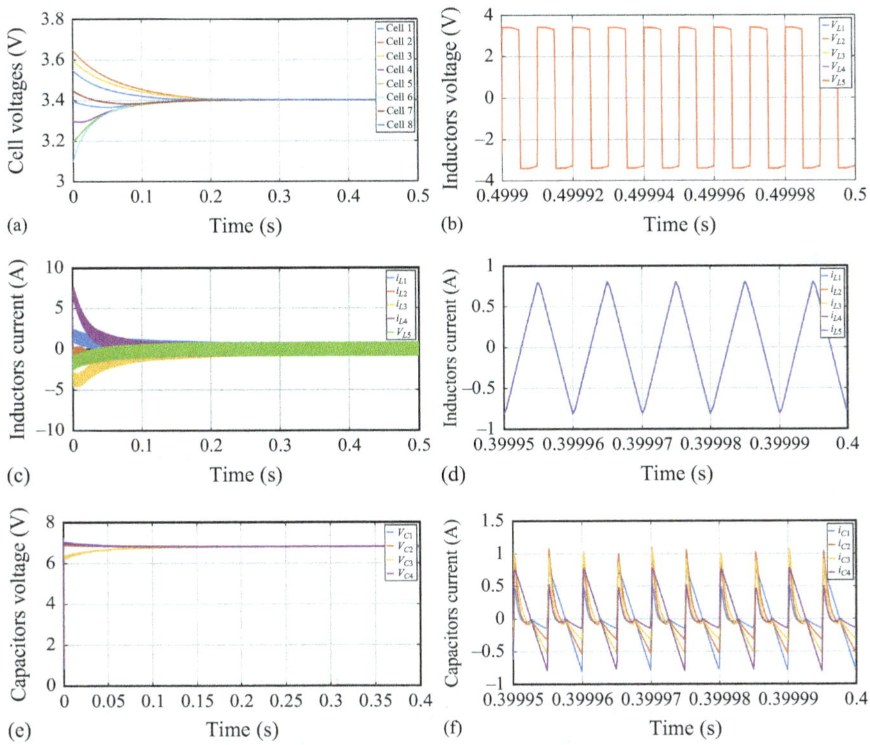


*Figure 7.8 Simulation results of Ćuk converter balancing circuit: (a) cell voltages vs time, (b) inductor voltage waveforms, (c) inductor current waveforms, (d) inductor current waveforms in steady state, (e) capacitor voltage waveforms, and (f) capacitor current waveforms in steady state*

the charging process and the positive currents belong to the cells which are discharging.

In order to see the inductor current waveforms in steady state, Figure 7.8(d) is presented. This figure shows that all currents have the same value, which means that all cells are equalized. If the inductor currents are not at the same level, that means some cells have higher voltages while the others have lower voltages, which means the battery cells are not equalized. At the end, capacitor voltage waveforms and capacitor current waveforms (in steady state) are shown in Figure 7.8(e) and (f), respectively. It can be observed that the average capacitor voltage values are twice the voltage of one cell. This can be explicated by having a KVL around the larger loop (through cell 1, $L1$, $C$, $L2$, and cell 2) in Figure 7.6. Therefore, the equation can be written as

$$-V_{\mathrm{cell}_1} + \bar{V}_{L_1} + \bar{V}_C - \bar{V}_{L_2} - V_{\mathrm{cell}_2} = 0 \tag{7.28}$$

However, as is known, the average inductor voltage is zero; therefore $\bar{V}_{L_1}$ and $\bar{V}_{L_2}$ will be zero. Thus, according to (7.28), the average capacitor voltage $\bar{V}_C$ can be rewritten as

$$\bar{V}_C = V_{\text{cell}_1} + V_{\text{cell}_2} \tag{7.29}$$

## Summary

A special Ćuk converter balancing circuit for lithium-based batteries is investigated. In an electric vehicle and a lot of battery appliances, batteries play an important role due to the fact that the cost depends on the batteries; thus the batteries with a longer life cycle are desired. Therefore, a battery management system is required to increase the life cycle of the batteries. In batteries, the cell voltages will differ due to the charging and discharging process; thus a balancing circuit is needed to ensure that all cells are at the same voltage level. Among existing cell balancing methods, Ćuk converter balancing has a feature of fast balancing time and fewer costs. In this chapter, eight cells in series in a battery pack are considered. This method uses only $N$ switches per $N$ cells. All switches are N-channel MOSFETs (low $R_{\text{DS(on)}}$), with body diodes, and they are triggered with PWM in synchronous patterns with 50% duty cycle. It is an advantage that the number of components is reduced, so reducing losses. The proposed circuit is tested with significant cell voltage differences to have voltage imbalances in order to verify the proposed topology. The simulation results are performed to testify the variability of the proposed circuit.

## References

[1] Mi C and Masrur MA. Hybrid electric vehicles: principles and applications with practical perspectives. Chichester, UK: John Wiley & Sons; 2017.

[2] Park YH, Choi SY, Choi YJ, *et al.* A novel charge equalizer with auxiliary circuit to control the allowable charging and discharging current of the Lithium-ion battery. In: 2017 IEEE 3rd International Future Energy Electronics Conference and ECCE Asia (IFEEC 2017-ECCE Asia). IEEE; 2017. p. 500–505.

[3] Lievre A, Sari A, Venet P, *et al.* Practical online estimation of lithium-ion battery apparent series resistance for mild hybrid vehicles. IEEE Transactions on Vehicular Technology. 2015;65(6):4505–4511.

[4] Lu L, Han X, Li J, *et al.* A review on the key issues for lithium-ion battery management in electric vehicles. Journal of Power Sources. 2013;226: 272–288.

[5] Uno M and Tanaka K. Single-switch cell voltage equalizer using multi-stacked buck-boost converters operating in discontinuous conduction mode

for series-connected energy storage cells. IEEE Transactions on Vehicular Technology. 2011;60(8):3635–3645.

[6] Emadi A, Lee YJ, and Rajashekara K. Power electronics and motor drives in electric, hybrid electric, and plug-in hybrid electric vehicles.IEEE Transactions on Industrial Electronics. 2008;55(6):2237–2245.

[7] Gallardo-Lozano J, Romero-Cadaval E, Milanes-Montero MI, *et al.* Battery equalization active methods. Journal of Power Sources. 2014;246:934–949.

[8] Uno M and Tanaka K. Accelerated charge-discharge cycling test and cycle life prediction model for supercapacitors in alternative battery applications. IEEE Transactions on Industrial Electronics. 2011;59(12):4704–4712.

[9] Kim Y, Samad NA, Oh KY, *et al.* Estimating state-of-charge imbalance of batteries using force measurements. In: 2016 American Control Conference (ACC). IEEE; 2016. p. 1500–1505.

[10] Lin X, Stefanopoulou AG, Li Y, *et al.* State of charge imbalance estimation for battery strings under reduced voltage sensing.IEEE Transactions on Control Systems Technology. 2014;23(3):1052–1062.

[11] Kim CH, Kim MY, and Moon GW. A modularized charge equalizer using a battery monitoring IC for series-connected Li-ion battery strings in electric vehicles. IEEE Transactions on Power Electronics. 2012;28(8):3779–3787.

[12] Park HS, Kim CH, Park KB, *et al.* Design of a charge equalizer based on battery modularization. IEEE Transactions on Vehicular Technology. 2009;58(7):3216–3223.

[13] Kutkut NH and Divan DM. Dynamic equalization techniques for series battery stacks. In: Proceedings of Intelec'96-International Telecommunications Energy Conference. IEEE; 1996. p. 514–521.

[14] Lindemark B. Individual cell voltage equalizers (ICE) for reliable battery performance. In: [Proceedings] Thirteenth International Telecommunications Energy Conference-INTELEC 91. IEEE; 1991. p. 196–201.

[15] Zhang X, Liu P, and Wang D. The design and implementation of smart battery management system balance technology. Journal of Convergence Information Technology. 2011;6(5):108–116.

[16] Moore SW and Schneider PJ. A review of cell equalization methods for lithium ion and lithium polymer battery systems.SAE Technical Paper; 2001.

[17] Speltino C, Stefanopoulou A, and Fiengo G. Cell equalization in battery stacks through state of charge estimation polling. In: Proceedings of the 2010 American Control Conference. IEEE; 2010. p. 5050–5055.

[18] Cao J, Schofield N, and Emadi A. Battery balancing methods: a comprehensive review. In: 2008 IEEE Vehicle Power and Propulsion Conference. IEEE; 2008. p. 1–6.

[19] Yuanmao Y, Cheng K, and Yeung Y. Zero-current switching switched-capacitor zero-voltage-gap automatic equalization system for series battery string. IEEE Transactions on Power Electronics. 2011;27(7):3234–3242.

[20] Sano K and Fujita H. Voltage-balancing circuit based on a resonant switched-capacitor converter for multilevel inverters. IEEE Transactions on Industry Applications. 2008;44(6):1768–1776.

[21] Park SH, Kim TS, Park JS, *et al.* A new battery equalizer based on buck-boost topology. In: 2007 7th International Conference on Power Electronics. IEEE; 2007. p. 962–965.

[22] Moo C, Hsieh Y, Tsai I, *et al.* Dynamic charge equalisation for series-connected batteries. IEE Proceedings-Electric Power Applications. 2003; 150(5):501–505.

[23] Phung TH, Crebier JC, Chureau A, *et al.* Optimized structure for next-to-next balancing of series-connected lithium-ion cells. In: 2011 Twenty-Sixth Annual IEEE Applied Power Electronics Conference and Exposition (APEC). IEEE; 2011. p. 1374–1381.

[24] Baek IK, Kim TH, Lim CS, *et al.* Modularized battery cell voltage equalization circuit using extended multi-winding transformer. In: 2012 IEEE Vehicle Power and Propulsion Conference. IEEE; 2012. p. 349–353.

[25] Fan S, Duan J, Sun L, *et al.* A fast modularized multiwinding transformer balancing topology for series-connected super capacitors. IEEE Transactions on Power Electronics. 2018;34(4):3255–3268.

[26] Oriti G, Julian AL, and Norgaard P. Battery management system with cell equalizer for multi-cell battery packs. In: 2014 IEEE Energy Conversion Congress and Exposition (ECCE). IEEE; 2014. p. 900–905.

[27] Lee WC, Drury D, and Mellor P. Comparison of passive cell balancing and active cell balancing for automotive batteries. In: 2011 IEEE Vehicle Power and Propulsion Conference. IEEE; 2011. p. 1–7.

[28] Hua CC and Fang YH. Battery charge equalization circuit with a multi-winding transformer. Journal of the Chinese Institute of Engineers. 2016; 39(5):556–564.

[29] Kim CH, Park HS, Kim CE, *et al.* Charge equalization converter with parallel primary winding for series connected Lithium-Ion battery strings in HEV. In: 2007 7th International Conference on Power Electronics. IEEE; 2007. p. 795–800.

[30] Lu X, Qian W, and Peng FZ. Modularized buck-boost + Cuk converter for high voltage series connected battery cells. In: 2012 Twenty-Seventh Annual IEEE Applied Power Electronics Conference and Exposition (APEC). IEEE; 2012. p. 2272–2278.

[31] Lee YS, Duh CY, Chen GT, *et al.* Battery equalization using bi-directional Cuk converter in DCVM operation. In: 2005 IEEE 36th Power Electronics Specialists Conference. IEEE; 2005. p. 765–771.

[32] Lee YS, Duh CY, Chen GT, *et al.* Battery equalization using bi-directional Cuk converter in DCVM operation. In: 2005 IEEE 36th Power Electronics Specialists Conference. IEEE; 2005. p. 765–771.

*Chapter 8*

# Active equalizer using coupled inductors

## 8.1 Introduction

This chapter develops a coupled inductor balancing method to overcome cell voltage variation among cells in series, for lithium-ion (Li-ion) batteries with applications in electric vehicles, and a lot of other today uses. In this book, we call the principle 'coupled inductor' balancing when the magnetic device has all the same number of turns. The term 'transformer' is used when there is a primary winding using a higher number of turns. Two variants will be proposed in the three different parts. The first variant proposes a circuit with two cells per winding. The second uses a Ćuk converter principle.

*Section 1*

# Coupled inductors with one winding per two cells

## 8.2 Introduction of one winding per two cells

An effective topology using one winding per two cells and one transistor per cell is proposed. For an example of eight cells in series, the developed balance circuit has four windings, so one magnetic circuit with one winding per two cells, and one controlled switch per cell as compared to the traditional inductor based equalizer that needs $N-1$ inductors and magnetic circuits for $N$ number of cells and more switches. The result is a substantially more efficient, cost-effective circuit, having only a low bill of materials. All switches are logic level N-channel MOSFETs and they are controlled by a pair of complementary signals in a synchronous trigger pattern. In the proposed topology, fewer components, a fast equalization, are achieved compared to a conventional battery management system technique based on the inductor balancing method. This scheme is suitable for fast equalization by the fact that the equalizing is obtained based on the leakage inductances which are substantially lower than balancing methods based on separate inductors. All windings have the same number of turns and are coupled with one magnetic core with an air gap. The theoretical derivation of the proposed circuit will be presented and numerical simulation relevant of the electrochemical storage devices was conducted to show the validity of the proposed balance circuit. A complete balancing circuit was built to verify the proposed circuit that could solve imbalance problems existing inside battery modules.

## 8.3 Coupled inductor balancing

The inductor and also the transformer balancing techniques are active methods, the operation principle is to move energy from a cell to another cell [1,2]. Due to the relatively high balancing current, they offer a smaller balancing execution time [3]. The transformer balancing may have even a smaller balancing execution time compared to inductors only. It also has a disadvantage such as high cost and magnetic losses. If all transistors are always switched at the secondary side, the leakage impedance determines the equalizing current for a voltage difference. In magnetic circuits, an equal voltage is obtained if all windings carry the same flux. This needs a special attention in magnetic cores with air gap. If, for example, four wires are wound 'in one hand' in one layer, a low leakage inductance is obtained and a lot of symmetry is maintained with respect to the air gap. The switched single inductor balancing method utilizes one inductor for transferring energy between the whole pack. The control system senses the voltage of the cells and selects the two cells which will be used for energy transfer, while the multi-switched inductor balancing needs $N$–1 inductor for balancing $N$ cells. They are featured by fast equalization time.

The conventional inductor based equalizer uses $N$–1 inductors and $2N$–2 number of switches in order to balance $N$ cells by storing energy in the inductor as shown in Figure 7.1 in Chapter 7. By controlling on-and-off switching, the energy flows between the cells.

In the first half period, the switches $S_1$, $S_2$, and $S_5$ are turned on in order to store energy in the inductors from the cells while the switches $S_3$, $S_4$, and $S_6$ are turned off. In the second half period, switches $S_1$, $S_2$, and $S_5$ are turned off while $S_3$, $S_4$, and $S_6$ will be turned on to transfer the stored energy in the inductors to the adjacent cells. The principle is to transfer the energy from a high voltage cell to the lower one. This method has a disadvantage of long balancing time because energy transfer is from the first cell to the last, and it also contains more inductors for a number of *N* cells in series and more switches that cause more power losses more components and not cost-effective.

## 8.4 The proposed coupled inductor balancing

Some superior performance can be expected as the coupling property can be achieved by magnetic devices and not by capacitors. The proposed circuit has fewer inductors and switches compared to the conventional method that uses 2*N*–2 switches and *N*–1 inductors. The proposed circuit uses one switch per cell, one winding per two cells, and one magnetic circuit per battery. This means fewer components, cost-effective, more efficient, and simple implementation.

The block diagram of the proposed system is shown in Figure 8.1 which consists of eight battery cells and eight logic level N-channel MOSFETs with body diodes and four coupled inductors. The operation principle of the circuit can be divided into two main states.

1. In stage one, the switches $S_1$, $S_3$, $S_5$, and $S_7$ are turned on to store energy in the magnetic, while the other switches are kept turned off. When switch $S_1$ is on and $S_2$ is off, the energy of cell 1 will be transferred to cell 2; therefore, the energy of cell 1 will be stored in the inductor $L_1$. The inductor $L_1$ can be expressed as

$$L_1 = M_{11} + L_{S_1} \tag{8.1}$$

where $M_{11}$ is the self-inductance of winding 1 and $L_{S1}$ is the leakage inductance of winding 1. Similarly, $M_{22}$, $M_{33}$, $M_{44}$, and $L_{S2}$, $L_{S3}$, $L_{S4}$ are mutual inductance and leakage inductance of winding 2, 3, 4, respectively.

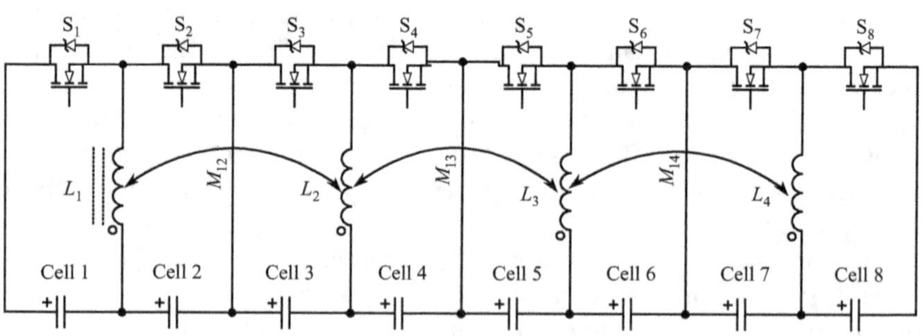


*Figure 8.1 The proposed circuit based on coupled inductor balancing*

$$\begin{bmatrix} L_1 \\ L_2 \\ L_3 \\ L_4 \end{bmatrix} = \begin{bmatrix} M_{11} \\ M_{22} \\ M_{33} \\ M_{44} \end{bmatrix} + \begin{bmatrix} L_{S_1} \\ L_{S_2} \\ L_{S_3} \\ L_{S_4} \end{bmatrix} \tag{8.2}$$

The inductor voltage of $V_{L_1}$ during ON interval can be expressed as

$$L_1 \frac{\mathrm{d}i_{L_1}}{\mathrm{d}t} + M_{12} \frac{\mathrm{d}i_{L_2}}{\mathrm{d}t} + M_{13} \frac{\mathrm{d}i_{L_3}}{\mathrm{d}t} + M_{14} \frac{\mathrm{d}i_{L_4}}{\mathrm{d}t} = V_{\mathrm{cell}_1} \tag{8.3}$$

where $M_{12}$ is the mutual inductance between $L_1$ and $L_2$, $M_{13}$ is the mutual inductance between $L_1$ and $L_3$, and $M_{14}$ is the mutual inductance between $L_1$ and $L_4$. Similarly, the inductor voltages $V_{L_2}$, $V_{L_3}$, and $V_{L_4}$ can be expressed as

$$L_k \frac{\mathrm{d}i_{L_k}}{\mathrm{d}t} + \sum_{j=1}^{4} M_{ij} \frac{\mathrm{d}i_{L_j}}{\mathrm{d}t} + M_{ij} \frac{\mathrm{d}i_{L_j}}{\mathrm{d}t} + M_{ij} \frac{\mathrm{d}i_{L_j}}{\mathrm{d}t} = V_{\mathrm{cell}_n} \tag{8.4}$$

where $k = 2, 3, 4$; $i = 2, 3, 4$; $j = 1, 2, 3, 4$; $i \neq j$; and $n = 3, 5, 7$. Similarly, the inductor voltages for $L_1$–$L_4$ can be rewritten as

$$\begin{bmatrix} V_{\mathrm{cell}_1} \\ V_{\mathrm{cell}_3} \\ V_{\mathrm{cell}_5} \\ V_{\mathrm{cell}_7} \end{bmatrix} = \begin{bmatrix} L_1 & M_{12} & M_{13} & M_{14} \\ M_{21} & L_2 & M_{23} & M_{24} \\ M_{31} & M_{32} & L_3 & M_{34} \\ M_{41} & M_{42} & M_{43} & L_3 \end{bmatrix} \times \frac{\mathrm{d}}{\mathrm{d}t} \begin{bmatrix} iL_1 \\ iL_2 \\ iL_3 \\ iL_4 \end{bmatrix} \tag{8.5}$$

If inductors are symmetrical, but with different leakage inductance due to circuit layout, (8.3) can be rewritten as

$$L_{S_1} \frac{\mathrm{d}i_{L_1}}{\mathrm{d}t} + M\left( \frac{\mathrm{d}i_{L_1}}{\mathrm{d}t} + \frac{\mathrm{d}i_{L_2}}{\mathrm{d}t} + \frac{\mathrm{d}i_{L_3}}{\mathrm{d}t} + \frac{\mathrm{d}i_{L_4}}{\mathrm{d}t} \right) = V_{\mathrm{cell}_1} \tag{8.6}$$

Note that the difference in 'leakage' can also come from the circuit: printed circuit board (PCB) and wiring. Indeed if the windings are made 'in one hand' the leakage of the transformer may be lower than the circuit.

In the next mode, when the switches $S_1$, $S_3$, $S_5$, and $S_7$ are turned off, the body diodes of switches $S_2$, $S_4$, $S_6$, and $S_8$ will be turned on, and the energy stored in the inductors will be transferred to cells 2, 4, 6, 8. Therefore, the inductor voltages $L_1$ can be expressed as

$$L_1 \frac{\mathrm{d}i_{L_1}}{\mathrm{d}t} + M_{12} \frac{\mathrm{d}i_{L_2}}{\mathrm{d}t} + M_{13} \frac{\mathrm{d}i_{L_3}}{\mathrm{d}t} + M_{14} \frac{\mathrm{d}i_{L_4}}{\mathrm{d}t} = -V_{\mathrm{cell}_2} \tag{8.7}$$

Similarly, the inductor voltages $L_2$–$L_4$ can be expressed as

$$L_k \frac{\mathrm{d}i_{L_k}}{\mathrm{d}t} + \sum_{j=1}^{4} M_{ij} \frac{\mathrm{d}i_{L_j}}{\mathrm{d}t} + M_{ij} \frac{\mathrm{d}i_{L_j}}{\mathrm{d}t} + M_{ij} \frac{\mathrm{d}i_{L_j}}{\mathrm{d}t} = -V_{\mathrm{cell}_n} \tag{8.8}$$

where $k = 2, 3, 4$; $i = 2, 3, 4$; $j = 1, 2, 3, 4$; $i \neq j$; and $n = 4, 6, 8$. Therefore, the inductor voltages for $L_1$–$L_4$ can be rewritten as

$$\begin{bmatrix} V_{\text{cell}_2} \\ V_{\text{cell}_4} \\ V_{\text{cell}_6} \\ V_{\text{cell}_8} \end{bmatrix} = \begin{bmatrix} L_1 & M_{12} & M_{13} & M_{14} \\ M_{21} & L_2 & M_{23} & M_{24} \\ M_{31} & M_{32} & L_3 & M_{34} \\ M_{41} & M_{42} & M_{43} & L_4 \end{bmatrix} \times \frac{-\mathrm{d}}{\mathrm{d}t} \begin{bmatrix} iL_1 \\ iL_2 \\ iL_3 \\ iL_4 \end{bmatrix} \tag{8.9}$$

According to (8.3), the inductor voltage of $L_1$ can be written as

$$V_{L_1} = L_1 \frac{\mathrm{d}i_{L(1+2+3+4)}}{\mathrm{d}t} = V_{\text{cell}_1} \tag{8.10}$$

$$V_{L_1} = L_{S_1} \frac{\mathrm{d}i_{L_1}}{\mathrm{d}t} + M \frac{\mathrm{d}i_{L(1+2+3+4)}}{\mathrm{d}t} = V_{\text{cell}_1} \tag{8.11}$$

$$\frac{\Delta i_{L(1+2+3+4)}}{DT} M + \frac{\Delta i_{L_1}}{DT} L_{S_1} = V_{\text{cell}_1} \tag{8.12}$$

For the second mode, when $S_1$ is off and $S_2$ is on, the inductor voltage of $L_1$ based on (8.7) can be expressed as

$$V_{L_1} = L_1 \frac{\mathrm{d}i_{L(1+2+3+4)}}{\mathrm{d}t} = -V_{\text{cell}_2} \tag{8.13}$$

$$V_{L_1} = L_{S_1} \frac{\mathrm{d}i_{L_1}}{\mathrm{d}t} + M \frac{\mathrm{d}i_{L(1+2+3+4)}}{\mathrm{d}t} = -V_{\text{cell}_2} \tag{8.14}$$

$$\frac{\Delta i_{L(1+2+3+4)}}{(1-D)T} M + \frac{\Delta i_{L_1}}{(1-D)T} L_{S_1} = -V_{\text{cell}_2} \tag{8.15}$$

2. In stage two, the switches $S_1$, $S_3$, $S_5$, $S_7$ are turned off, while the switches $S_2$, $S_4$, $S_6$, $S_8$ are turned on to transfer the stored energy to the adjacent cells. In the first stage, the switch $S_2$ is turned on and $S_1$ is turned off and the energy of cell 1 will be stored in the inductor $L_1$.

$$\begin{bmatrix} V_{\text{cell}_2} \\ V_{\text{cell}_4} \\ V_{\text{cell}_6} \\ V_{\text{cell}_8} \end{bmatrix} = \begin{bmatrix} L_1 & M_{12} & M_{13} & M_{14} \\ M_{21} & L_2 & M_{23} & M_{24} \\ M_{31} & M_{32} & L_3 & M_{34} \\ M_{41} & M_{42} & M_{43} & L_4 \end{bmatrix} \times \frac{\mathrm{d}}{\mathrm{d}t} \begin{bmatrix} iL_1 \\ iL_2 \\ iL_3 \\ iL_4 \end{bmatrix} \tag{8.16}$$

In the next mode, when the switch $S_2$ is off and the body diode of MOSFET $S_1$ is on, the energy of the inductor $L_1$ will be transferred to cell 1. As discussed in the previous state, the principles are similar to stage one, expect that in this mode cell 2 acts as input and the energy is transferred in the opposite direction. The principle for the other cells is the same.

$$\begin{bmatrix} V_{\text{cell}_1} \\ V_{\text{cell}_3} \\ V_{\text{cell}_5} \\ V_{\text{cell}_7} \end{bmatrix} = \begin{bmatrix} L_1 & M_{12} & M_{13} & M_{14} \\ M_{21} & L_2 & M_{23} & M_{24} \\ M_{31} & M_{32} & L_3 & M_{34} \\ M_{41} & M_{42} & M_{43} & L_4 \end{bmatrix} \times \frac{-\mathrm{d}}{\mathrm{d}t} \begin{bmatrix} iL_1 \\ iL_2 \\ iL_3 \\ iL_4 \end{bmatrix} \tag{8.17}$$

## 8.5 Simulation results

To verify the assumption, the model of eight-battery-cell is built in MATLAB®. In the simulation, eight capacitors with the value of 50 mF are used to model eight battery cells. The initial voltages are 3 V–3.6 V in order to perform significant voltage differences of the cells in a battery pack. All the switches are N-channel MOSFETs with body diode. An N-channel MOSFET is chosen due to the lower $R_{DS(on)}$ in both simulation and experimental set-ups. The switches are controlled by a pair of complementary signals in a synchronous trigger pattern with the switching frequency of 45 kHz and a duty ratio of 45%, which is generated by the pulse generator. Note that the 45% is a non-ideal case but enough. A duty ratio of 49% would give a bit faster equalizing between even and odd cells. The inductors are modelled with coupled inductor block in Simulink® with a leakage inductance of 100 nH, a magnetizing inductance of 20 μH, and a series resistance of 1 mΩ. The simulation parameters are shown in Table 8.1.

In order to compare between the conventional inductor-based balancing circuit and the proposed coupled inductor balancing circuit, the conventional inductor balancing with four cells is investigated in the simulation. Figure 8.2(a) shows the cell voltages of the conventional circuit [4,5]. It can be noticed that it has longer equalization time compared to the proposed coupled inductor balancing circuit. In Figure 8.2(b), the simulation result of the cell voltages of the proposed coupled inductor balancing circuit with eight cells is presented. In the proposed topology, eight cells are considered. Due to the fact that, by increasing the number of cells, the equalization time will increase, so the proposed circuit has fast balancing time. Therefore, the proposed circuit is capable of balancing more than four cells in series with fast equalization time.

As can be seen in the figure, the proposed circuit performs fast equalizing time compared with the conventional circuit. Figure 8.2(c) shows the simulation result of the coupled inductor currents. It is visible that the $L_1$–$L_2$ currents are at the same values in steady state, which indicates no imbalances in the cells. In Figure 8.2(d), the inductor voltages of $L_1$–$L_4$ are expressed. It is known that the average inductor voltage is zero in the period which is clear in the figure. Figure 8.2(e) and (f) shows the current waveforms of the coupled inductors in transient and current waveforms of the cells in steady state, respectively.

*Table 8.1 Simulation parameters*

| Parameter | Values |
|---|---|
| Equivalent capacitors | 50 mF |
| Initial capacitor voltages | 3 V–3.6 V |
| Switching frequency | 45 kHz |
| Duty ratio | 45% |
| Magnetizing inductance | 20 μH |
| Leakage inductance | 100 nH |
| Series resistance | 1 mΩ |

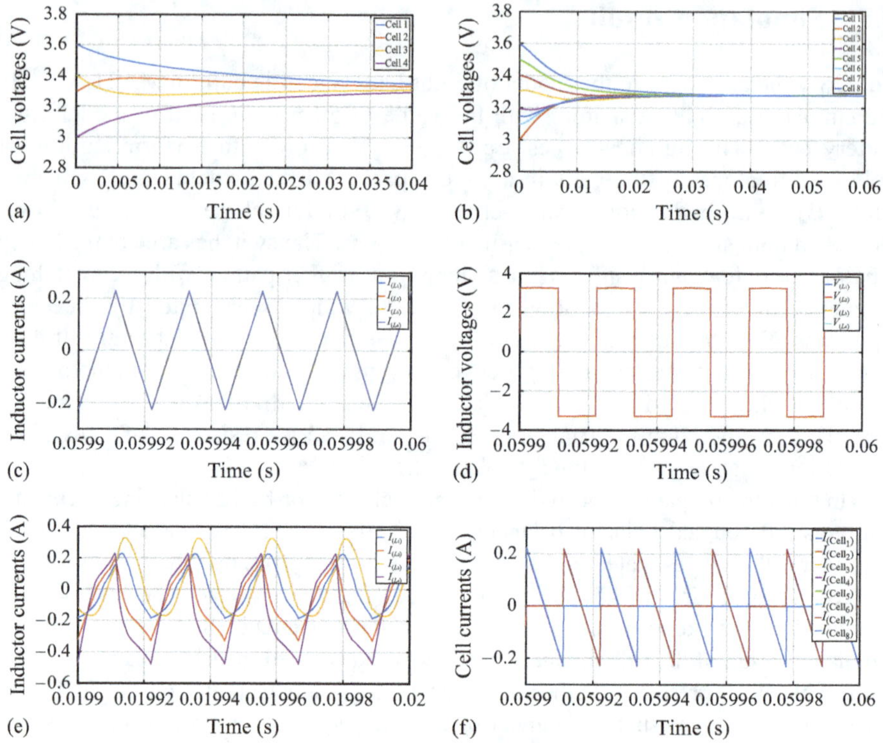


*Figure 8.2 Simulation results of the proposed coupled inductor balancing: (a) cell voltages vs time (conventional circuit), (b) cell voltages vs time (proposed circuit), (c) inductor currents in steady state, (d) inductor voltages in steady state, (e) inductor currents in transient, and (f) cell current waveforms*

## 8.6 Experimental results

A prototype of the coupled inductor equalizer is implemented and tested. All the switches are logic level N-channel MOSFETs (with body diodes), and they are controlled by a pair of complementary signals in a synchronous trigger pattern generated by a microcontroller. The odd switches are controlled by a non-inverting MOSFET driver, which is triggered by a microcontroller, and the even switches are controlled by an inverting MOSFET driver. In Figure 8.3(a), the control block diagram of the proposed balancing circuit is presented. The coupled inductors are made with a winding ratio '1', and they are all coupled with one magnetic core with an air gap. In order to test the proposed topology, a prototype of the coupled inductor balancing circuit is made to examine the feasibility of the system. This circuit equalizes eight cells in series using only four coupled inductors. As can be seen in Figure 8.3(a), the gate-to-source voltage to the odd switches is applied to a non-inverting MOSFET driver and the gate-to-source voltage to the even switches is applied to the inverting driver.

When the odd switches are triggered, the energy will be stored in the inductor. In the next stage, when the even switches are triggered, while the odd switches are turned off, and the energy stored in the inductors will circulate through the cells. The cells with lower voltage will absorb the energy, and they will be charged to the nominal voltage. The gate-to-source voltage (*V*gs) to the MOSFETs showing the implemented 45% duty ratio is presented in Figure 8.3(b). The switching frequency generated by microcontroller is 45.5 kHz with 45% duty cycle and 5% dead time, which has a lot of safety to avoid shoot-through between switches. The control circuit is isolated with an optocoupler at the input for safety. When the optocoupler is triggered, the circuit is enabled.

All switches are logic level N-channel MOSFETs with lower $R_{DS(on)}$ to decrease losses. The coupled inductor values are 10 μH for each inductor, and it is measured by an LCR meter. The leakage inductance is in the order of 50 nH (inductor and circuit). The circuit inductance can be limited by a ground layer plane. In order to limit the inductor current, a resettable fuse (positive temperature coefficient (PTC)) is implemented in the circuit in series with each inductor. The PTC has been chosen in such a way that it limits the current above 2 A. The output is filtered with output filter capacitors with a value of 100 μF. The schematic and the photograph of the PCB set-up of the practical experiment is presented in Figures 8.4 and 8.5, respectively. It can be seen in the figure that an optocoupler is implemented to enable or disable the whole circuit if it is desired. By doing this, the user can control the circuit when balancing is not required.

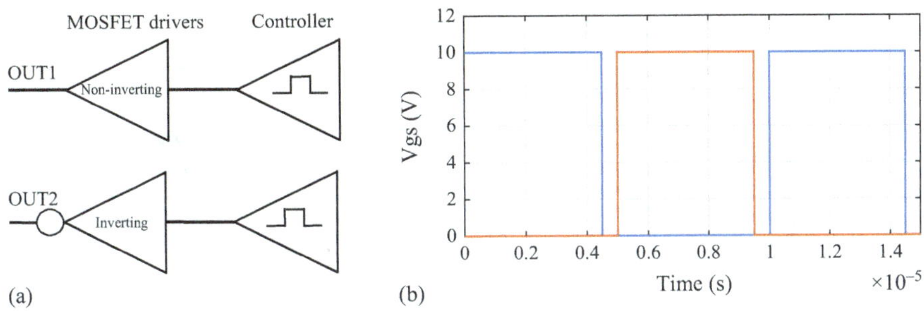


*Figure 8.3 (a) Control block diagram and (b) MOSFETs gate-to-source voltages*

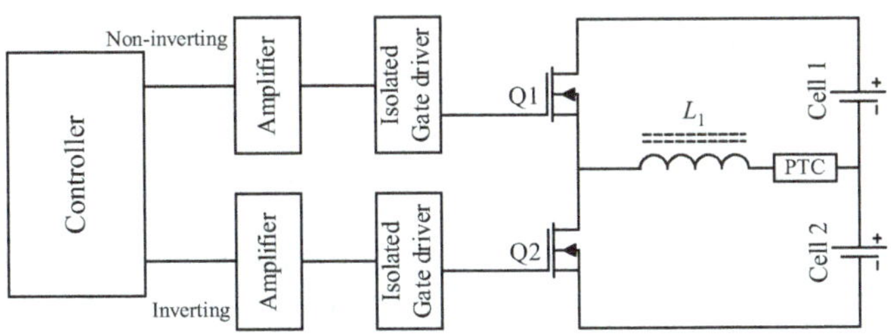


*Figure 8.4 The schematic of the proposed circuit*

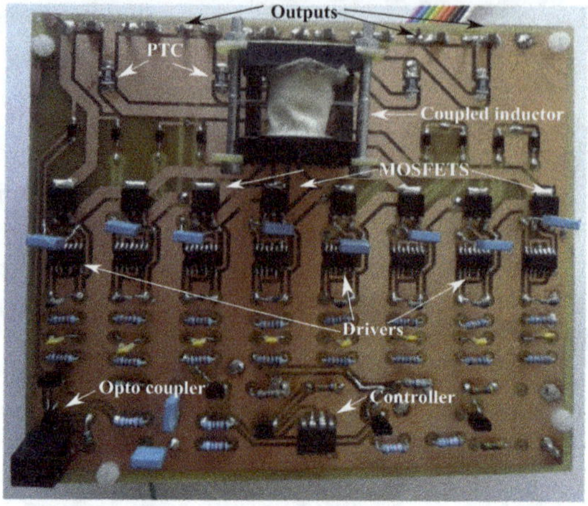


*Figure 8.5 The photograph of the PCB set-up of the practical experiment*

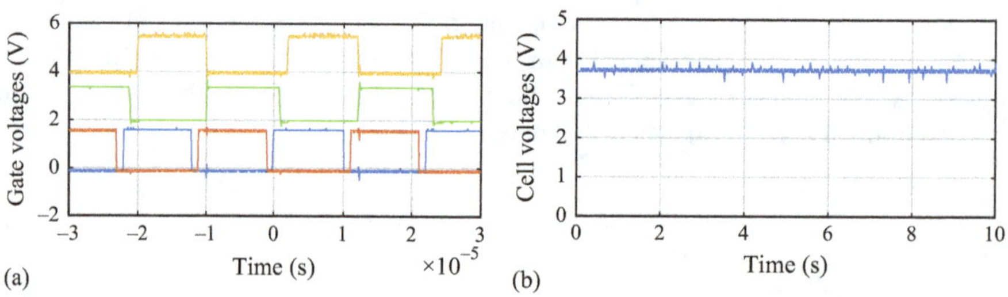


*Figure 8.6 (a) The driver output voltage, gate–source voltage of the switches, and (b) the input voltage to the cells (output voltage of the circuit)*

The output voltage of the driver (inverting and non-inverting), the gate-to-source voltage to the odd switches, and gate-to-source voltage of the even switches is presented in Figure 8.6(a). The blue line is the non-inverting output of the driver, while the red is inverting output. The orange waveform is the gate-to-source voltage to the odd switches, while the green is the gate-to-source voltage of the even switches. The input voltages to the cells, for example cell 1, are presented in Figure 8.6(b). To verify the feasibility of the system, an imbalanced cell was considered to ensure how the circuit will equalize it. In Figure 8.7(a), the charging process of cells with the proposed inductor balancing is presented. It can be noticed that all cells are equalized and fully charged. The cell-equalizing process with different cell voltage levels is presented in Figure 8.7(b).

It has been known that the average voltage of the inductor is almost equal to zero. In Figure 8.8(a), the inductor voltage value $L_1$ is presented to verify this concept. In order to test between the balanced and imbalanced cell inductor

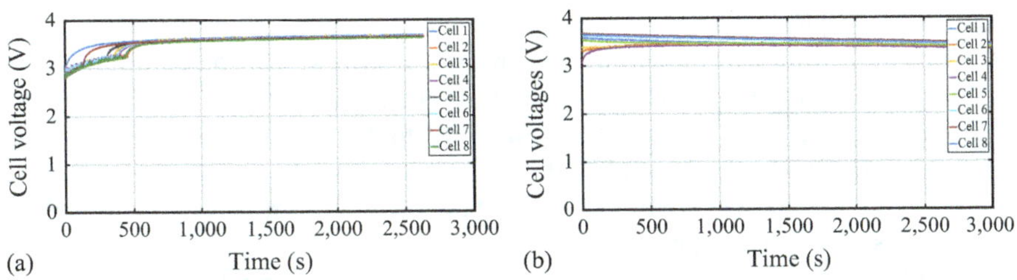


*Figure 8.7 (a) Cells 1–8 charging and (b) cells 1–8 equalizing results*

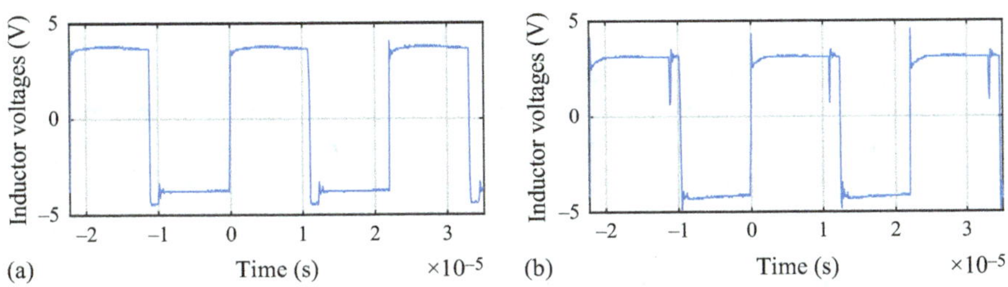


*Figure 8.8 Inductor voltage of ($L_1$) with cell 1 (a) 0.8 V and (b) 1.2 V imbalanced*

*Table 8.2 Parameters values*

| | |
|---|---|
| $L_1$, $L_2$, $L_3$, $L_4$ | 10 μH |
| MOSFET ($R_{DS(on)}$) | 104 mΩ max |
| Switching frequency | 45.5 kHz |
| Duty cycle | 45% |
| Cell voltages | 3.3 V nominal |
| Dead time | 5% |
| Number of cells | 8 cells |
| PTC | 2 A |
| Output filter capacitors | 100 μF |

voltages, cell 1 with 1.2 V imbalance is considered and the inductor voltage $L_1$ is shown in Figure 8.8(b). All circuit parameter values of the proposed topology are presented in Table 8.2.

## 8.7 Some note about winding the coupled inductor

The core needs some air gap, as if the even cells have a different average voltage than the odd cells, and some DC mmf will be present in the magnetic core, so that core needs some air gap. Too high currents are avoided by the NTC resistors. The presence of an air gap means that the flux/winding is not equal everywhere, as close to the air gap, the flux is lower than close to the yoke (or far away from the air gap). If the component is wound using four wires in one hand, so like a 'ribbon', each winding has on average almost the same distance from the air gap, which results in a good symmetry. If the windings would be wound in four different rooms of a coil-former, this would not be the case, and the cells connected to the windings close to the air gap would equalize on a lower EMF. Winding with four wires in one hand is still rather easy (wires cut to the right length and stripped before winding may go faster). The coil former needs at least eight pins for eight cells, and most of coil formers have at least that number. The realized coupled inductor had 10 μH magnetizing inductance.

## Summary

In this part, the concept of inductor-based balancing method is investigated. The proposed topology with a coupled inductor circuit is proposed as shown in Figure 8.1. This balancing method improves the balancing time compared to the conventional inductor circuit. The conventional inductor circuit based on separate inductors has a slow balancing speed because the energy is transferred cell by cell. This problem can be solved by using coupled inductors, which is much faster. Moreover, the coupled inductor of two cells per winding has the advantages of having two times less windings than cells to equalize. It is also bidirectional: the high-voltage cells can be discharged whereas the low-voltage charged by the same circuit. All the inductors are coupled with only one magnetic core with an air gap and they are made with a 1:1:1:1 winding ratio. The switches are logic level N-channel MOSFET and they are triggered by a microcontroller with a frequency of 45.5 kHz and duty cycle of 45% to avoid shoot-through between switches. The circuit is tested with different cell imbalances (0.8 V and 1.2 V), respectively, to testify the viability of the system. In order to protect the charging currents, a PTC thermistor protection is implemented for each inductor to limit the balancing current up to 2 A. The experimental and simulation results are performed to show the feasibility of the proposed topology.

# *Section 2*
# Ćuk converter balancing by using coupled inductors

## 8.8 Introduction of the Ćuk type with coupled inductors

In this section, a Ćuk converter balancing method by using a coupled inductor for lithium-based batteries is investigated. The proposed circuit is an active balancing circuit that will equalize eight battery cells in series. The conventional Ćuk converter has 2*N*–2 switches to balance *N* cells, while the proposed circuit requires only *N* switches for *N* cells in series. Also, the proposed developed topology uses coupled inductors instead of separate inductors, unlike the basic Ćuk converter balancing method. Since the normal Ćuk balancing transfers the energy among two adjacent cells, it requires a proportionately long equalization time particularly for long string battery packs, but the coupled inductor Ćuk converter type overcomes this problem. The switches are all N-channel MOSFETs to achieve a lower drain-source on-resistance, $R_{DS(on)}$, and less voltage drop as compared to the P-channels. The switches are triggered by complementary signals. The coupled inductor is made in such a way to hold the same magnetizing inductance. It can be done by using five wires in one hand. The circuit contains five inductors, one magnetic core, with five windings for eight cells, and one capacitor for two cells. Therefore, the overall circuitry and complexity of the circuit are reduced, resulting in more cost-effective and easy-to-implement solution. The system also does not demand complicated control for battery equalizing. The experimental circuit was implemented and the simulation results are made to confirm the validity of the proposed system. Note that the Ćuk contains inductors towards the cells and has capacitors at the level of the transistors, so that some inductance in the wiring to the cell is less important and has no direct detrimental effect in losses or too much ringing.

## 8.9 The conventional Ćuk converter balancing

The conventional Ćuk converter balancing circuit is shown in Figure 8.9. The Ćuk converter has non-dissipative currents, a bidirectional energy transferring capacity, and high efficiency [6–8]. The conventional Ćuk converter requires 2*N*–2 switches, and *N*–1 capacitors in order to balance *N* number of battery cells [8]. The operational principle is divided into two stages, First, the switches Q1, Q3, and Q5 will be turned on, while the other switches are turned off. In the next stage, switches Q2, Q4, and Q6 will be turned on while the other switches will be turned off. This circuit equalizes four cells in a battery string connected in series.

To investigate the difference between conventional and the proposed Ćuk converter balancing circuit, the circuit with four cells are simulated with MATLAB software and the simulation results are presented. The operational principle is simplified with two cells to investigate the circuit. In the first mode, Q1 is turned on while Q2 is turned off, and the cell voltage equations can be written as

$$V_{\mathrm{cell}_1} = L_1 \frac{\mathrm{d}i_{L_1}}{\mathrm{d}t} \tag{8.18}$$

$$V_{\mathrm{cell}_2} = -L_2 \frac{\mathrm{d}i_{L_2}}{\mathrm{d}t} + V_{C_1} \tag{8.19}$$

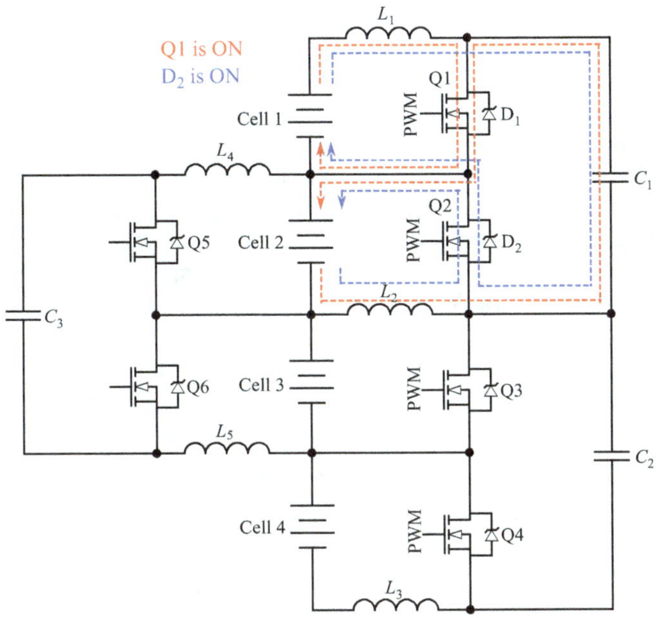


*Figure 8.9 Conventional Ćuk converter balancing circuit*

In the next mode, Q1 and Q5 are turned off and the body diode of Q2 is turned on, the cell voltage equations can be expressed as

$$V_{\text{cell}_1} = L_1 \frac{\mathrm{d}i_{L_1}}{\mathrm{d}t} + V_{C_1} \tag{8.20}$$

$$V_{\text{cell}_2} = -L_2 \frac{\mathrm{d}i_{L_2}}{\mathrm{d}t} \tag{8.21}$$

By applying volt–second balance across $L_1$, $L_2$

$$V_{\text{cell}_1} DT + (V_{\text{cell}_1} - V_{C_1})(1 - D)T = 0 \tag{8.22}$$

$$(-V_{\text{cell}_2} + V_{C_1})DT + (-V_{\text{cell}_2})(1 - D)T = 0 \tag{8.23}$$

The expression for average inductor current can be obtained from charge balance of $C_1$ as

$$i_{L_1}(1 - D)T_s - iL_2 DT_s = 0 \tag{8.24}$$

where $D$ is duty cycle and $T_s$ is switching frequency.

## 8.10 The simulation result of the conventional Ćuk balancing

The simulation results of the conventional Ćuk converter with MATLAB are expressed in this section. The switches are N-channel MOSFETs with body diodes and they are triggered by pulse width modulation (PWM), created by a pulse generator with a synchronous pattern. The switching frequency is 50 kHz with the duty ratio of 50%. The batteries are modelled with the capacitors with the value of 0.5 F. The inductors are 10 μH and the capacitors are 100 μF. The simulation parameters are shown in Table 8.3.

In Figure 8.10(a), the cell voltages of the conventional Ćuk balancing circuit after equalization are presented. It has been found that it is not converging at the end. The inductor voltages in steady state are shown in Figure 8.10(b). It appears

*Table 8.3 Simulation parameters*

| Parameter | Values |
|---|---|
| Equivalent capacitors | 500 mF |
| Initial capacitor voltages | 3 V–3.7 V |
| Switching frequency | 50 kHz |
| Ćuk inductors | 10 μH |
| Ćuk capacitors | 100 μF |

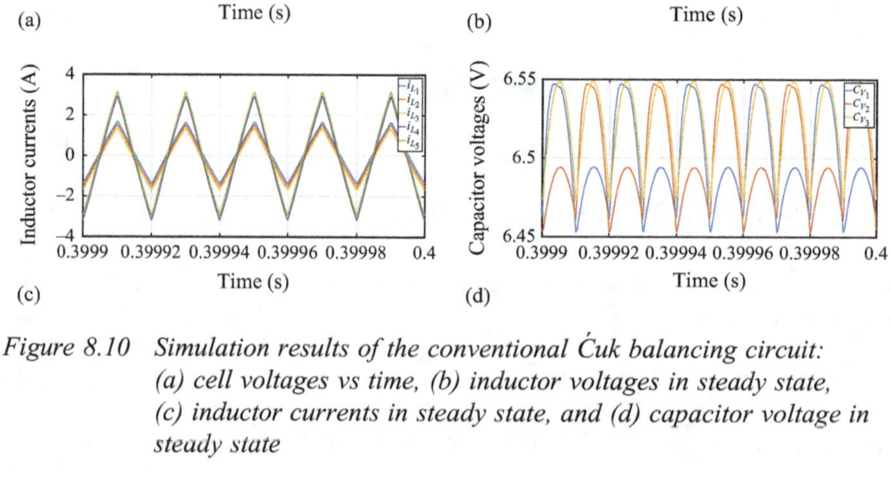


*Figure 8.10 Simulation results of the conventional Ćuk balancing circuit: (a) cell voltages vs time, (b) inductor voltages in steady state, (c) inductor currents in steady state, and (d) capacitor voltage in steady state*

that the inductor voltages are not at the same level of voltage. They should be at the same voltage, which means that they are not equalized well. Figure 8.10(c) and (d) shows the inductor currents and capacitor voltages, respectively. In Figure 8.10(c), it can be noticed that the inductor currents are not at the same level, which means that they are not balanced cells.

## 8.11 The proposed Ćuk converter balancing circuit

In this section, the proposed Ćuk converter balancing circuit with coupled inductors is investigated. The schematic of the proposed circuit is shown in Figure 8.11(a) and (b).

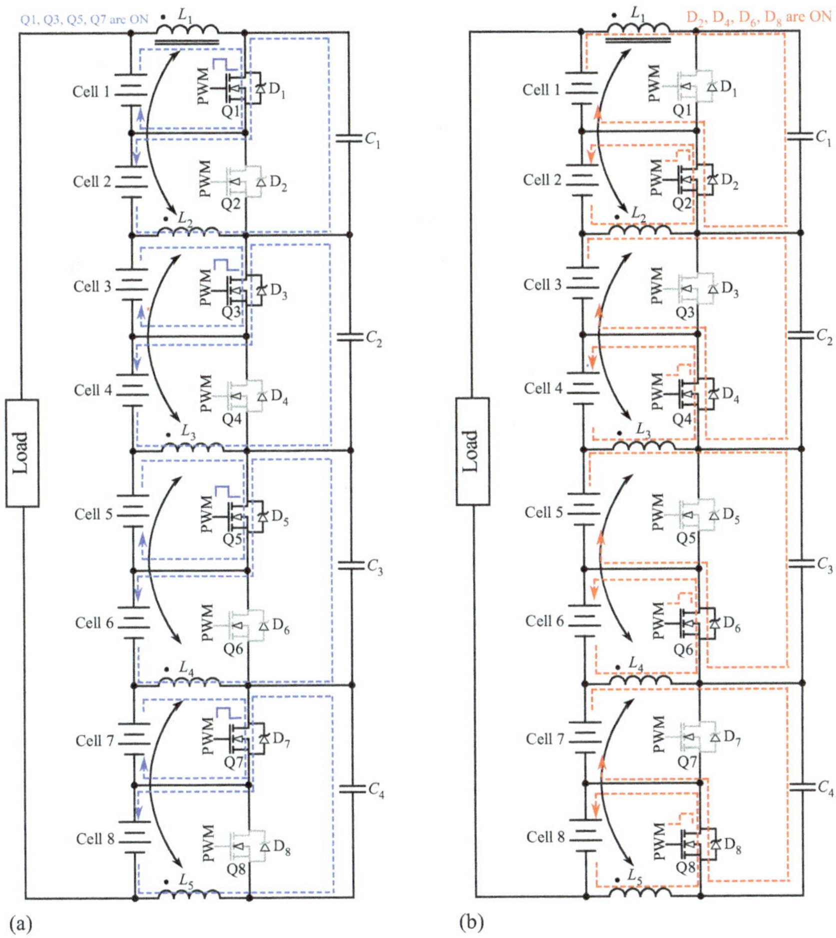


*Figure 8.11 The operational principle of the proposed Ćuk converter balancing with coupled inductors: (a) first mode, (b) second mode*

In some applications, improved performance can be achieved by magnetic devices and not by capacitors, and in this circuit a coupled inductor is used instead of inductors. The circuit will equalize eight battery cells in series using one magnetic core and five coupled inductors. The conventional Ćuk converter requires 2*N*–2 switch per *N* cells, while the proposed circuit uses *N* switches per *N*–1 cells. Furthermore, the conventional Ćuk circuit requires *N* capacitors per *N* cells, while the proposed circuit requires only *N*/2 capacitors per *N* cells; thus the number of capacitors is reduced by a factor of 2. The proposed Ćuk converter balancing circuit with coupled inductors uses one magnetic core with five coupled inductors for eight cells. The inductors are made in such a way to have the same leakage inductances. The operational principle can be split into two modes. The operational principle of the first mode is presented in Figure 8.11(a). The blue lines indicate the current direction. In addition, the operational principle of the second mode is presented in Figure 8.11(b). The red lines indicate the current direction.

In the first mode, the MOSFETs Q1, Q3, Q5, and Q7 will be turned on while the other switches are turned off. The energy from even cells (cells 1, 3, 5, 7) will be transferred to $L_{1,2,3,4,5}$; therefore, the equivalent equation can be derived as

$$V_{\text{cell}_1} = L_1 \frac{\mathrm{d}i_{L_1}}{\mathrm{d}t} + M_{12}\frac{\mathrm{d}i_{L_2}}{\mathrm{d}t} + M_{13}\frac{\mathrm{d}i_{L_3}}{\mathrm{d}t} + M_{14}\frac{\mathrm{d}i_{L_4}}{\mathrm{d}t} + M_{15}\frac{\mathrm{d}i_{L_5}}{\mathrm{d}t} \tag{8.25}$$

where $M$ is the mutual inductance, between $L_1$, $L_2$, $L_3$, and $L_4$. The cell voltage $V_{\text{cell2}}$ can be expressed as

$$V_{\text{cell}_2} = -L_2 \frac{\mathrm{d}i_{L_2}}{\mathrm{d}t} - M_{21}\frac{\mathrm{d}i_{L_1}}{\mathrm{d}t} - M_{23}\frac{\mathrm{d}i_{L_3}}{\mathrm{d}t} - M_{24}\frac{\mathrm{d}i_{L_4}}{\mathrm{d}t} - M_{25}\frac{\mathrm{d}i_{L_5}}{\mathrm{d}t} + V_{C_1} \tag{8.26}$$

The cell voltage $V_{\text{cell}_{(3,5,7)}}$ can be written as

$$V_{\text{cell}_n} = L_k \frac{\mathrm{d}i_{L_k}}{\mathrm{d}t} + M_{ij}\frac{\mathrm{d}i_{L_j}}{\mathrm{d}t} + M_{ij}\frac{\mathrm{d}i_{L_j}}{\mathrm{d}t} + M_{ij}\frac{\mathrm{d}i_{L_j}}{\mathrm{d}t} + M_{ij}\frac{\mathrm{d}i_{L_j}}{\mathrm{d}t} \tag{8.27}$$

where $n = 3, 5, 7$; $k = 2, 3, 4$; $i = 2, 3, 4$; $j = 1, 2, 3, 4, 5$; and $i \neq j$.

Similarly, the cell voltage $V_{\text{cell}_{(4,6,8)}}$

$$V_{\text{cell}_n} = -L_k \frac{\mathrm{d}i_{L_k}}{\mathrm{d}t} - M_{ij}\frac{\mathrm{d}i_{L_j}}{\mathrm{d}t} - M_{ij}\frac{\mathrm{d}i_{L_j}}{\mathrm{d}t} - M_{ij}\frac{\mathrm{d}i_{L_j}}{\mathrm{d}t} - M_{ij}\frac{\mathrm{d}i_{L_j}}{\mathrm{d}t} + V_{C_m} \tag{8.28}$$

where $n = 4, 6, 8$; $k = 3, 4, 5$; $i = 3, 4, 5$; $j = 1, 2, 3, 4, 5$; $m = 2, 3, 4$; and $i \neq j$; therefore, the cell voltages (cells 1, 3, 5, 7) can be rewritten as

$$\begin{bmatrix} V_{\text{cell}_1} \\ V_{\text{cell}_3} \\ V_{\text{cell}_5} \\ V_{\text{cell}_7} \end{bmatrix} = \begin{bmatrix} L_1 & M_{12} & M_{13} & M_{14} & M_{15} \\ M_{21} & L_2 & M_{23} & M_{24} & M_{25} \\ M_{31} & M_{32} & L_3 & M_{34} & M_{35} \\ M_{41} & M_{42} & M_{43} & L_4 & M_{45} \end{bmatrix} \times \frac{\mathrm{d}}{\mathrm{d}t} \begin{bmatrix} iL_1 \\ iL_2 \\ iL_3 \\ iL_4 \\ iL_5 \end{bmatrix} \tag{8.29}$$

Similarly, the cell voltage (cells 2, 4, 6, 8) can be written as

$$\begin{bmatrix} V_{\mathrm{cell}_2} \\ V_{\mathrm{cell}_4} \\ V_{\mathrm{cell}_6} \\ V_{\mathrm{cell}_8} \end{bmatrix} = \begin{bmatrix} M_{21} & L_2 & M_{23} & M_{24} & M_{25} \\ M_{31} & M_{32} & L_3 & M_{34} & M_{35} \\ M_{41} & M_{42} & M_{43} & L_4 & M_{45} \\ M_{51} & M_{52} & M_{53} & M_{54} & L_5 \end{bmatrix} \times \frac{-\mathrm{d}}{\mathrm{d}t} \begin{bmatrix} iL_1 \\ iL_2 \\ iL_3 \\ iL_4 \\ iL_5 \end{bmatrix} + \begin{bmatrix} V_{C_1} \\ V_{C_2} \\ V_{C_3} \\ V_{C_4} \end{bmatrix} \tag{8.30}$$

In the second mode, the MOSFETs Q2, Q4, Q6, and Q8 will be turned on, while the other switches, the MOSFETs Q1, Q3, Q5, and Q7 are turned off. The equivalent equation for $V_{\mathrm{cell}_2}$ can be expressed as

$$V_{\mathrm{cell}_1} = L_1 \frac{\mathrm{d}i_{L_1}}{\mathrm{d}t} + M_{12} \frac{\mathrm{d}i_{L_2}}{\mathrm{d}t} + M_{13} \frac{\mathrm{d}i_{L_3}}{\mathrm{d}t} + M_{14} \frac{\mathrm{d}i_{L_4}}{\mathrm{d}t} + M_{15} \frac{\mathrm{d}i_{L_5}}{\mathrm{d}t} + V_{C_1} \tag{8.31}$$

Similarly, the cell voltage $V_{\mathrm{cell}_2}$ can be expressed as

$$V_{\mathrm{cell}_2} = -L_2 \frac{\mathrm{d}i_{L_2}}{\mathrm{d}t} - M_{21} \frac{\mathrm{d}i_{L_1}}{\mathrm{d}t} - M_{23} \frac{\mathrm{d}i_{L_3}}{\mathrm{d}t} - M_{24} \frac{\mathrm{d}i_{L_4}}{\mathrm{d}t} - M_{25} \frac{\mathrm{d}i_{L_5}}{\mathrm{d}t} \tag{8.32}$$

The cell voltage $V_{\mathrm{cell}_{(3,5,7)}}$ can be written as

$$V_{\mathrm{cell}_n} = L_k \frac{\mathrm{d}i_{L_k}}{\mathrm{d}t} + M_{ij} \frac{\mathrm{d}i_{L_j}}{\mathrm{d}t} + M_{ij} \frac{\mathrm{d}i_{L_j}}{\mathrm{d}t} + M_{ij} \frac{\mathrm{d}i_{L_j}}{\mathrm{d}t} + M_{ij} \frac{\mathrm{d}i_{L_j}}{\mathrm{d}t} + V_{C_m} \tag{8.33}$$

where $n = 3, 5, 7$; $k = 2, 3, 4$; $i = 2, 3, 4$; $j = 1, 2, 3, 4, 5$; $m = 2, 3, 4$; and $i \neq j$.

Similarly, the cell voltage $V_{\mathrm{cell}_{(4,6,8)}}$ can be written as

$$V_{\mathrm{cell}_n} = -L_k \frac{\mathrm{d}i_{L_k}}{\mathrm{d}t} - M_{ij} \frac{\mathrm{d}i_{L_j}}{\mathrm{d}t} - M_{ij} \frac{\mathrm{d}i_{L_j}}{\mathrm{d}t} - M_{ij} \frac{\mathrm{d}i_{L_j}}{\mathrm{d}t} - M_{ij} \frac{\mathrm{d}i_{L_j}}{\mathrm{d}t} \tag{8.34}$$

where $n = 4, 6, 8$; $k = 3, 4, 5$; $i = 3, 4, 5$; $j = 1, 2, 3, 4, 5$; and $i \neq j$.

Therefore, the cell voltages $V_{\mathrm{cell}_{(1,3,5,7)}}$ can be rewritten as

$$\begin{bmatrix} V_{\mathrm{cell}_1} \\ V_{\mathrm{cell}_3} \\ V_{\mathrm{cell}_5} \\ V_{\mathrm{cell}_7} \end{bmatrix} = \begin{bmatrix} L_1 & M_{12} & M_{13} & M_{14} & M_{15} \\ M_{21} & L_2 & M_{23} & M_{24} & M_{25} \\ M_{31} & M_{32} & L_3 & M_{34} & M_{35} \\ M_{41} & M_{42} & M_{43} & L_4 & M_{45} \end{bmatrix} \times \frac{\mathrm{d}}{\mathrm{d}t} \begin{bmatrix} iL_1 \\ iL_2 \\ iL_3 \\ iL_4 \\ iL_5 \end{bmatrix} + \begin{bmatrix} V_{C_1} \\ V_{C_2} \\ V_{C_3} \\ V_{C_4} \end{bmatrix} \tag{8.35}$$

Similarly, the cell voltage $V_{\mathrm{cell}_{(2,4,6,8)}}$ can be written as

$$\begin{bmatrix} V_{\mathrm{cell}_2} \\ V_{\mathrm{cell}_4} \\ V_{\mathrm{cell}_6} \\ V_{\mathrm{cell}_8} \end{bmatrix} = \begin{bmatrix} M_{21} & L_2 & M_{23} & M_{24} & M_{25} \\ M_{31} & M_{32} & L_3 & M_{34} & M_{35} \\ M_{41} & M_{42} & M_{43} & L_4 & M_{45} \\ M_{51} & M_{52} & M_{53} & M_{54} & L_5 \end{bmatrix} \times \frac{-\mathrm{d}}{\mathrm{d}t} \begin{bmatrix} iL_1 \\ iL_2 \\ iL_3 \\ iL_4 \\ iL_5 \end{bmatrix} \tag{8.36}$$

The average capacitor voltage values ($C_1$–$C_3$) are twice a voltage of one cell. This can be explained by having a KVL around the larger loop (through cell 1, $L_1$, $C_1$, $L_2$, and cell 2) in Figure 8.11. Therefore, the KVL equation can be written as

$$V_{C_1} = V_{\text{cell}_1} + V_{\text{cell}_2} \tag{8.37}$$

Similarly the average capacitor voltages for $C_2$, and $C_3$ can be driven as

$$V_{C_2} = V_{\text{cell}_3} + V_{\text{cell}_4} \tag{8.38}$$

$$V_{C_3} = V_{\text{cell}_5} + V_{\text{cell}_6} \tag{8.39}$$

In order to write the state-space equation of the proposed circuit, the capacitor currents equation should be written. Therefore, the capacitor currents in $C_1$–$C_4$ in the first stage, when the switches Q1, Q3, Q5, and Q7 are turned on, can be written as

$$C_1\frac{dv_{C_1}}{dt} = -iL_2 \quad C_2\frac{dv_{C_2}}{dt} = -iL_3 \quad C_3\frac{dv_{C_3}}{dt} = -iL_4 \quad C_4\frac{dv_{C_4}}{dt} = -iL_5 \tag{8.40}$$

The average capacitor voltage values ($C_1$–$C_3$) are twice a voltage of one cell. This can be explained by having a KVL around the larger loop (through cell 1, $L_1$, $C_4$, $L_4$, and cell 2) in Figure 8.11(b), Therefore, the KVL equation can be written as

$$V_{C_1} = V_{\text{cell}_1} + V_{\text{cell}_2} \tag{8.41}$$

Similarly, the average capacitor voltages for $C_2$ and $C_3$ can be driven as

$$V_{C_2} = V_{\text{cell}_3} + V_{\text{cell}_4} \tag{8.42}$$

$$V_{C_3} = V_{\text{cell}_5} + V_{\text{cell}_6} \tag{8.43}$$

Similarly, the capacitor currents $C_1$–$C_4$ during the second stage, when the switches Q2, Q4, Q6, and Q8 are turned on, can be written as

$$C_1\frac{dv_{C_1}}{dt} = i_{L_1} \quad C_2\frac{dv_{C_2}}{dt} = i_{L_2} \quad C_3\frac{dv_{C_3}}{dt} = i_{L_3} \quad C_4\frac{dv_{C_4}}{dt} = i_{L_4} \tag{8.44}$$

The above equations can be used to solve the differential equations, or to make a state space model [9].

## 8.12 The simulation results of the proposed Ćuk (four cells)

The simulation results of the proposed Ćuk converter with coupled inductors are expressed in this section. First, the proposed circuit is tested with four numbers of cells in MATLAB to compare the conventional and proposed circuit. In the next section, the proposed Ćuk balancing circuit is tested with eight cells. The switches are N-channel MOSFETs with body diodes, and they are triggered by PWM, created by pulse generator with a synchronous pattern. The switching frequency is 50 kHz with a duty ratio of 50%. As compared to the basic Ćuk converter circuit which uses one switch and one diode, the proposed circuit uses two switches instead of a diode.

This means less voltage drop across it and more symmetry voltages for both odd cells and even cells. The batteries are modelled as equivalent cells using the capacitors with the value of 0.5 F, and the initial capacitor voltages are chosen in such a way to present different voltage levels. The voltages are chosen from 3 V to 3.7 V in order to have different imbalanced cells. The magnetizing inductance for the coupled inductors is 10 μH and the leakage inductance of 1 μH. All capacitors are 100 μF with no series resistance. The simulation results of the proposed Ćuk converter with coupled inductors and four cells in series are presented in Figures 8.12 and 8.13.

The cell voltages of cells 1–4, after balancing, are shown in Figure 8.12(a). The circuit has a faster balancing time as compared to the conventional circuit presented in the previous section and the cell voltages are converged at the end. Figure 8.12(b) shows the inductor voltages of the proposed Ćuk balancing circuit with four cells in steady state. It is known that the inductor average voltage in a period in steady state is about zero. Thus, it can be seen that they are at the same levels, which means all cells are balanced, unlike the conventional circuits which are not at the same voltage levels. In Figure 8.13(a), the inductor currents of the proposed Ćuk balancing circuit with four cells in steady state are presented. It can be found that all inductor currents are the same, which means that those are balanced cells. Figure 8.13(b) shows the capacitor voltages in steady state. It can be seen in the figure that, according to (8.37) and (8.42), the average voltages across capacitors are twice the voltage of one cell.

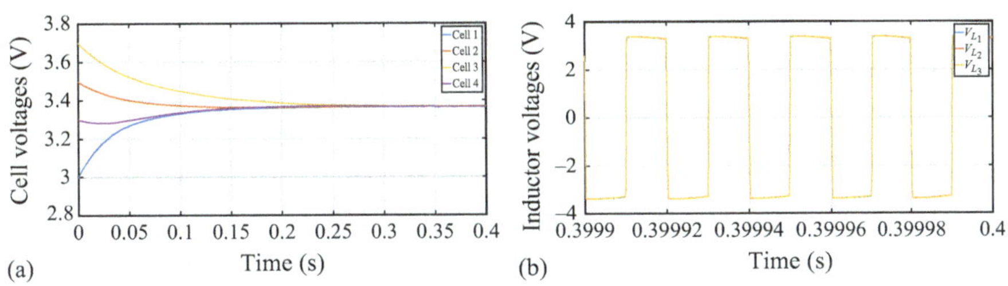


*Figure 8.12 (a) Cell voltage waveforms vs time and (b) inductor voltage waveforms vs time of the proposed Ćuk converter balancing circuit with four cells in series*

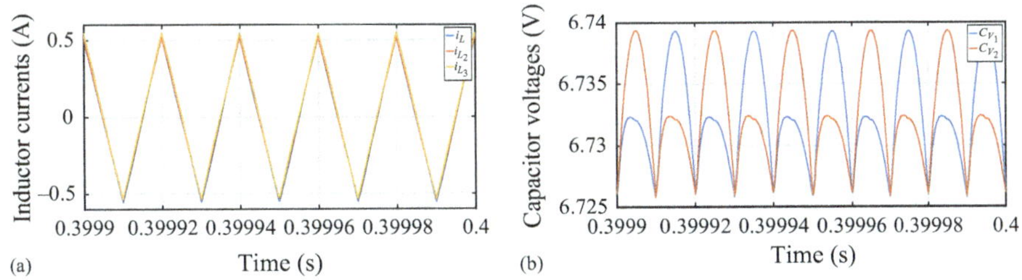


*Figure 8.13 (a) Inductor current waveforms and (b) capacitor voltage waveforms of the proposed Ćuk converter balancing circuit with four cells in series*

## 8.13 The simulation results of the proposed Ćuk balancing (eight cells)

In this section, the simulation results of the proposed Ćuk converter with coupled inductors are presented. Similarly to the previous simulation, all simulations parameters are the same, except the number of cells, which are increased to eight cells. By doing this, the circuit will be tested if the cells are increased from four to eight. The cell voltages of cells 1=8 are shown in Figure 8.14(a). The circuit has a faster balancing time as compared to the conventional circuit presented, and the cell voltages are also converged at the end like the proposed circuit with four cells. The simulation results are presented to verify the validity of the circuit. As discussed before, the inductor average voltage in a period in steady state is near zero. Similarly to the previous section, it can be seen that all inductor voltages are at the same voltage level, resulting in all cells to be balanced. The inductor voltages of the proposed Ćuk balancing circuit with four cells in steady state are presented in Figure 8.14(b).

Figure 8.15(a) shows the inductor currents of the proposed Ćuk balancing circuit with eight cells in transient. The figure shows that after the balancing time is done, all inductor currents reach the same level in transient. The inductor currents of the proposed Ćuk balancing circuit with eight cells in steady state, in a closer view, are presented in Figure 8.15(b). It can be noticed that all inductor currents are the same, which means that those are balanced cells. In Figure 8.16(a) and (b), the capacitor voltages in transient and steady state are presented, respectively.

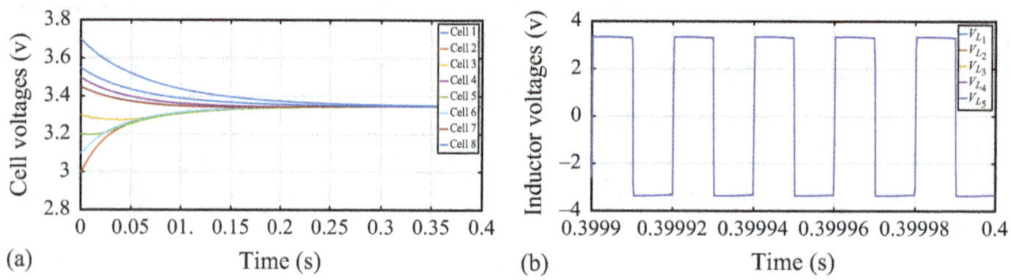


*Figure 8.14 (a) Cell voltage and (b) inductor voltage ($L_1$–$L_5$) waveforms vs time of the proposed Ćuk converter balancing circuit with eight cells*

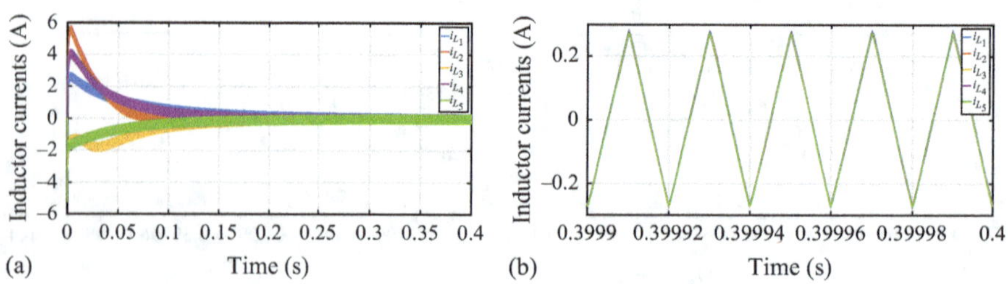


*Figure 8.15 Inductor current waveforms of the proposed Ćuk converter balancing circuit eight cells in (a) transient and (b) steady state*

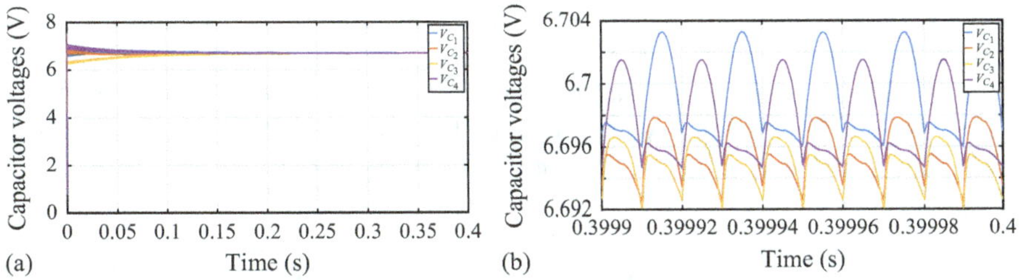


*Figure 8.16 Capacitor voltages of the proposed Ćuk balancing with eight cells in (a) transient and (b) steady state*

*Table 8.4 MOSFET characteristics*

| Parameters | Values |
|---|---|
| $V_{\mathrm{DSS}}$ | 60 V |
| $R_{\mathrm{DS(on)}}$ | 89 mΩ |
| $I_D$ | 12 A |
| $V_{\mathrm{gs}}$ | ±15 |
| $V_{\mathrm{gs(th)}}$ | 1.6 V (min) |
| $Q_{\mathrm{g}}$ | 7.4 nC (typ.) |
| $t_r$ | 104 ns |
| $t_f$ | 40 ns |

## 8.14 Experimental results

In this section, the experimental results of the proposed Ćuk converter are presented. In order to verify the proposed balancing circuit, a prototype of the Ćuk converter balancing circuit with coupled inductors is implemented and tested with eight cells in series. The circuit will equalize eight battery cells, but the number of cells can increase or decrease depending on the desired application. The switches are logic level N-channel MOSFETs with body diodes, and they are controlled by a pair of complementary signals in a synchronous trigger pattern which are generated by a microcontroller.

The N-channel MOSFETs are used in order to achieve lower drain–source onresistance, $R_{\mathrm{DS(on)}}$, and less voltage drop as compared to the P-channel types. The MOSFETs Q1, Q3, Q5, and Q7 are controlled by a non-inverting MOSFET driver, which is triggered by a microcontroller, while the switches Q2, Q4, Q6, and Q8 are controlled by an inverting MOSFET driver. The MOSFET characteristics are shown in Table 8.4. The proposed Ćuk converter circuit contains five coupled inductors and one magnetic core with an air gap. The ferrite and corresponding coil former of E 25/13/7 (EF 25) is chosen. The windings are made with five wires in one hand to obtain more symmetry in the leakage inductance of the coupled inductor winding. The magnetizing inductance of 9.89 μH is obtained by tuning the air-gap thickness of the ferrite core. The capacitors are 100 μF multilayer ceramic capacitors in order to achieve lower equivalent series resistance.

The driver is an inverting and non-inverting MOSFET driver controlled by a microcontroller. The frequency of 48 kHz with a duty ratio of 45% is applied to the driver. The simplified driver schematic of the proposed circuit with two cells is presented in Figure 8.17.

In the first mode, the odd MOSFETs will be triggered by a non-inverting driver which is generated by a microcontroller. Then, even MOSFETs are triggered by an inverting driver. The duty cycle of 5% is applied to the driver in order to avoid shoot-through between MOSFETs. The input signals to the drivers which are generated by a PWM controller are first amplified, and then they are applied to the drivers.

The batteries are lithium polymer battery (Li-poly) with the nominal voltage of 3.7 V and a minimum capacity of 1000 mAh, and they are connected in a series configuration. The cut-off discharge voltage of the battery is 3 V and the maximum charge voltage is 4.2 V. The cell voltages are measured individually and the data are collected by Arduino Nano. The gate-to-source voltages to the MOSFETs were generated by the microcontroller as it can be seen in Figure 8.18. The blue

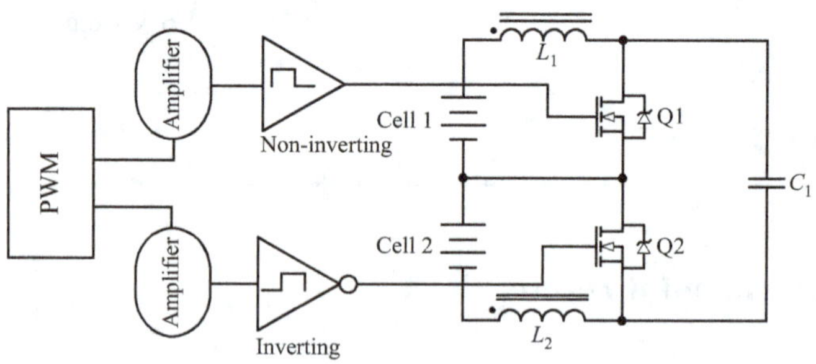


*Figure 8.17 The control flowchart of the proposed Ćuk converter balancing circuit*

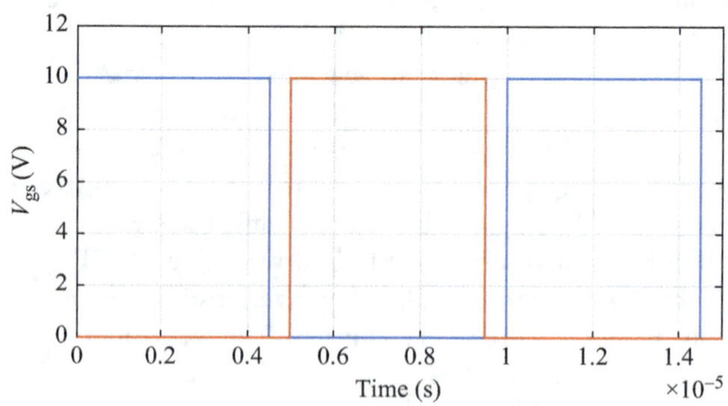


*Figure 8.18 The gate-to-source voltages to the MOSFETs*

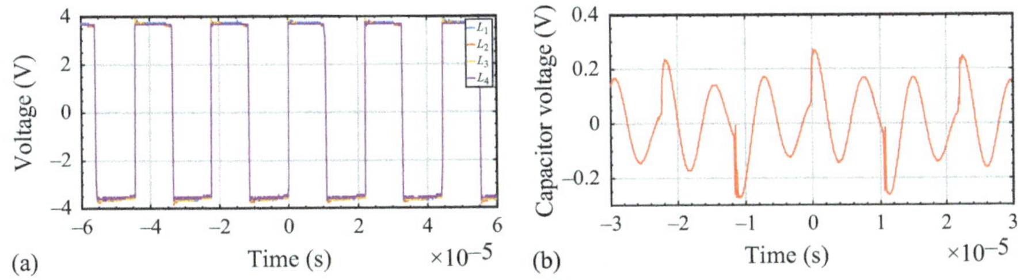


*Figure 8.19 Experimental results of the (a) inductor voltage and (b) capacitor voltage waveforms of the proposed Ćuk converter balancing circuit*

signal is applied to the MOSFETs Q1, Q3, Q5, and Q7, while the red signal is applied to the MOSFETs Q2, Q4, Q6, and Q8. The inductor current waveforms are presented in Figure 8.19(a). As can be noticed, the average inductor voltage is zero and all inductors have the same voltage levels, which means that all cells are balanced. The capacitor voltage waveform $C_1$ is presented in Figure 8.19(b). For clarity, only the capacitor voltage waveform of one capacitor $C_1$ is presented.

To verify the proposed Ćuk balancing circuit, the circuit was tested with eight lithium-ion battery cells in series. In order to have cell imbalances, the battery cells with different voltages were tested. In Figure 8.20(a), the cell voltage waveforms of the proposed circuit are shown. The cell voltages vary from 2.9 V to 3.7 V. The cell voltage waveforms with more voltage differences and one cell fully charged are shown in Figure 8.20(b). The cell voltages vary from 2.8 V to 4 V in order to have significant voltage differences. In Figure 8.20(c), the charging and equalizing voltage waveforms of the proposed circuit are presented. It can be seen that all cells are charged and balanced from the beginning to the end and they are at the same voltage level. The discharging voltage waveforms of the cells with nominal voltages without the balancing circuit are shown in Figure 8.20(d). It can be seen that each cell has a different discharging point, and they are not converged at the end. In Figure 8.20(e), the discharging waveforms of the cells with fully charged cells without a balancing circuit are presented. It can be noticed that they are not converged at the end, because the balancing circuit is not presented. In order to verify the circuit, the battery cells were discharged with the proposed circuit. The discharging waveforms of the cells with the existing one of the balancing circuit are shown in Figure 8.20(f). To verify the proposed circuit, the battery cells with different voltage values were tested. The cell voltages may vary from 2 V to 4 V in order to apply significant voltage differences. The photograph of the experimental set-up of the proposed Ćuk converter balancing circuit is shown in Figure 8.21. In this case, five windings are needed but can still be implemented in a small core such as an EF25. The number of five wires is on the limit to be wound 'in one hand' so as a kind of 'ribbon'. Also an air gap has to be implemented.

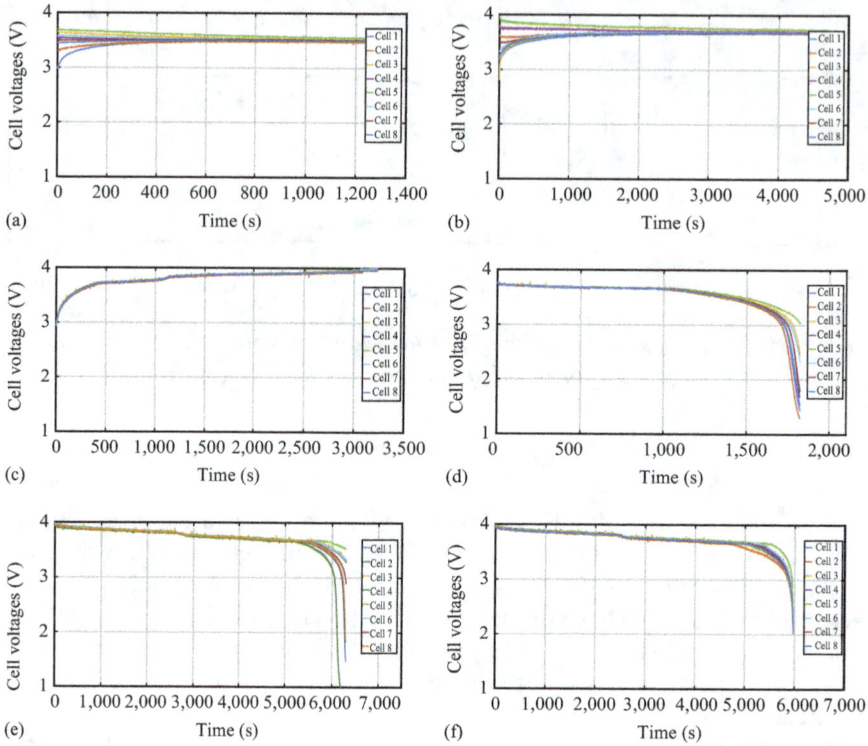


*Figure 8.20 The experimental results of the proposed Ćuk converter balancing: (a) cell voltage waveforms of the proposed circuit after balancing, (b) cell voltage waveforms after balancing with one cell fully charged, (c) charging and equalizing cell voltage waveforms with balancing circuit, (d) discharging cell voltage waveforms with nominal voltages without balancing circuit, (e) discharging waveforms of the cells with fully charged cells without balancing circuit, and (f) discharging waveforms of the cells with the existing ones of the balancing circuit*

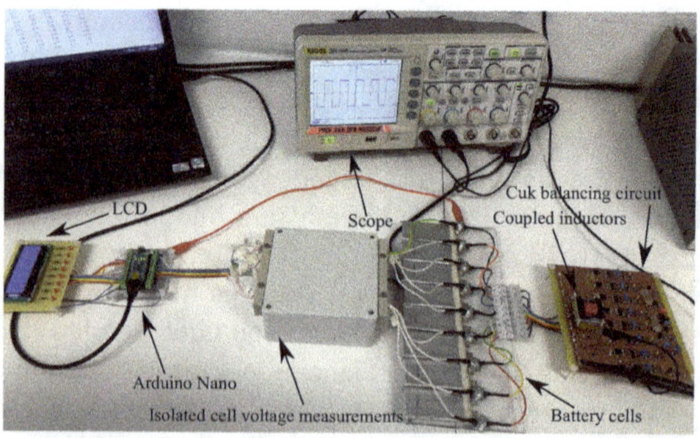


*Figure 8.21 The photograph of the experimental set-up of the proposed Ćuk converter balancing circuit*

## Summary

A Ćuk converter balancing technique, by using a coupled inductor for lithium-ion batteries in ultralight electric vehicles, is investigated in this part. The proposed circuit equalizes eight Li-ion battery cells in series. The conventional Ćuk converter balancing circuit requires $2N-1$ switches in order to equalize $N$ cells, but the proposed circuit requires only $N$ switches for $N$ cells in a battery pack. A conventional Ćuk converter uses one switch and one diode, but the proposed circuit uses two MOSFETs as switch per cell, therefore, resulting in less voltage drop. Furthermore, the proposed circuit uses coupled inductors instead of uncoupled inductors, unlike the conventional type. The fact of using coupled inductors needs less magnetic cores, but also fewer windings compared to separate inductors, and also the symmetry is maintained. There is always a voltage available to power the MOSFETs, even without a bootstrap. The coupling permits a faster equalizing than capacitors alone. As the leakage is lower than the magnetizing, a lower impedance is created. As no separate voltage measurement is used, it is less sensitive to false readings. The gate drivers can be made in such a way that a virtual zero consumption is obtained at rest. The proposed circuit will equalize eight cells in a battery pack in series. The number of cells can be decreased or increased depending on the desired application. Experimental and simulation results are performed in order to show the feasibility of the proposed system.

## References

[1] Imtiaz AM, Khan FH, and Kamath H. A low-cost time shared cell balancing technique for future lithium-ion battery storage system featuring regenerative energy distribution. In: 2011 Twenty-Sixth Annual IEEE Applied Power Electronics Conference and Exposition (APEC). IEEE; 2011. p. 792–799.

[2] Einhorn M, Roessler W, and Fleig J. Improved performance of serially connected Li-ion batteries with active cell balancing in electric vehicles. *IEEE Transactions on Vehicular Technology*. 2011;60(6):2448–2457.

[3] Moghaddam AF and Van den Bossche A. An active cell equalization technique for lithium ion batteries based on inductor balancing. In: 2018 9th International Conference on Mechanical and Aerospace Engineering (ICMAE). IEEE; 2018. p. 274–278.

[4] Chen Y, Liu X, Cui Y, *et al.* A multiwinding transformer cell-to-cell active equalization method for lithium-ion batteries with reduced number of driving circuits. *IEEE Transactions on Power Electronics*. 2015;31(7):4916–4929.

[5] Li S, Mi C, and Zhang M. A high efficiency low cost direct battery balancing circuit using a multi-winding transformer with reduced switch count. In: 2012 Twenty-Seventh Annual IEEE Applied Power Electronics Conference and Exposition (APEC). IEEE; 2012. p. 2128–2133.

[6] Wang Z, Chen C, Zhang Y, *et al.* A resonant switched-capacitor balancing strategy for series battery pack. In: 2014 IEEE Conference and Expo Transportation Electrification Asia-Pacific (ITEC Asia-Pacific). IEEE; 2014. p. 1–5.

[7] Cui X, Shen W, Zhang Y, *et al.* A fast multi-switched inductor balancing system based on a fuzzy logic controller for lithium-ion battery packs in electric vehicles. *Energies*. 2017;10(7):1034.

[8] Li S, Mi CC, and Zhang M. A high-efficiency active battery-balancing circuit using multiwinding transformer. *IEEE Transactions on Industry Applications*. 2012;49(1):198–207.

[9] Farzan Moghaddam A, and Van den Bossche A. A Ćuk converter cell balancing technique by using coupled inductors for lithium-based batteries. *Energies*. 2019;12(15):2881.

# Chapter 9
# Active transformer-based equalizer

## 9.1 Introduction

The concept of an active cell balancing technique, based on multi-winding transformer balancing for lithium-ion (Li-ion) batteries is presented. In Li-ion-based batteries, voltage differences exist between cells due to the charging and discharging process; thus, a battery management system is required to ensure that all cells are equally charged or discharged. An equalizing method is essential to achieve the best performance. Several cell balancing methods have been presented. Among them, the multi-winding transformer balancing method has fast equalization time and is easy to control. The proposed method uses MOSFETs as a switch in the output and a single phase full bridge at the input. Suppressing diodes results in lower losses and has a faster response. Be aware that the output transistor conducts when the input is also on, so it is a forward configuration. Fast equalization, easy to control, high efficiency and easy isolation are its features. Unlike the forward structure, the proposed method does not use a demagnetizing winding or resistor-capacitor-diode circuit in order to demagnetize the primary winding. The simulation results are performed to verify the feasibility of the system.

The transformer balancing is an active method to move energy from a cell to another cell. It has fast equalization time compared to other methods. It works based on Faraday's law, and the same electromotive forces (EMFs) are induced in different windings which have the same number of turns. The multi-winding transformer topology has a magnetic core with primary and multi secondary windings one for each cell. The advantages of the transformer balancing method are fast balancing time, easy to control because it reduces the voltage sensors and simplifies the control strategy, and also direct energy transfer is possible between non-adjacent cells [1–5]. In Figure 9.1, the conventional multi-winding transformer balancing is presented. It requires one diode for each secondary and the primary absorbs power from the total battery pack throughout Q1. The primary one is controlled by a MOSFET triggered in a synchronous pattern. It has a simple circuit and is easy to control due to one switch and does not need an air gap; so the voltages are uniform, but still diodes give voltage drop losses in the secondary and cannot discharge rapidly a too high cell.

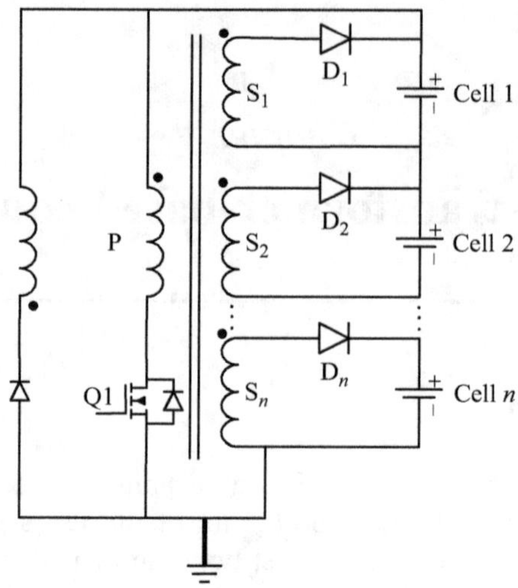


*Figure 9.1 The conventional multi-winding transformer, forward type*

*Section 1*

# Full-bridge multi-winding transformer balancing circuit

## 9.2 The proposed multi-winding transformer balancing

The proposed multi-winding transformer balancing circuit is shown in Figure 9.2. On the primary side, it possesses a full-bridge inverter that uses four MOSFETs with body diodes as switches which are connected to the primary winding. All switches are N-channel MOSFET and they are controlled by a pair of complementary signals in synchronous pattern with a 50% duty ratio. The full-bridge inverter is fed from the total battery pack. Note that it is not essential to use a full bridge, and a forward converter using Q1 and Q2 and where Q3 and Q4 are replaced by two diodes gives about the same result. The full bridge allows a zero-voltage switch and has no diode losses.

The inverter switches have a zero-voltage switching turn on that reduces switching losses. On the secondary side, it has a single MOSFET with a body diode for each secondary. On the other hand, the multiple transformers balancing as in [6] uses several multiple core transformers, one core for each cell. The multi-transformer solution is more expensive than the multi-windings transformer scheme. Considering again the proposed scheme, the switches are turned on with an almost 50% duty cycle.

The primary winding has $N_p$ turns and the secondary windings have $N_s$ turns to transfer energy from the higher cells to the lower cells. The secondary turns ratio depends on the voltage of the whole battery pack. In this part, eight cells are considered; thus the turn-ratio is $a=N_p/N_s$. The operational principle can be divided into two modes. In the first mode, the switches Q1 and Q2 are turned on while the switches Q3 and Q4 are kept turned off; thus, the energy is transferred to each secondary windings.

On the secondary side, all switches are turned on simultaneously with the same duty ratio of 50%; thus, the energy from a higher cell is transferred to the lower one. The voltages of these switches are zero because they have the signal with the same duty ratio. The current from the higher cell has positive values (discharging) and the current from the lower cell has a negative value (charging). Hence, the

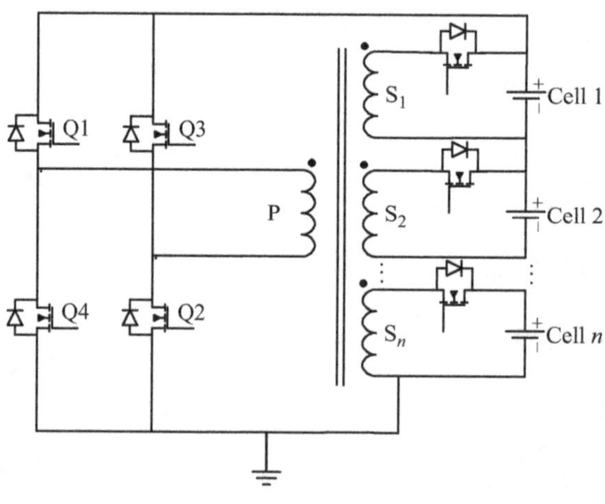


*Figure 9.2 The proposed multi-winding transformer balancing circuit*

higher voltage cells are discharging and the lower cells charging. In the second mode, the switches Q1 and Q2 are kept turned off while the switches Q3 and Q4 are turned on. During this negative cycle, the body diode does not conduct and the switches are off; therefore, the discharging and charging process is done.

The selection of MOSFETs instead of a diode on the secondary side can significantly influence on the efficiency. The equalizing process can be simplified as the step response of an $R_L$ circuit. The value of $R_{eq}$ is the equivalent series resistance of the MOSFET $R_{DS(on)}$ and inductor series resistance $R_L$. $R_{eq}=R_{DS(on)}+R_L$. When the MOSFET is turned on, the inductance is loaded by cells EMFs; thus, the voltage loop and current can be expressed as, where $V$ is the voltage difference of the EMF of the transformer and the EMF of the battery, and take $L$ as the equivalent leakage inductance of the transformer and the circuit inductance.

## 9.3 Simulation results

In this section, the simulation results are presented to verify the feasibility of the proposed circuit. The proposed model is simulated in MATLAB®. The simulation parameters are shown in Table 9.1. The battery cells are modelled with the capacitor with the value of 50 mF. Throughout the book, the same values are used, so that a comparison is possible. The switches are MOSFETs with a body diode. The switches are triggered with synchronous pulse with the duty ratio of 50%. The switching frequency is 50 kHz generated by a pulse generator.

The transformer is a multi-winding transformer with eight secondary ones. The battery voltages are chosen to be from 3.1 V to 3.6 V inappropriately to present the significant voltage difference of the cells. In Figure 9.3(a), the cell voltage waveforms of the proposed circuit are presented. The cell voltages are chosen from 3.1 V to 3.6 V to model a lithium-ion battery with a significant voltage difference to investigate the equalization time. In Figure 9.3(b), the current waveforms of the cells are presented. It can be seen that in the steady state the equalization current is zero which indicates that all cells are equalized. The primary and secondary voltages of the multi-winding transformer are presented in Figure 9.3(c). The primary voltage is the voltage of the whole battery pack and the secondary voltages are eight times less than the primary. In this chapter, the nominal voltage of each cell is considered to be 3.6 V; thus, the total battery pack voltage with eight cells can be driven as 28.8 V. In Figure 9.3(d), the primary and secondary currents are shown. The figure indicated that, due to the cell voltage differences, the secondary currents have a positive value for the higher cells and negative value for the lower cells.

*Table 9.1 Simulation parameters*

| Parameter | Values |
|---|---|
| Equivalent capacitors | 50 mF |
| Initial capacitor voltages | 3.1 V–3.6 V |
| Switching frequency | 50 kHz |

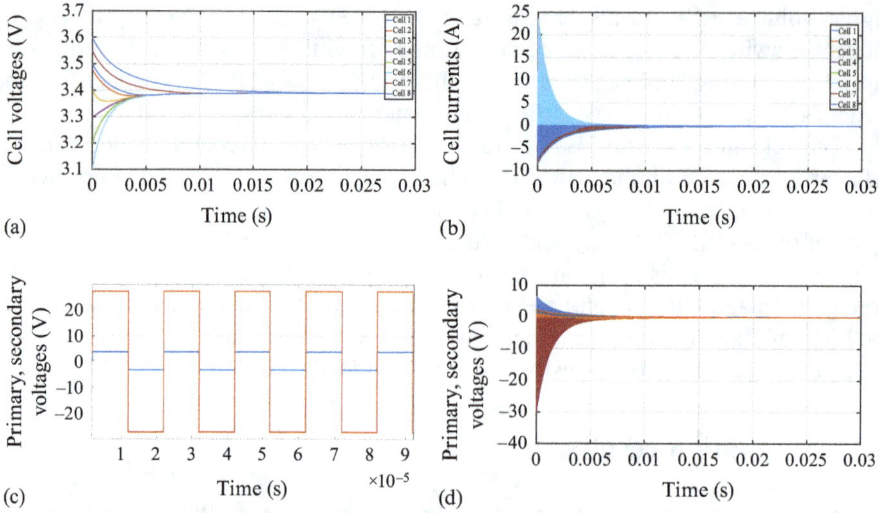

*Figure 9.3 The simulation results of transformer balancing: (a) the cell voltages waveform, (b) the cell current waveforms, (c) the primary and secondary voltages, and (d) the primary and secondary currents*


## Summary

The concept of active balancing technique by using a multi-winding transformer balancing circuit (with forward structure) is presented. It equalizes eight cells in a series battery pack. As compared to the conventional method with flyback structure that uses diodes in the secondary side with an air gap, this scheme with forward structure needs no air gap, thus, reduces the effect of the unequal voltage of the secondary windings. On the primary side, it has a full-bridge inverter and no air gap is needed. It has zero-voltage switching turned on that reduces switching losses. The proposed circuit has fast equalization time and more efficient. The primary can start from a separate voltage or from a series connection of battery packs, and the primaries of transformers can be put in parallel. The circuit has a drawback that as many secondaries are needed as number of cells. The simulation results verify the feasibility of the proposed circuit.

*Section 2*

# A single-transformer balancing circuit with two cells/winding

## 9.4 Introduction

The concept of an active cell balancing technique, by using a multiple outputs double forward converter for lithium-ion (Li-ion) batteries, is investigated. The principle controls two times more cells than secondaries. With four secondaries, it equalizes eight cells in a series. Four secondaries can reasonably be wound 'in one hand' with the same back EMF. This means, a low pin count on the transformer and a low bill of materials. The bridge uses four N-channel MOSFETs as switches, which means low switching losses. This scheme is applied for equalizing the minimum voltage among the cells of the lithium-ion battery. However to operate, it takes energy from the whole battery, so also reducing the higher cells. It can be active as well in an almost full charged battery as close to discharge. It utilizes a multi-winding transformer based on a double forward converter structure. Conventional schemes using a multi-winding transformer for electric vehicles and battery appliances require as much number of secondaries per cell. This scheme requires one secondary for two adjacent cells; thus, the number of secondaries is reduced by a factor of 2. Also, the redistribution of charge from a high cell to a low cell does not require many switching components and little intelligence to determine low cell voltage detection. The basic principle of this method is to use the overall battery pack voltage as a reference to supply individual cells, using a forward converter containing a transformer with a well-chosen winding ratio. The experimental and simulation results are performed to verify the feasibility of the proposed system.

## 9.5 Full-bridge double forward transformer balancing with two cells/winding

The concept of a multi-winding transformer with four secondaries for a battery with eight cells by a double forward converter will be explained and a proof of concept will be given through practical test results. The multi-winding transformer has an advantage that direct energy transfer is possible between non-adjacent battery cells. The multi-winding transformer topology has a magnetic core with primary and multi secondary windings for each cell. The advantage of this method can be expressed as simple, easy to control, and low cost. The conventional multi-winding transformer balancing method requires one secondary and one switch per cell, so the number of secondaries and switches are two times more compared to the proposed circuit shown in Figure 9.5.

This section proposes a direct balance circuit using a multi-winding transformer. Energy transfers from the total package to the lower voltage cells. The main suitability of this section is that the number of secondary windings is reduced by a factor of 2 by using a double forward converter. It results in a cost-effective, more efficient, and less complicated circuit. Also, the number of switches is reduced, resulting in a simpler control and fewer losses. The basic principle of this circuit is to use the overall battery pack voltage as a reference to supply

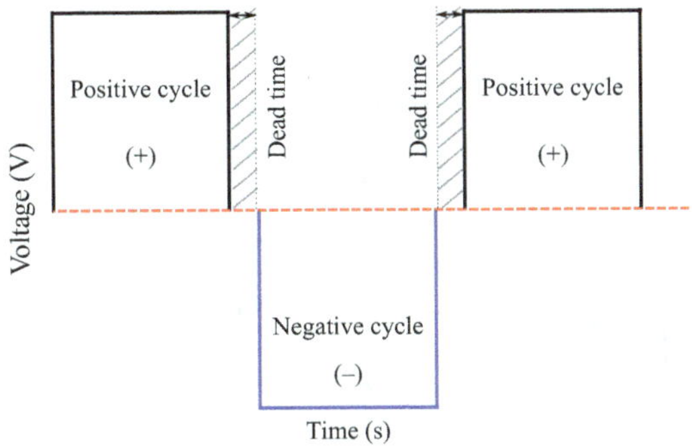


*Figure 9.4 Balancing principle, output voltage of the double forward converter*

individual cells, using a forward converter containing a transformer with a well-chosen winding ratio.

When a cell drops below the voltage level of accompanying pack cells, an extra charge will be supplied to this cell from the entire battery pack. The transformer has a single primary winding and four secondary windings. In each secondary winding, an equal induced voltage is assumed. When an imbalance occurs, a current will flow proportional to the extent of the imbalance. On the primary side, this current will allow the detection of imbalance and a battery pack by-pass when the imbalance is critical. It has the advantage that each cell has not to be compared with a reference voltage, but that the principle relies on a transformer ratio.

The operational principle of this circuit is based on a double forward converter. The double forward aspect indicates the usage of both up and down sides of a primary signal for balancing. This means, a positive half cycle is used to balance one-half of the cells, while the negative half cycle is used to balance the other half, as displayed in Figure 9.4.

Assume that cell 1 is at a lower voltage, while other cells are at a nominal voltage. Since all the cells surrounding cell 1 are at a higher voltage, the transformed reference voltage will also be higher, allowing the balancing of cell 1. By doing this, the number of secondary windings can be limited to 4. This means that the secondaries can easily be 'wound in one hand', where the symmetry allows having an equal EMF in the windings. The forward transformer needs no air gap, so the EMF is quite equal among the windings. The positive–negative can be adjusted by the duty ratio if required. The transformer can be loaded with an asymmetric DC current if sufficient dead time (DT) is given between each half period.

## 9.6 Operational principle

In Figure 9.5, the implementation of the balancing circuit is schematically represented. Assume that the MOSFETs Q3 and Q4 are conducting; thus the primary one receives a positive voltage. Through a rectifying Schottky diode, $D_2$, and smoothing capacitor,

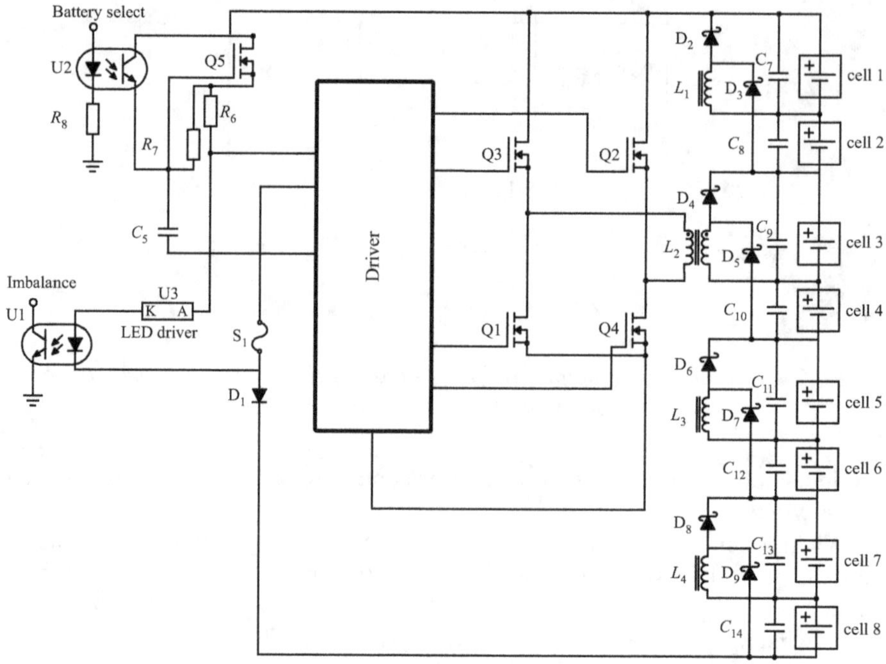


*Figure 9.5 The simplified schematic of the proposed balancing circuit*

$C_7$, cell 1 receives a DC voltage of the same polarity as the primary voltage. However, cell 2 will receive no voltage on its terminals since diode $D_3$ is reverse-biased.

When the MOSFETs Q1 and Q2 are conducting, the primary one receives a negative voltage and cell 2 receives a voltage of the opposite polarity. The MOSFETs Q1 to Q4 build up the full-bridge converter, supplying the signal to the primary.

They are driven by a self-oscillating full-bridge driver integrated circuit (IC). This IC receives its $V$cc from the battery pack voltage through the resistor $R_6$. This only occurs when MOSFET Q5 is conducting with a sufficiently high gate-source voltage $V_{gs}$. This is achieved through the presence of resistor $R_7$. The gate signal for the MOSFET Q5 originates from the optocoupler U2. There is a 'bank select' to reduce the number of simultaneous signals. When receiving a bank select signal, the optocoupler U2 allows a collector-emitter current to flow and to pull up the gate of MOSFET Q5, thereby supplying the $V$cc of the driver.

As it is not preferable to disable the balancing in a short absence of a bank select signal (multiplexing), $C_5$ holds up the gate of Q5 for a limited amount of time, sufficient for the downtime of the bank select. When the bank is critically imbalanced, or a surge current can flow, the polymer positive temperature coefficient (PPTC) resettable fuse $S_1$, in the form of a PPTC, will trip and primary current will be limited by the 'resettable fuse' and the LED driver/constant current regulator. When the voltage drop across the PPTC is too high, the full-bridge driver will go into under-voltage lockout (typically: *V*ccuv-=9 V).

An imbalance (too low cell) indication occurs through optocoupler monitoring. When a significant imbalance occurs, the LED driver supplies enough current to the optocoupler U1 to notify the microcontroller that an imbalance has occurred. The pack could then be decoupled from the load to prevent further unbalancing. When the pack is charged again, it might recover and be used again. The capacitors $C_7$ to $C_{14}$ have a purpose.

They operate as smoothing capacitors as the inductance of wiring and circuit needs to be partly compensated by these capacitors, thus reducing the inductive drop and optimizing the increase of balancing current per voltage imbalance. The wiring inductance acts similarly to an increased transformer leakage. The diode $D_1$ has the purpose to reduce possible circulating current. The selection of this added voltage drop follows from the proposed transformer winding ratio.

To evaluate the balancing current, consider the following configuration. In principle, the secondary balancing current is governed by the voltage imbalance and the leakage inductance of the transformer, by the following relation:

$$i_{\text{balancing}}(t) = \frac{1/n \times \left[\left(\sum_{i=1}^{8} V_{\text{bi}}\right) - V_{\text{D}_1}\right] - V_{\text{Schottky}} - V_{b'}}{L_{\sigma 2}} \times t \tag{9.1}$$

where $0 \leq t \leq t_{\text{on}}$

where $n$ is the primary to secondary turns ratio, $i_{\text{balancing}}$ is the current of considered unbalanced cell, $V_{\text{bi}}$ is the voltage of cell $i$, $V_{b'}$ is the voltage of considered unbalanced cell, $V_{\text{D}_1}$ is the voltage drop of diode $D_1$, $V_{\text{Schottky}}$ is the voltage drop of Schottky diode, $L_{\sigma 2}$ is the transformer leakage referred to the secondary, $t_{\text{on}}$ is the balancing current rise time.

The average balancing current is obtained by taking into account the flyback or recovery time of the leakage inductance. The driver is provided with an internal dead time, which is explained in the following section. As such

$$t_{\text{on}} = T/2 - DT \tag{9.2}$$

with $T$ is the period of the input signal.

The average balancing current is then given by

$$I_{\text{balancing, average}} = \frac{\frac{i_{\text{balancing}}(t_{\text{on}})}{2}\left(T/2 - DT + t_{\text{off}}\right)}{T} \tag{9.3}$$

In order to obtain the flyback or recovery time of the leakage inductance, the recovery voltage during the DT must be known.

$$V_{\text{off}} = -\left(V_{b'} + V_{\text{Schottky}}\right) \tag{9.4}$$

The recovery time is governed by the rate of current decrease and the peak current

$$t_{\text{off}} = L_{\sigma 2} \times i_{\text{balancing}}(t_{\text{on}})/V_{\text{off}} \tag{9.5}$$

In the proof of concept, the balancing current is measured in the function of the imbalance.

## 9.7 Practical implementation

In this section, an implementation of the proposed circuit is done and the results are presented. One of the most critical elements is the transformer ratio. The winding ratio is determined out of the battery pack voltage according to the following reasoning. Assume that all cells are at nominal voltage, being 3.2 V. This causes a total battery pack voltage of $V_{\text{pack, nom}} = 25.6$ V. Due to the limited forward voltage drop of the Schottky diodes, represented in Figure 9.6, the primary to secondary winding ratio, $n$, can be selected as (7.5).

The peak voltage over a secondary winding at nominal battery cell voltage, neglecting primary side voltage drops and leakage inductance, can be expressed as

$$V_{\text{secondary, peak}} = \frac{V_{\text{pack,nom}}}{n} \tag{9.6}$$

This gives a Schottky forward voltage drop of 0.213 V. Therefore, according to Figure 9.6, a circulating current can flow, but only at increased temperatures. This could result in some positive feedback during some time. The lowest voltage cell has the largest current and will have the hottest diode. At a full-charge cell voltage

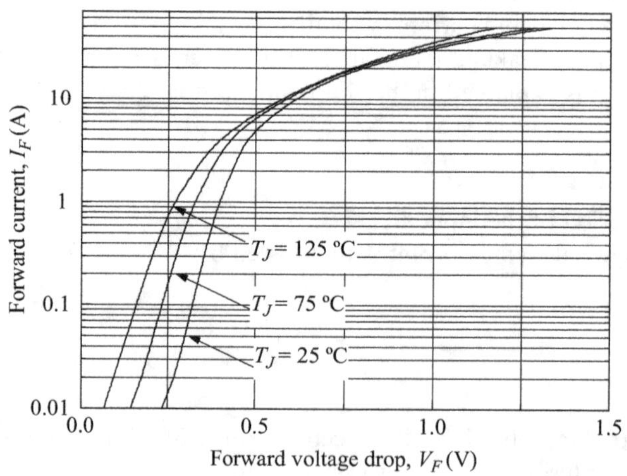


*Figure 9.6 Forward characteristics of a Schottky diode*

of 3.65 V, and thus a 29.2 V pack voltage, the forward voltage drop over the Schottky amounts to 0.243 V, which could lead to more significant circulating current. To reduce the circulating current, an extra voltage drop is introduced in the form of diode $D_1$in Figure 9.5. Assuming a 0.6-V diode voltage drop, therefore, the peak voltage over the secondary winding can be written as

$$V_{\text{secondary,peak}} = \frac{\left(V_{\text{pack}} - V_{\text{diode}}\right)}{n} \tag{9.7}$$

By choosing another type of diode, or more diodes, the precise effect can be tuned. So, by doing this, no circulating current is present, both at full charge and nominal voltage. Only at elevated temperatures, a small circulating current appears. In the above-simplified argumentation for the transformer winding ratio, all primary side voltage drops, other than the diode drop, and all losses were neglected. Instead of attempting to model these, they will be evaluated in the experimental results section by experimental validation.

In the selection of the full-bridge driver, a possible saturation of the transformer core has to be taken into account. In a balanced pack, there is no risk for saturation, since the peak flux density stays well below 0.3 T. In an unbalanced cell situation, there is a flux imbalance between positive and negative half-cycles, with using a 50% duty ratio ($\delta$). The unbalanced cell will require a charge and thus a current in order to balance, while another cell will not require this balancing charge. Assume the positive half-cycle is the balancing cycle. During this positive half-cycle, a voltage drop and thus a flux drop occurs due to surrounding circuit components. The subsequent negative half-cycle will not have this flux drop and might, over a certain time, saturate the core. Thus, it needs a transformer core reset. A solution for this was found in a self-oscillating full-bridge driver IC with an internal DT of 1 μs. Depending on the extent of the imbalance and the corresponding balancing current, the core can demagnetize during this DT.

As an argumentation for this selection, a simplified estimation of the required recovery time is given through a calculation example, performed on a transformer equivalent L-scheme, shown in Figure 9.7 based on measured and estimated values. For a matter of simplicity, only one secondary is considered, and values are converted to the primary. Note that this is a rough estimation simply to give an order of magnitude. The inductances were measured at the operating frequency. Since the

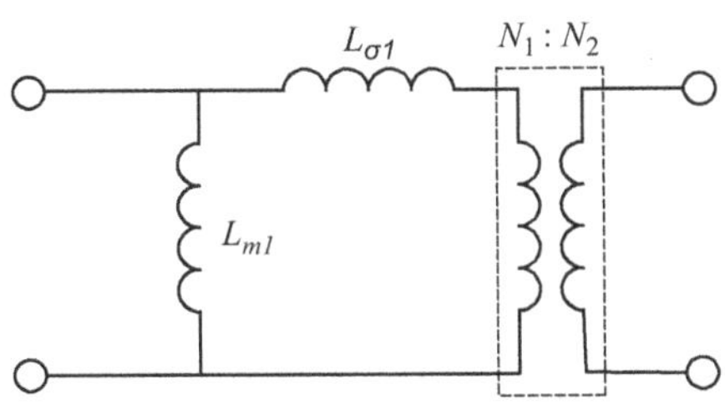


*Figure 9.7 The equivalent L model of the transformer, one secondary considered*

*Table 9.2 Calculation parameters*

| Parameters | Values |
|---|---|
| $f$ | 67 kHz |
| $\delta$ | 50% |
| $DT$ | 1 μs |
| $L_{\sigma p} = (L_{\sigma 1} + L'_{\sigma 1}) + L_{\text{Layout}}(N_1/N_2)^2$ | 6.4 μH |
| $R_1$ | 100 mΩ |
| $R_2$ | 50 mΩ |
| $R_{\text{DS(on)}}$ | 400 mΩ |

leakage inductance (5.8 μH) at the primary side is small compared to the magnetizing inductance (377 μH), and the primary is wound close to the core, the use of a simplified L-scheme is allowed. In a symmetric T-scheme, primary and secondary leakages would have to be appropriately attributed to each winding. The required values are represented in Table 9.2.

$L_{\sigma p}$ is the total leakage (transformer and circuit) measured at the primary side, referred to the primary. The printed circuit board (PCB) layout inductance is measured using an LCR meter, by subtracting the leakage inductance, measured at the primary from the inductance measured with a shorted Schottky diode and output capacitor. $R_1$ is a conservative estimate of the primary resistance, located on the left-hand side of the magnetizing inductance in the L-scheme. The DC resistance is estimated as follows:

$$R = MLT \times 15 \times \rho_{Cu}/A \tag{9.8}$$

where $MLT$ is the mean length of turn, $\rho$ is the resistivity of copper (25 °C), $A$ is wire cross-section area.

However, there is also skin-effect to take into account for the AC resistance. The skin depth is can be expressed as

$$\delta = \sqrt{\frac{2\rho}{(2\pi f \mu_r \mu_0)}} \tag{9.9}$$

Since there is also a proximity effect and a DC component, a conservative estimate of 100 mΩ is assumed. Also, $R_2$ was obtained following a similar reason. $R_{\text{DS(on)}}$ is the total drain-source on-state resistance of the two-switched MOSFETs. This value was obtained from the data sheet taking into account the temperature of 120 °C and amounted to 400 mΩ. As a reference, a severe imbalance is considered, with a peak primary current of 2 A ($I_{\text{peak}}$).

This amounts to a peak balancing current of 15 A, which corresponds to a critically unbalanced pack. In reality, this current is very unlikely, due to the limiting influence of the PPTC and the imbalance that would have to be present. And if it happens, the PPTC protects. In the proof of concept, a realistic balancing curve is presented.

In this worst-case, a simplified calculation to estimate the required DT foreseen in the driver is given below. The flux linkage leakage field corresponding to a

2-A $I_{\text{peak}}$, neglecting magnetizing current can be expressed as

$$\psi_{\sigma 1} = I_{\text{peak}}\, L_{\sigma p} \tag{9.10}$$

At an input voltage ($V_{\text{in}}$) of 25 V, and following the previous argumentation regarding balancing current, this gives an unbalanced cell of about 2 V. The required flyback time for this leakage flux linkage is

$$\Delta t_{\text{flyback}} = \frac{\psi_{\sigma 1}}{V_{\text{off}}\, n} \tag{9.11}$$

The DC voltage drop at the output, compared to a no-load situation, corresponding with this leakage is

$$\Delta V_L = \frac{\psi_{\sigma 1}}{[(T/2 - DT\, n)]} \tag{9.12}$$

The primary resistive voltage drop is estimated as follows:

$$\Delta V_R = \left(R_1 + R_{\text{DS(on)}}\right) I_{\text{peak}}/2 \tag{9.13}$$

The factor 2 originates from the simplification of a linearly increasing current with a peak value of 2 A (normally an exponential behaviour with $\tau = L/R$ is expected). To obtain the voltage drop over the half-cycle, this peak drop can be halved. The recovery time required for this in order to obtain $V$s-balance is

$$\Delta t_R = \frac{\Delta V_R (T/2 - DT)}{V_{\text{off}}\, n} \tag{9.14}$$

According to the previous argumentation, the magnetizing current has 0.34 μs time to equalize a DC magnetizing component in the one microsecond dead time. From this rough estimate, a $DT = 1$ μs seems sufficient.

## 9.8 Simulation results

To test the proposed circuit, the simulation results of the proposed circuit with MATLAB software are expressed. The switches are N-channel MOSFETs with body diodes and they are triggered by pulse width modulation (PWM), created by a pulse generator with a synchronous pattern. The switching frequency is 67 kHz with a duty ratio of 45%. The batteries are modelled with the capacitors, $C = 50$ mF.

In order to apply cell imbalances, the battery pack with different cell voltages was considered. The cell voltages vary from 3.1 V to 3.6 V to apply the significant voltage differences. The simulation parameters are presented in Table 9.3. In Figure 9.8, the cell balancing simulation results of the proposed transformer balancing circuit are presented. It can be seen in the figure that all cells are equalized and converged at the end. Note that the convergence goes faster for a too low cell, rather than a too high cell. The reason is that the too high cells are only discharged by the transistor bridge at the primary side. During

*Table 9.3 Simulation parameters*

| Parameter | Values |
|---|---|
| Equivalent capacitors | 50 mF |
| Initial capacitor voltages | 3.1 V–3.6 V |
| Switching frequency | 67 kHz |

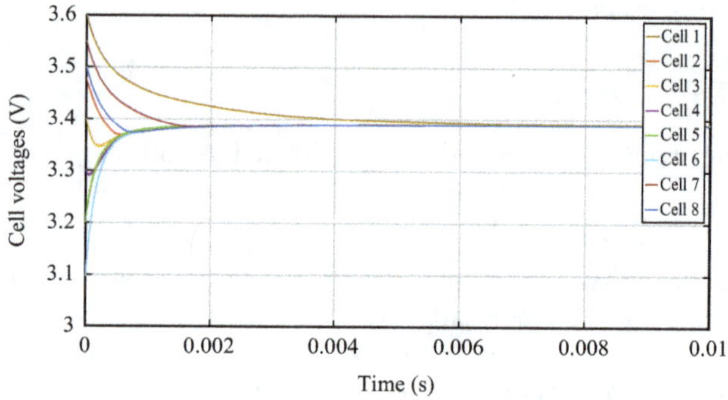


*Figure 9.8 The simulation results of the proposed transformer balancing circuit*

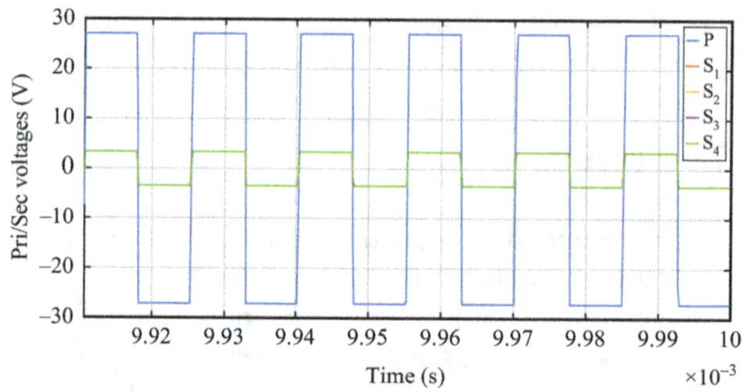


*Figure 9.9 The transformer primary and secondary voltage waveforms*

the first test, the battery cells were not connected to the load, but in the next test, the battery cells were connected to the load. The transformer primary and secondary voltage waveforms in steady state are shown in Figure 9.9. It can be noticed that all secondary voltages are at the same level, which means all cells are balanced. The transformer has four secondary windings ($S_1$–$S_4$) but they all overlap.

## 9.9 Experimental results

To evaluate the properties of the proposed circuit concept, a testing PCB set-up with two layers was constructed. In order to assure that the induced voltages in each of the secondaries would be equal and the transformer is correctly wound, the magnetizing inductance is measured for the primary and each of the secondaries. If it is done correctly, the ratio of both would have to be $N^2$. In Table 9.4, the magnetizing inductance ratio for each of the secondary windings is presented. The photograph of the experimental set-up of the proposed transformer balancing circuit is presented in Figure 9.10. In the previous sections, some calculations were driven to estimate the amount of DT that would be required to avoid saturation. However, the voltage drops obtained in these calculations are not accurate.

Therefore, the characteristic of cell voltage vs balancing current is measured. Since it is not preferable to purposely unbalance battery cells before usage, the test is done by supplying the circuit with 25.6 V using a DC source. The imbalance is simulated by attaching different loads to the circuit terminals in the form of resistors. The result is shown in Figure 9.11.

*Table 9.4 Magnetizing inductance*

| Parameters | Values (μH) |
|---|---|
| $L_{m1}$ | 56.20 |
| $L_{m2}$ | 56.21 |
| $L_{m3}$ | 56.21 |
| $L_{m4}$ | 56.20 |

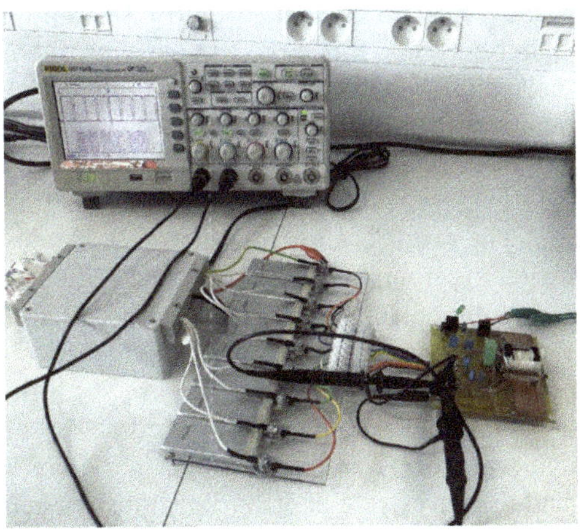

*Figure 9.10 The photograph of experimental set-up of the proposed transformer balancing circuit*

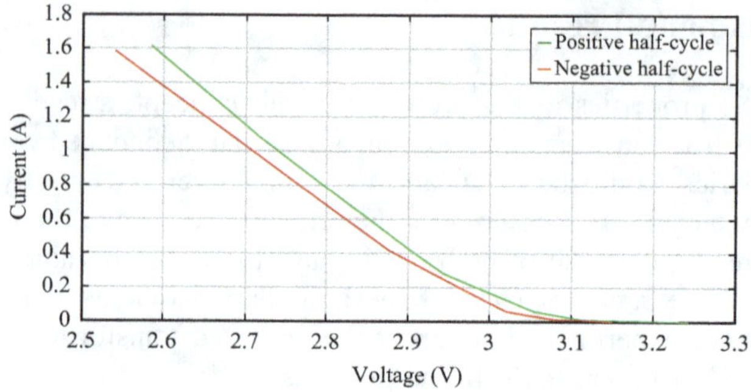


*Figure 9.11 Voltage vs balancing current (experimental result)*

It can be observed that at small imbalances, the balancing current is limited. A cell at 3.15 V, with the overall pack voltage being 25.6 V, gives a balancing current of 0.32 mA to that cell in a positive half-cycle. Therefore, there is little risk of unbalancing the pack outside of the final voltage regions. There is a small, but noticeable difference, in the characteristic between positive and negative half-cycle balancing. As the duty ratio is almost symmetrical, this difference is most likely caused by a difference in impedance, caused by an asymmetry in the PCB or the measuring set-up. The balancing characteristic can be defined by a balancing amperage per voltage imbalance for higher currents. For the positive half-cycle, this amounts to 3.33 A/$V_{imbalance}$, whereas for the negative half-cycle, this amounts to 2.57 A/$V_{imbalance}$. To emphasize the importance of the design on the balancing characteristic, these values are compared to a previous design, where less attention was given to the inductance of tracks and wiring. As described in the previous section, the layout inductance was measured for each of the designs. In the testing phase of the design, the primary current is limited by the PPTC in order to avoid damage. In further stages, the trip current of the PPTC can be selected higher to allow balancing at higher currents. Since the trip current of the PPTC is 0.2 A, the balancing current is normally limited to 1.5 A (23 °C, $t_{trip} = 4$ s). In this design, this corresponds to an imbalance of about 0.6 V. To evaluate the current limiting performance of the PPTC, the secondary current is measured at an imbalance that should trip the PPTC. However, 10 s already allows supporting the weaker cells a bit during acceleration. The result is shown in Figure 9.12.

To protect the switching components and Schottky diodes from over-current, the PPTC should be over-dimensioned at a sufficiently low trip current, compared to the maximum ratings, due to its thermal inertia. The PWM signals applied to the gate to source of the transistors are shown in Figure 9.13(a). The blue signal is applied to Q3 and Q4, while the red signal is applied to Q1 and Q2. The primary and secondary voltage waveforms are represented in Figure 9.13(b), with a frequency of 67.6 kHz and the duty cycle of 50%. The transient behaviour is due to the various parasitic capacities around the circuit charging up when the MOSFETs turn off. When the MOSFETs turn back on, these discharge-causing currents spike. The behaviour at the fall side of the balancing cycle assures volt–second balance, as

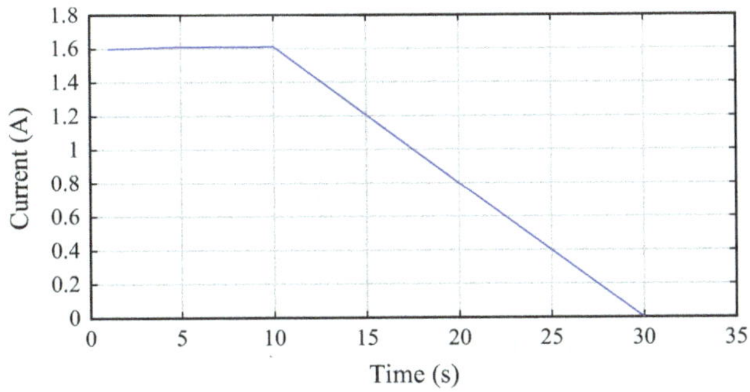


*Figure 9.12 Current limiting performance PTC*

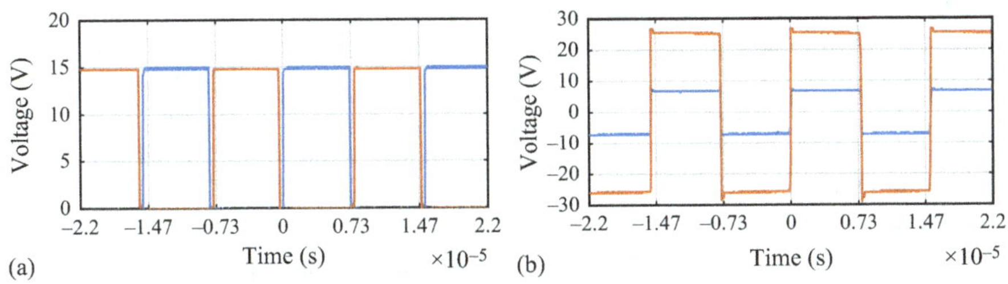


*Figure 9.13 (a) Gate-to-source voltages to the MOSFETs and (b) primary and secondary voltage waveforms*

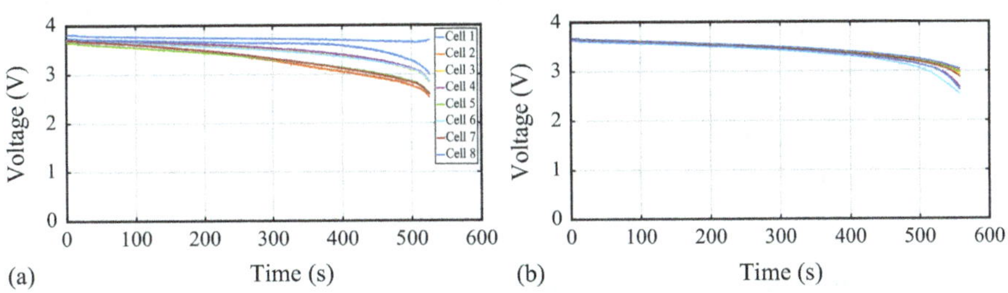


*Figure 9.14 The discharging process of the proposed transformer balancing circuit: (a) discharging without a balancing circuit and (b) discharging with a balancing circuit*

explained before. Note that at this imbalance the PPTC has already increased in resistance, thereby limiting the balancing current. To verify the proposed circuit, the battery cells with different voltage values were tested. The cell voltages vary from 2.7 V to 3.65 V to have imbalanced cells. First, the discharging process of the cells without a balancing circuit was investigated. The discharging results of the cell are presented in Figure 9.14(a). As it can be seen in the figure, the cell voltages

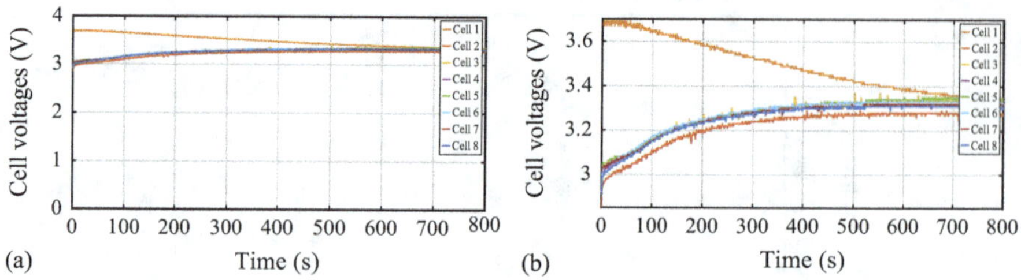


*Figure 9.15 The cell voltage waveforms of the transformer balancing circuit: (a) cell voltage vs time and (b) cell voltage vs time in closer view*

are not converged and there are significant voltage differences because the balancing circuit is not present. Then the discharging progress of the cells with the balancing circuit was examined. In Figure 9.14(b), the discharging results of the cells with the proposed circuit are presented. It can be seen that the cell voltages are converged due to the balancing circuit.

To verify the proposed circuit, the circuit was tested with cell voltages from 2.7 V to 3.65 V to apply voltage differences. The cell voltage waveforms of the battery cells of the proposed circuit after balancing are presented in Figure 9.15(a). In Figure 9.15(b), the cell voltage waveforms of the battery cells of the proposed circuit in closer view are presented. The voltage differences are significant, but they are equalized at the end.

## Summary

In this section, a multi-winding transformer cell balancing based on a double forward converter is proposed. Battery equalization is very important to enhance the battery life cycle and to maintain the total storage capacity and providing safe operation. As long as the multi-winding transformer is sufficiently symmetrical, the circuit can achieve favourable balance voltage. This method is specially implemented for a lithium-ion battery string of eight cells. The energy is transferred by a double forward converter. Compared to the common methods, this scheme reduces the required number of secondaries by a factor of 2, resulting in a smaller size and lower cost. The proposed circuit also uses a full-bridge driver that controls four N-channel MOSFETs as switches, which means two times fewer transistors than cells, resulting in less switching losses. The experimental results and simulation have shown that the proposed equalizer demonstrates a good and competitive overall performance in terms of the equalization speed, efficiency, and complexity of implementation. This method may be also suitable for electrical vehicles, hybrid electric vehicle applications, and also grid-connected battery applications and diverse appliances.

*Section 3*

# Multiple-secondaries transformer 'flyback principle'

## 9.10 Introduction of multiple-secondaries transformer using the flyback principle

Here, a flyback balancing method will be investigated. It also shows the advantages of the easy isolation, simple control, and small size [7]. A flyback transformer is a coupled inductor with a gapped core, but where a primary is added. During each cycle, when the input voltage is applied to the primary winding, energy is stored in the core. Next, it will be transferred to the secondary winding to provide energy to the load. Flyback transformers need a gapped-core construction, to allow an energy storage without saturating the core. This energy storage aspect distinguishes the flyback from other topologies such as forward-mode where energy transfers immediately from primary to secondary.

## 9.11 The proposed flyback converter balancing

The flyback is similar to a buck–boost converter structure, but with a galvanic separation. The core of the flyback needs an air gap to store energy. The schematic of the proposed flyback converter balancing is shown in Figure 9.16. The primary side is connected to the complete battery pack, and every battery cell is connected to secondary windings through a diode. The secondaries are in such a way that the diodes conduct when the primary transistor is off. The flyback secondary winding polarities are opposite regarding to the primary winding. When the main switch, which is a MOSFET, is turned on, the energy will be stored in the transformer as a flux. The stored energy cannot be transferred to the secondary, because the diode is reverse biased.

In the next mode, when the switch is turned off, the energy will be transferred to the secondary windings. The flyback principle is widely used in power supplies,

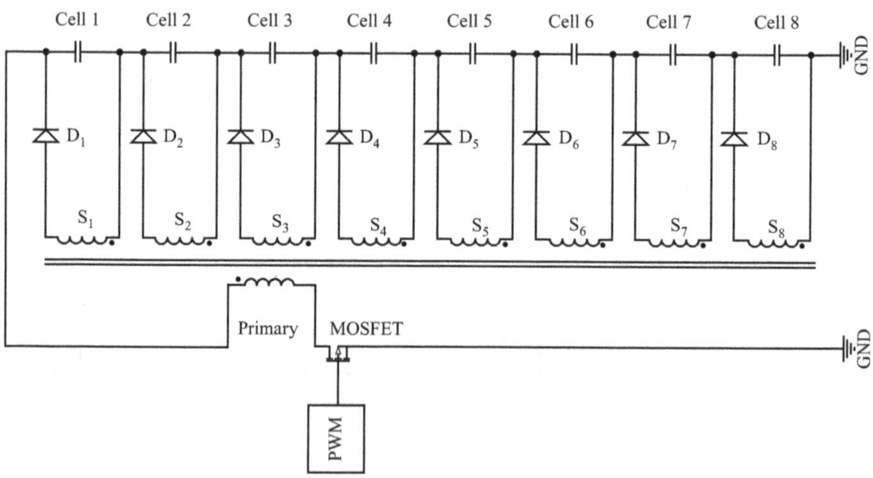


*Figure 9.16 Schematic of the proposed flyback balancing circuit*

so the typical low-cost ICs can be used. The primary side and secondary side are isolated, and it can provide multi-outputs. In order to analyse the circuit, let us consider the operational principle of the circuit in two modes. In mode one, when the MOSFET is ON, the primary inductor voltage is

$$V_{L_p} = V_{\text{Battery}} \tag{9.15}$$

$$T_{(\text{on})} = D \times T_p \tag{9.16}$$

where $T_p$ is the period, and $D$ is duty cycle. The primary peak current will be calculated as

$$i_{Lp} = \frac{V_{\text{Battery}}}{L} \times T_{(\text{on})} \tag{9.17}$$

In the next mode, during OFF time, primary voltage is

$$V_{L_p} = \frac{-V_{\text{cell}_1}}{n} = \frac{-V_{\text{cell}_2}}{n} = \cdots = \frac{-V_{\text{cell}_\text{m}}}{n} \tag{9.18}$$

where $n$ is the winding ratio, and $m = 1, 2, \ldots, 8$ is the cell number.

$$T_{(\text{off})} = (1 - D) \times T_p \tag{9.19}$$

Also, the primary current, during off, can be written as

$$i_{L_p} = 0 \tag{9.20}$$

By applying the volt–second rule, which means that the average voltage across inductor during steady state is zero, the transfer function can be driven as

$$T_{(\text{on})}\, V_{L_p} = V_{\text{Battery}}\, D\, T_p \quad \text{during(on)} \tag{9.21}$$

$$T_{(\text{off})}\, V_{L_p} = -\frac{V_{\text{cell}_m}}{n} (1 - D)\, T_p \quad \text{during(off)} \tag{9.22}$$

$$V_{\text{cell}_m} = n \frac{D}{1 - D} V_{\text{Battery}} \tag{9.23}$$

In order to calculate the primary current, assume one secondary for simplicity. The output voltage is 3.6 V and assumes 2-A output current. The output power can be calculated as

$$P_{\text{out}} = V_{\text{out}}\, I_{\text{out}} \tag{9.24}$$

If the diodes and switches are ideal, the efficiency can be considered 100% for simplicity, which means $P_{\text{in}} = P_{\text{out}}$. Then the average primary current can be expressed as

$$I_{p,\text{ave}} = \frac{P_{\text{in}}}{V_{\text{Battery}}} \tag{9.25}$$

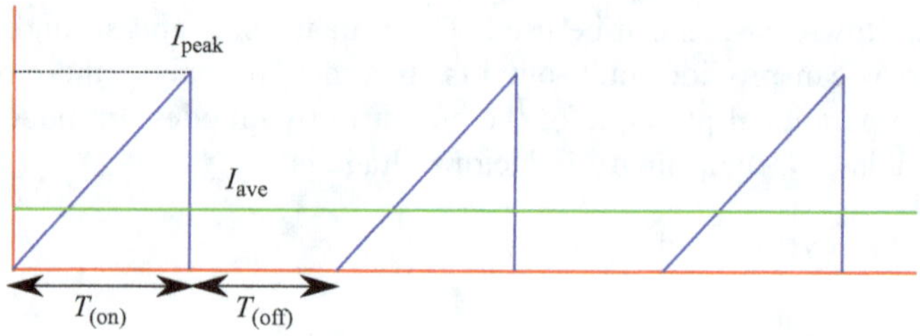


*Figure 9.17 Expected peak and average primary current waveform*

When designing a flyback transformer, the worst case condition of maximum load at the minimum input voltage creates the highest (peak) primary current requirement. Select a transformer that has a peak primary current ($I_{peak}$) or primary saturation current ($I_{sat}$) that is sufficient above the expected primary current peak for the application. If the application's peak primary current exceeds the transformer rating, core saturation will saturate and cause the primary inductance to drop and might overheat the transistor, or the required control is not obtained.

The peak primary current ($I_{peak}$) or primary saturation current ($I_{sat}$) of a flyback transformer should be well designed. The primary peak current of a flyback is a sawtooth waveform as shown in Figure 9.17; therefore, the area is the area of square divided by two. The average current of primary can be expressed as

$$I_{p,\mathrm{ave}} = \frac{1}{T_p}\int_0^{Tp} I(t)\mathrm{d}t = \frac{\mathrm{Area}}{T_p} \tag{9.26}$$

$$\frac{0.5 \times I_{peak} \times D \times T_p}{T_p} = 0.5 \times D \times I_{peak}$$

$$\text{where} \quad D = 0.5, \quad I_{p,\mathrm{ave}} = 0.25 \times I_{peak}$$

Now, based on the previous equation, the primary peak current can be written as

$$I_{peak} = 4 \times I_{p,\mathrm{ave}}$$

In order to find primary inductance, the general equation of inductor voltage can be used as

$$V = L\frac{\mathrm{d}i}{\mathrm{d}t} \Rightarrow L_p = \frac{V}{\mathrm{d}i}\mathrm{d}t \tag{9.27}$$

$$L_p = \frac{V_{\mathrm{Battery}}}{I_{peak}} T_{(\mathrm{on})} \tag{9.28}$$

$$T_{(\mathrm{on})} = \frac{D}{f} \tag{9.29}$$

The induced voltage during (on) and (off) interval can be written as

$$V_{L_p} = V_{\text{Battery}} - I\,R_{\text{L}} - I\,R_{\text{on}} = 0 \tag{9.30}$$

$$V_{L_p} = \frac{-V_{\text{cell}_m}}{n} - I\,R_{\text{L}} + \frac{V_D}{n} = 0 \tag{9.31}$$

Flyback transformers are commonly used for output current below about 10 A and output power below about 100 W. Here, the ability to regulate the multiple output voltages with a single control is used. It can be designed just to equalize, or to be a trickle charging. The diode drop is a disadvantage compared to a full transistorized control but needs less control circuit. Some Schottky diodes have low drops of 0.26 V at 100 °C 1N5820, DMT6007LFG, 0.1 V difference between 20 °C and 100 °C, or synchronous rectifying, or MOSFETs designed as diodes. Also synchronous rectifier circuits or 'MOSFET ideal diode' circuits can be used, but this is again circuitry.

## 9.12 Simulation results

In this section, the simulation results are performed to test the proposed flyback converter circuit. In order to simulate the circuit, the multi-winding transformer is chosen with a turn ratio of $a = 1{:}8$. The transformer has one primary and eight secondaries, which are in parallel with each cell. The total battery pack voltage is applied to the primary side and the each secondary is connected to one cell via a diode. The switch is a MOSFET, which is controlled by PWM signal with a frequency of 50 kHz and 50% duty cycle. The simulation parameters are shown in Table 9.5.

In the first stage, when the switch is turned on, the energy will be stored in the transformer. It cannot be transferred to the secondaries due to diodes which are reverse biased. In the next stage, when the switch is turned off, the stored energy will be transferred to the secondaries. This energy will charge each battery cells, but the lower voltage cells will get the major part.

In Figure 9.18, the gate-to-source voltages of the MOSFET are presented. The batteries are modelled with the capacitors with a value of 500 mF and voltage differences from 3.1 V to 3.6 V in order to apply significant voltage

*Table 9.5 Simulation parameters*

| Parameters | Values |
|---|---|
| Equivalent capacitors | 500 mF |
| Initial capacitor voltages | 3.1 V–3.6 V |
| Switching frequency | 50 kHz |

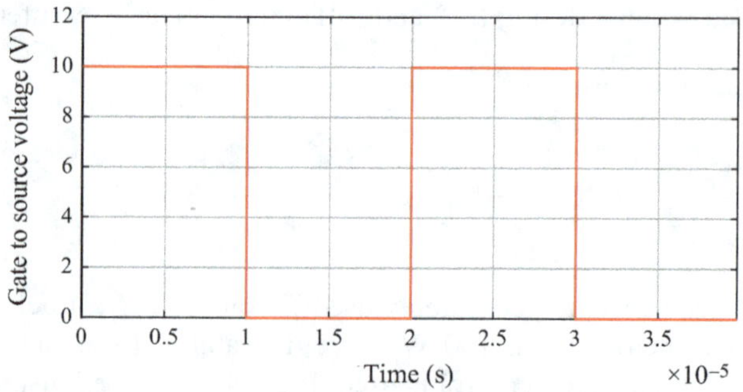


*Figure 9.18 Gate-to-source voltage (Vgs) to the MOSFET*

imbalances. Note that the 500 mF values are used throughout the book, not only to compare, but also to avoid unnecessary long simulations. The cell voltages of the battery, with imbalances after equalization, are presented in Figure 9.19(a). As can be seen in the figure, the equivalent cells are balanced fast in 0.02 s. The primary voltage and secondary voltages are presented in Figure 9.19(b). The primary voltage is the summation of eight battery cells, which is 3.6 V × 8 = 28.8 V and 3.6 V for the secondary side corresponding to the battery cells. Figure 9.19(c) presents the primary current waveform of the flyback converter circuit in steady state. It can be noticed that, during $T_{(on)}$, the primary current is rising, and during $T_{(off)}$, when the switch is not conducting, the primary current is zero. The current waveforms of the cells in steady sate are presented in Figure 9.19(d). It can be seen that all currents are in the same value which means the cells are equalized and are at the same voltage level. The cell current waveforms during transient are shown in Figure 9.19(e). The cell current waveforms at the beginning of the period are not equal, due to the reason that the cells are not balanced yet. Then after transient time, in steady state, the cells are all equalized; thus the currents will be the same for each cell. Note that the too high cells equalize faster in the beginning than at the end. This is as some energy is drawn from the pack in the beginning to charge the low cells. At the end, the high-voltage cells reduce their charge, mainly by the losses of the converter.

In Figure 9.19(f), the voltage waveforms across MOSFET during ON and OFF time are presented. It can be noticed that during ON time when the switch is turned on, the voltage is zero, while in OFF time, the voltage is approximately two times that primary voltage. The switch should be chosen in such a way that it should hold twice the input voltage. Although the solution using a flyback converter seems simple, having one primary and eight secondaries, winding is not easy. The coil former would need 2 + 2 × 8 = 18 pins, and it is difficult to wind and to have an equal EMF in all the secondary windings (same average distance to the air gap). A practical solution uses two transformers in parallel; hence each transformer needs only ten pins, one primary and four secondaries wound in one hand. But it means that there is still room for improving the topology or converter type.

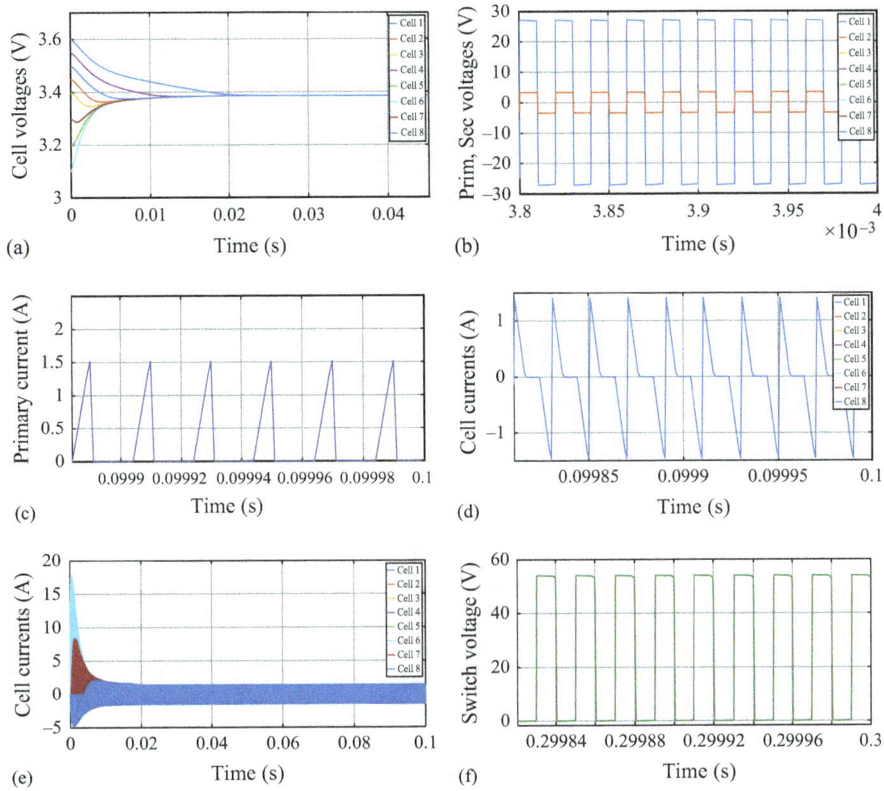


*Figure 9.19 Simulation results of the flyback balancing circuit: (a) cell voltages vs time, (b) primary and secondary voltage waveforms, (c) primary current waveform, (d) cell current waveforms in steady state, (e) cell current waveforms, and (f) switch voltage waveforms*

## Summary

The concept of active balancing by using a flyback converter structure is investigated in this chapter. This proposed circuit equalizes eight cells in series in a battery pack. Flyback balancing has fast equalization time and easy to control. The correct implementation of the transformer can be more troublesome. It requires only one switch for $N$ cells and $N$ diodes at the output of the secondaries for each cell. The switch is an N-channel MOSFET with body diode which is triggered by a pulse width modulation (PWM) signal with a switching frequency of 50 kHz and 50% duty cycle. The batteries are modelled with capacitors with a value of 5 F, from 3.1 V to 3.6 V. The transformer has eight secondaries with a saturable core. Unlike the forward converter that the energy from primary will transfer to the secondary, in flyback, the energy will store in the transformer during ON-time and then during OFF-time, it will be transferred to the secondaries. %The advantages of flyback as compared to the forward converter are that it will not require a reset winding in the primary side, and also extra diode and inductor at the secondary side. Therefore, fewer components are easy to implement. The diodes are considered to be ideal diodes. The simulation results are presented to verify the feasibility of the circuit. In practice having more than one transformer is needed to avoid having eight secondaries that should have the same EMF in the presence of an air gap, which makes the transformer design more difficult.

## References

[1] Li S, Mi CC, and Zhang M. A high-efficiency active battery-balancing circuit using multiwinding transformer. *IEEE Transactions on Industry Applications*. 2012;49(1):198–207.
[2] Kim MY, Kim CH, Cho SY, *et al.* A cell selective charge equalizer using multi-output converter with auxiliary transformer. In: 8th International Conference on Power Electronics-ECCE Asia. IEEE; 2011. p. 310–317.
[3] Shang Y, Xia B, Zhang C, *et al.* A modularization method for battery equalizers using multiwinding transformers. *IEEE Transactions on Vehicular Technology*. 2017;66(10):8710–8722.
[4] Lim CS, Kim RY, and Hyun DS. Battery voltage sensorless charge equalizer using the multi-winding transformer. In: 2012 IEEE Vehicle Power and Propulsion Conference. IEEE; 2012. p. 789–793.
[5] Baek IK, Kim TH, Lim CS, *et al.* Modularized battery cell voltage equalization circuit using extended multi-winding transformer. In: 2012 IEEE Vehicle Power and Propulsion Conference. IEEE; 2012. p. 349–353.
[6] Cadar DV, Petreus DM, and Patarau TM. An energy converter method for battery cell balancing. In: 33rd International Spring Seminar on Electronics Technology, ISSE 2010. IEEE; 2010. p. 290–293.
[7] Kutkut NH, Wiegman HL, Divan DM, *et al.* Design considerations for charge equalization of an electric vehicle battery system. *IEEE Transactions on Industry Applications*. 1999;35(1):28–35.

# Chapter 10
# Buck–forward converter multipack cell equalizer

## 10.1 Introduction

In this chapter, the concept of a forward balancing technique fed by a buck converter for lithium-based batteries in electric vehicle applications is investigated. The proposed active topology equalizes eight cells in a series in a battery pack, by using a forward converter for each battery pack, and the whole battery packs, using a buck converter. The battery bank consists of four battery packs which are in series. Therefore, the proposed system will equalize 32 cells in series as an example. In this chapter, the proposed circuit employs a single transistor used in a zero-voltage switch (ZVS) for the forward converter. In practice, it means that a capacitor in parallel with the switch at the same time helps to obtain a demagnetizing of the transformer without excessive voltage on the transistor.

The ZVS is useful to obtain a low electromagnetic interference and reduces ringing. To overcome the problem of many pins on a coil former, the transformer secondary windings are made by using a hairpin type of winding, on a ring core. It permits, e.g. having eight secondaries and uniform output voltages. Each secondary winding is made by two 'hairpin' turns using two zero-ohm resistor* in series. The proposed topology has less components and circuitry, and it can equalize multiple battery packs by using a single-buck converter and several forward converters for each battery pack. Experimental and simulation results are performed to verify the viability of the proposed topology.

The forward balancing technique has a fast balancing time and is easy to control. It has one magnetic core with one primary winding and based on desired application, it can have multiple secondaries [1–3]. When the switch is on, the energy is transferred to the secondaries, which is different compared to the flyback balancing. In the forward balancing type, the core of the transformer could be saturated, if the supplementary circuit to avoid over-voltage during the demagnetizing time is not well designed. The energy stored in the transformer should be

*Zero-ohm resistors are the ones with a continuous wire inside, but with a shape of resistor. We use also the fact that the wires do not touch and that the wires have not to be stripped and automated insertion can be considered.

allowed to reset in order to avoid the transformer saturation. To limit the voltage, an extra reset winding, a resistor-capacitor-diode (RCD) snubber circuit or a two-switch forward type [4,5] may be used. An extra demagnetizing winding results in more complexity. Similarly, an RCD circuit has a low efficiency, because the magnetic energy will be dissipated through a resistor. The two-switch type also requires more switches and provides more losses.

## 10.2 Buck–forward converter multipack cell equalizer

In this section, a non-dissipative ZVS is used for the forward converter with a single-transistor design. The chapter is focused on the minimum and maximum voltage equalizing which can be easily done using dissipative techniques. The circuit requires only one MOSFET as a switch and one capacitor in order to reset the transformer and has no extra demagnetizing winding. The conventional forward converter circuit has two diodes and one inductor at the output.

In addition, to overcome the problem of non-uniform voltage of the secondaries, in the proposed topology, the transformer secondary windings are made by using hairpins winding, on a ring core. The secondary windings are made by two zero-ohm resistors in series for each secondary winding as shown in Figure 10.1. The zero-ohm can be bent, manipulated automatic inserted as normal resistors, and would be easier to implement than many secondary windings. The wires need no stripping, and the body separates the wires so that no short circuit happens. The total battery consists of four packs, and each pack includes eight cells in series. The forward converters equalize each individual pack and buck converter equalizes whole packs because all forward converters are fed by the buck converter from total battery packs. In the proposed forward converter, in the output, a diode and no separate inductor is used for each cell. Often forward converters contain an inductor, as a part of filtering, but the term 'forward' refers to the fact that the diode conducts when the transistor conducts, so the proposed circuit is also a 'forward' converter.

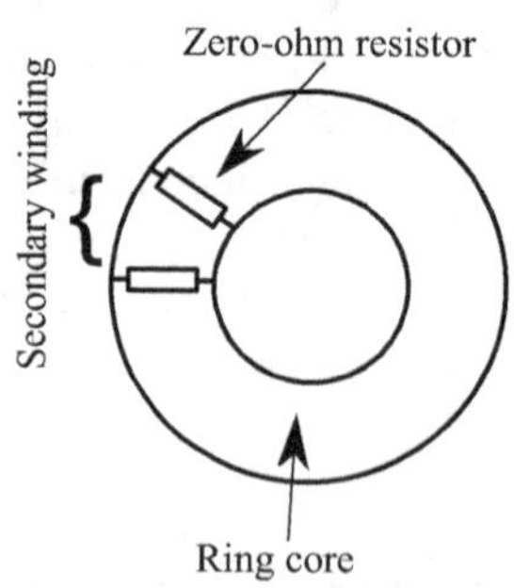


*Figure 10.1 The secondary winding by two zero-ohm resistors*

## 10.3 The conventional forward balancing circuit

Forward-mode transformers, also known as forward-converter transformers, are used to provide circuit isolation and voltage transformation in forward-mode DC–DC converters. The forward mode includes simple forward converters as well as half-bridge, full-bridge types. Forward-mode transformers are attractive because they allow you to achieve high efficiencies. The core is not used for energy storage in forward-mode transformers. Instead, the primary and secondary conduct simultaneously (and directly) when the switch is on, and energy is processed directly through the transformer.

The conventional forward balancing circuit has a fast balancing time, a simpler structure, and high efficiency as compared to multi-winding transformer. It has one magnetic core with one primary winding and multiple secondaries. The forward converter can be controlled with one switch; thus the number of component is decreased. They are mainly used for low voltage and high current application. Even though the forward converter has magnificent features, the main disadvantage is that the core of the transformer could be saturated, if the accessory circuit is not well designed.

The energy stored in the magnetizing inductor should be reset in order to avoid the transformer saturation when the switch is off [6].

A supplement circuit can be implemented by either an extra reset winding (lossless), RCD snubber circuit (loss-reset), or a two-switch forward type. Additional demagnetizing winding results in bulky size and more complexity. Comparably, an RCD circuit has a low efficiency, for the reason that the magnetic energy will be dissipated through a resistor. In Figure 10.2(a) and (b), the schematic of a forward converter with extra reset winding and a forward converter with RCD circuit is presented, respectively.

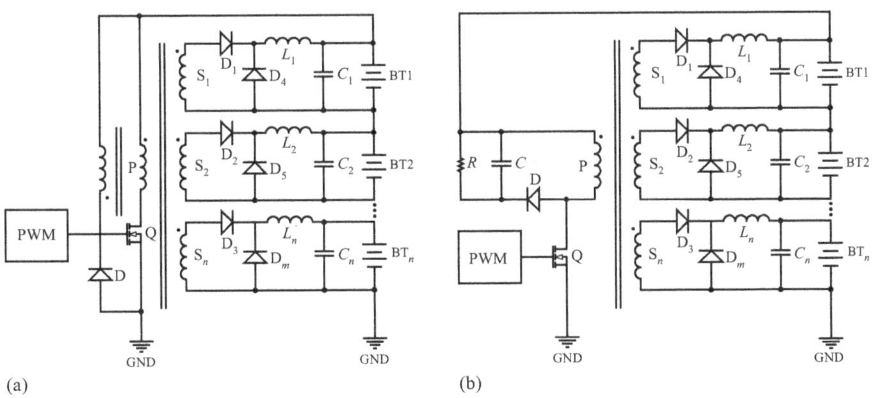


*Figure 10.2 Schematics of the classic forward converter with (a) reset winding or (b) RCD circuit*

The extra reset winding circuit recovers the magnetizing energy. When the switch is on, the energy is transferred to the secondaries. The reset winding diode is reverse biased. When the switch is off, a negative voltage will be induced in reset winding; thus, the core will be reset through diode which is forward biased.

The RCD circuit consists of a reset resistor, capacitor, and diode. The voltages of the reset capacitor and resistor decrease, because the capacitor voltage will be discharged through resistor; therefore, the efficiency of this circuit is low. When the switch is turned on, the higher cell with the lowest voltage will be charged. The energy comes from the total pack.

Similarly, a two-switch forward converter type requires one more switch and two extra diodes resulting in more components. In Figure 10.3, the schematic of a two-switch forward is shown. When the switches Q1 and Q2 are turned on, the energy will be induced to the primary and this energy will be transferred to the secondaries. In the next mode, when the switches are turned off, the voltage polarity of the primary and secondary will be reversed; therefore, the diodes $D_1$ and $D_2$ will be forward biased and will reset the transformer core. So, the magnetizing energy is recovered. It has some advantages and disadvantages. The advantages of single switch converter are simple construction and operation, low input capacitor ripple current, and lower current on the secondary diodes. However, the disadvantages are requirement of a high transistor rating (twice the input voltage), requirement for an active snubber circuit for resetting the transformer core, higher conduction losses, and bigger transformer.

In order to investigate the difference between a conventional forward converter and the proposed forward converter with ZVS in the next section, the simulation results of two-switch forward converter are presented.

The switching frequency is 50 kHz. The battery cells are modelled with the capacitors in series with the value of 500 mF. The cell voltages are chosen from

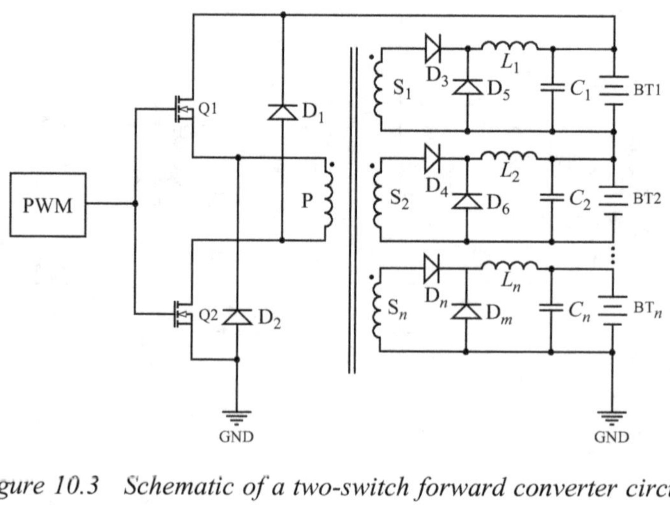


*Figure 10.3 Schematic of a two-switch forward converter circuit*

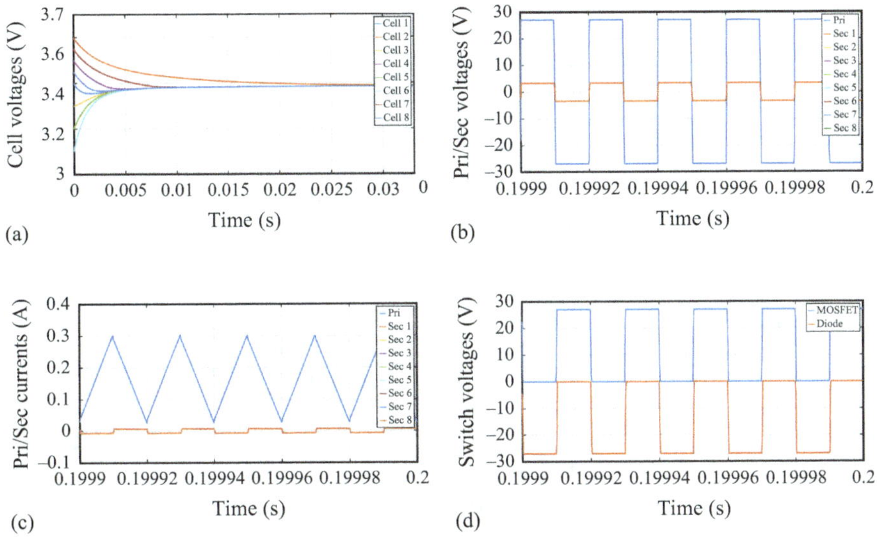


*Figure 10.4 Simulation results of the conventional forward converter: (a) cell voltages vs time of the circuit, (b) primary/secondary voltages vs time, (c) primary current vs time of the circuit in steady state, and (d) MOSFET and diode voltages vs time of in steady state*

3.1 V to 3.6 V to present voltage differences. The switch is a MOSFET with a body diode, and it is controlled with pulse width modulation (PWM) generated by a pulse generator. The transformer is a multi-winding transformer with one primary and eight secondaries. In Figure 10.4(a), the cell voltages after balancing are presented. The primary and secondary voltages of a transformer are presented in Figure 10.4(b). The primary current is shown in Figure 10.4(c). The MOSFET and diode voltages are shown in Figure 10.4(d). The two-switch type seems simple but needs a high side driver, which is less easy to implement.

## 10.4 The proposed ZVS forward converter circuit

In this section, the proposed ZVS forward balancing topology for lithium-based batteries will be expressed. The schematic of the forward circuit is presented in Figure 10.5. It consists of eight cells in a series for a battery pack. The forward converter balancing method poses fast balancing time, simpler structure, and higher efficiency. In the proposed circuit, a capacitor is implemented in parallel with the switch in order to limit the voltage in the part of the period when the transformer is demagnetized. The nominal voltage of each cell is 3.3 V and the total battery pack consists of eight cells. Therefore, the total battery voltage is 26.4 V. The voltage drop across each Schottky diode is considered 0.3 V. The operational principle can be divided into two stages.

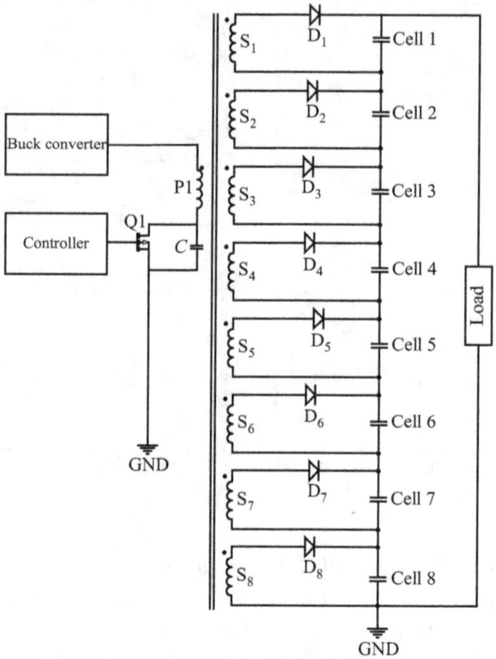


*Figure 10.5 Schematic of the proposed forward converter balancing circuit*

In stage one, when the switch is turned on, the voltage will be transferred to the secondaries through the diodes. The primary and secondary voltages can be expressed as

$$V_{\text{Primary}} = V_{\text{buck (output)}}, \quad V_{\text{secondary}} = V_{\text{buck (output)}} \frac{N_s}{N_p} \tag{10.1}$$

where $N$ is the turns ratio $N_s/N_p$, which can be calculated as

$$V_{\text{cell}_{(1-8)}} = N V_{\text{buck (output)}} - V_{\text{diode}} \tag{10.2}$$

$$N = \frac{V_{\text{cell}_{(1-8)}} + V_{\text{diode}}}{V_{\text{buck (output)}}} \tag{10.3}$$

For the forward converter, energy is transferred from the source to the load while switch is closed. Therefore, magnetizing inductance $L_m$ is not a parameter that includes in the input–output relationship and it is considered to be large.

In stage two, when the switch is turned off, the diode will be in reverse biased; therefore, the energy cannot be transferred to the secondaries. The primary and secondary voltages can be written as

$$V_{\text{Primary}} = -V_{\text{buck (output)}}, \quad V_{\text{secondary}} = -V_{\text{buck (output)}} N \tag{10.4}$$

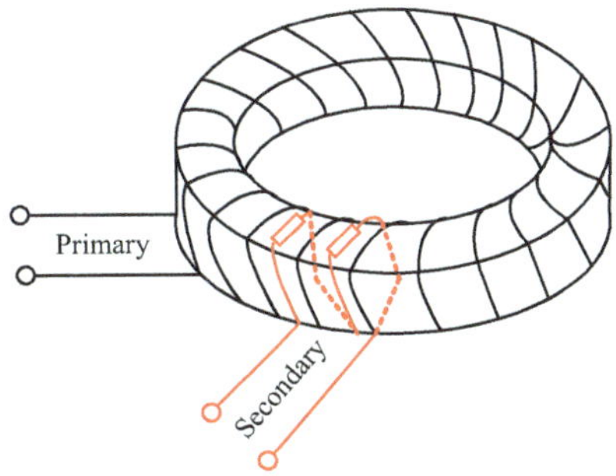


*Figure 10.6 Schematic of the core with primary and one secondary*

The voltage through the magnetizing inductance $L_m$ is $V_{\text{buck (output)}}$; therefore, the average current through $L_m$ can be written as

$$\Delta i_{L_m} = \frac{V_{\text{buck (output)}} DT}{L_m} \tag{10.5}$$

It can be noted in (10.5) that the magnetizing current is increasing linearly in $L_m$ when the switch is closed. The switch current can be expressed as

$$i_{(\text{switch})} = i_1 + i_{L_m} \tag{10.6}$$

The transformer is a ferrite ring core (dimension R25×14.8×10) with one primary and eight secondaries. In order to avoid non-uniform voltages of the secondaries the proposed transformer topology uses hairpin winding, which is made with zero-ohm resistors as secondaries to allow an automated insertion and to keep distance in order to avoid short circuit.

Each two hairpins (zero-ohm resistors in series) form one secondary which are connected in series on PCB. Totally 16 hairpins were used in order to implement eight secondaries as shown in Figure 10.6. A magnetizing inductance of the transformer at the primary with the value of 1 mH was measured with an LCR meter. The desired voltage at output is 3.6 V, which is 3.3 V nominal voltage of the battery plus 0.3 V is the diode voltage drop (Schottky diode) and the primary voltage is 40 V delivered from buck converter. The 40 V has been chosen to be compatible with a lot of configurations that use at least 16 cells. The secondary winding has two turns; therefore, the primary turn can be calculated as

$$\frac{N_1}{N_2} = \frac{V_1}{V_2} \tag{10.7}$$

## 10.5 The buck converter

In order to equalize the battery packs, a buck converter is added to the circuit to balance the whole system. Each battery pack will be equalized by a separate forward converter fed by a buck converter coming from the total pack. The schematic of the buck converter and forward converters is presented in Figure 10.7.

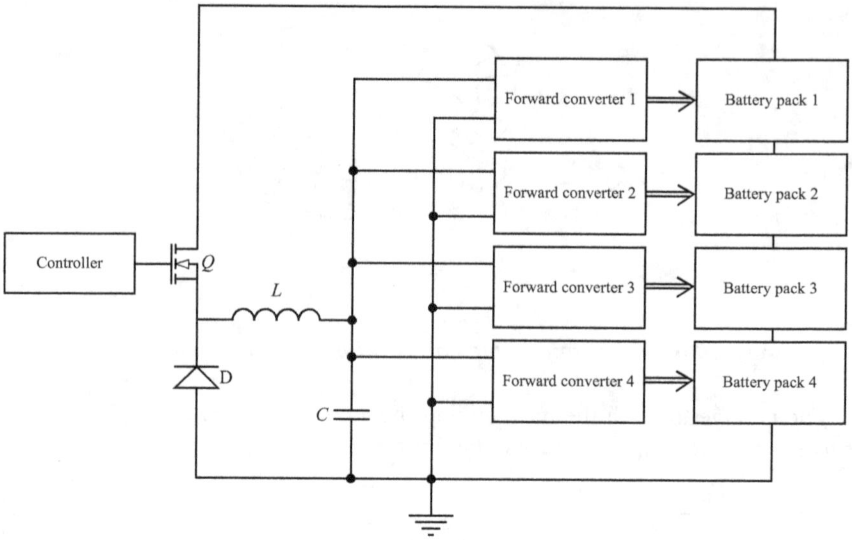


*Figure 10.7 Schematic of the buck converter and forward converters*

The duty cycle of the buck converter is approximately

$$D = \frac{V_{(\text{battery pack1}-4)}}{V_{(\text{total battery pack})}} \tag{10.8}$$

The inductor voltage can be calculated as

$$V_L = V_{(\text{total battery pack})} - V_{(\text{battery pack }1-4)} \tag{10.9}$$

by choosing $\Delta i_L$, at 1% of the output current, which is 1 A, the inductance value can be calculated as

$$\Delta t = \frac{D}{f} \quad \text{and} \quad L = \frac{V_L}{\Delta i_L}\Delta t \tag{10.10}$$

The buck output capacitor with 1% of the full voltage ripple can be estimated as

$$C = \frac{\Delta i_L \times T}{8 \times \Delta V} \tag{10.11}$$

The best is to use a film or ceramic capacitors and avoid electrolytic capacitors. Electrolytic capacitors need to be chosen much larger as their impedance at that frequency is mainly determined by their equivalent series resistance $R_{\text{ESR}}$. If we design the circuit well, no electrolytic capacitor is needed in the proposed BMS. So this means that a long life can be obtained, which would not be limited compared to the battery itself. The buck converter switch is an N-channel MOSFET controlled by PWM with a frequency of 45 kHz and a duty cycle of 37.87%. The inductor is made by an EE Ferrite core transformers with an air gap of 4.96 mH.

The input voltage is the voltage of the total battery pack which is 105.6 V in the example, and the output voltage of 40 V. The real circuit uses a current control to equalize the packs evenly.

## 10.6 Simulation results

In this section, the simulation results are performed to verify the viability of the proposed topology. The battery cells are modelled with the capacitors in series with the value of 0.5 F with different initial voltages, in order to model the cells. The cell voltages are chosen from 3.1 V to 3.6 V to present significant voltage differences. The switch is MOSFET with a body diode, and it is controlled with PWM generated by a pulse generator. The switching frequency is chosen to be 50 kHz, with 50% duty cycle. The diodes are considered to be ideal diodes. The transformer is a multi-winding transformer with one primary and eight secondaries. The simulation parameters are shown in Table 10.1.

The forward converter balancing has a fast balancing time, a simpler structure; however, the main disadvantage is that the core of the transformer could generate a high voltage at turn-off, so some additional circuit or method is needed. A supplement circuit can be implemented by either an extra reset winding or RCD clamp circuit. The idea of additional demagnetizing winding results in a bulky size and a more complex implementation. On the other hand, an RCD circuit retains low efficiency, due to the fact that the magnetic energy will be dissipated through a resistor. Thus, the proposed forward converter employs neither extra reset winding nor an RCD snubber circuit. The proposed topology applies ZVS topology, which requires a capacitor in parallel with the switch in order to limit the voltage. The gate-to-source voltages of the MOSFETs for forward and buck converter are presented in Figure 10.8(a) and (b), respectively.

The transistor has a duty of 50% and the transformer primary voltage is 40 V and secondary is 3.6 V. The buck has a controlled duty cycle of 36.63% in order to deliver 40 V for the forward converters. The buck should fed four forward converters which are connected to battery cells. Each pack consists of eight cells. The number of packs can be extended to the desired application.

In Figure 10.9(a), the cell voltages of the circuit with imbalances are presented. As it can be seen, the voltage imbalances are chosen from 3.1 V to 3.6 V, in order to verify the balancing time. The figure shows that all equivalent cells are equalized

*Table 10.1 Simulation parameters*

| Parameter | Values |
|---|---|
| Equivalent capacitors | 500 mF |
| Initial capacitor voltages | 3.1 V–3.6 V |
| Switching frequency | 50 kHz |
| Duty ratio | 50% |

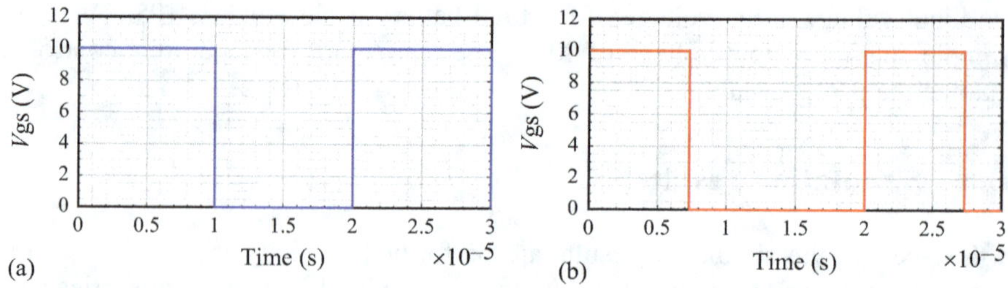


*Figure 10.8 Gate-to-source voltage of the (a) forward and (b) buck converter*

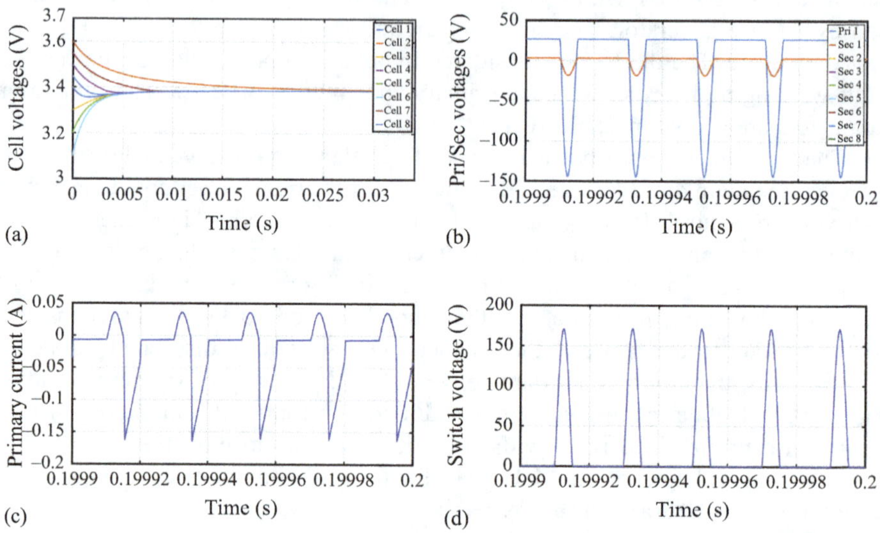


*Figure 10.9 Simulation result of the proposed buck–forward circuit: (a) cell voltages vs time of the circuit, (b) primary and secondary voltages vs time, (c) primary current vs time of the circuit in steady state, and (d) switch voltage vs time of the circuit in steady state*

around 0.025 s (with 500 mF/cell). The primary and secondary voltages of the transformer in steady state are shown in Figure 10.9(b). The figure presents that the ZVS (quasi resonant) is achieved. The primary current waveform in steady state is presented in Figure 10.9(c). The switch voltage in steady state is shown in Figure 10.9(d). The drain-to-source voltage rating of the MOSFET should be at least two times higher than the primary voltage. To equalize total battery pack, a buck converter is designed. The buck converter will receive its input voltage from the battery pack which is 105.6 V when all packs are fully charged. The output of the buck feeds each forward converter to equalize every pack at the same time. In Figure 10.10, the battery pack (1–4) output voltages after balancing is shown. Each pack has different voltage imbalances.

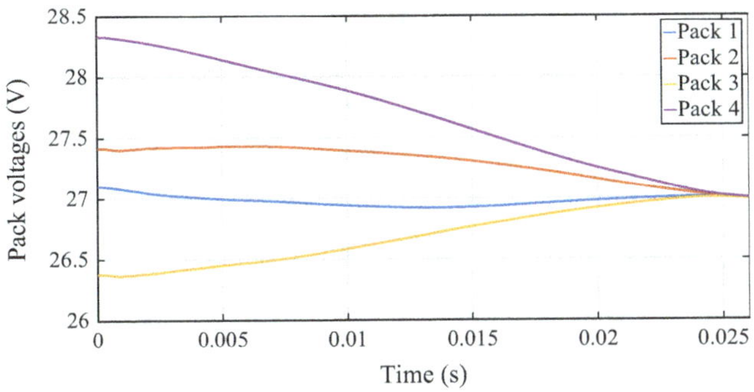


*Figure 10.10 Battery pack (1–4) equalized voltages (with 500 mF/cell)*

*Table 10.2 Secondary voltages*

| **Secondaries** | **Values (V)** |
|---|---|
| $S_1$ | 3.611 |
| $S_2$ | 3.611 |
| $S_3$ | 3.612 |
| $S_4$ | 3.612 |
| $S_5$ | 3.611 |
| $S_6$ | 3.612 |
| $S_7$ | 3.611 |
| $S_8$ | 3.612 |

## 10.7 Experimental results

In this section, the experimental results of the proposed forward-buck converter are presented. The forward converter is implemented with a ferrite ring core transformer with 22 turns primary winding and 2 turns secondaries. The secondary windings are made in such a way that they have same output voltages. To avoid non-uniform output voltage, the secondary windings are made with two hairpins made by two zero-ohm resistors in series for each winding. The hairpins are soldered on the PCB and are connected at the bottom layer of the PCB, which will result in less leakage inductance and easy inserted in manufacturing. A magnetizing inductance of 1.06 mH was obtained. The secondary output voltages are presented in Table 10.2. As it can be seen in the table, the output voltages are almost the same.

The switch for the forward converter is a surface-mount device (SMD) N-channel MOSFET with body diode. An N-channel MOSFET is chosen, in order to have less $R_{DS(on)}$ and it is controlled by a microcontroller with a frequency of 45 kHz and duty cycle of 45%. A capacitor is implemented in parallel with the switch to reset the transformer by ZVS technique. Using the ZVS technique,

*Table 10.3 MOSFET characteristics*

| Parameters | Values |
|---|---|
| $V_{DD}$ | 200 V |
| $R_{DS(on)}$ | 64 mΩ |
| $I_D$ | 24 A |
| $V_{gs}$ | ±20 |
| $V_{gs(th)}$ | 3 V (min) |
| $Q_g$ | 25 nC (typ) |
| $t_r$ | 22.4 ns |
| $t_f$ | 14.8 ns |

*Table 10.4 Schottky diode characteristics*

| Parameters | Values |
|---|---|
| $I_{F(AV)}$ | 4 A |
| $V_{RRM}$ | 20 V |
| $I_{FSM}$ | 150 A |
| $V_F$ | 0.3 V |
| $T_{J,max}$ | 150°C |

an extra reset winding is unnecessary, resulting in less circuitry and easy to implement. The capacitor value can be calculated as

$$\omega_0 = 2\pi f = \frac{1}{\sqrt{LC}} \tag{10.12}$$

To calculate the capacitor, the resonance *T*/2 should be smaller than *T*/2 of the switch. With the switching frequency of 46 kHz and magnetizing inductance value of 1.01 mH, the 2.96 nF capacitor was calculated. However, taking into account the parasitic capacitance of the transistor, a value of 2 nF was chosen to perform the best result.

$$C < \left(\frac{T}{4\pi}\right)^2 \frac{1}{L} \tag{10.13}$$

When the switch is turned on, the primary will receive the energy from buck converter, and it will be delivered to the secondaries through the SMD diodes. Unlike the conventional forward converter that uses an extra diode and inductor as an output, this method requires only one diode and capacitor as an output, resulting in less components, which makes it more efficient. The N-channel MOSFET characteristics are shown in Table 10.3.

All diodes are a surface-mounted Schottky barrier rectifier in order to have less voltage drop across it, resulting in less losses and more efficiency. The diodes can be cooled via PCB surface by increasing the copper area for each diode. The diode characteristics are shown in Table 10.4. The output capacitors are SMD multi-layer ceramic capacitor with a value of 100 μF, a voltage rating of 6.3 V DC, and an operating temperature range of −55 °C to 85 °C. As was discussed before, the forward converters are fed by a buck converter. It operates based on current control topology. In order to control the buck converter, an LED driver constant current control IC (HV9861A) was chosen. The controller is a fast average current control, which is programmable with constant off-time switching. The IC is equipped with a current limit comparator for hiccup-mode output short circuit, and also internal over-temperature protection is provided. The internally regulated voltage ($V_{DD}$) is 7.5 V. The IC can be powered from 15 V to 450 V supply. The output current can be programmed by an internal 270 mV reference and a shunt. The IC employs a special scheme, achieving fast and very accurate control of average current in the buck inductor through sensing the switching current only. Therefore, no compensation of the current-control loop is required. The timing resistor connected to $R_T$ pin determines the off-time of the gate driver. The equation governing the off-time according to the data sheet can be expressed as

$$T_{(\mathrm{off})(\mu\mathrm{s})} = \frac{R_{T(\mathrm{k}\Omega)}}{25} + 0.3 \tag{10.14}$$

The desired off-time is 20 μs, according to a maximum switching frequency of 50 kHz. Thus, a value of 492.5 kΩ for $R_T$ was obtained. Therefore, finally the nearest value of 500 kΩ was chosen for the resistor. The inductor value can be calculated as

$$L = \frac{V_o(\max)\, T_{(\mathrm{off})}}{I_o} \tag{10.15}$$

The desired output current is 1.5 A DC, 3 A (peak). Therefore, the inductance value can be obtained as 270 μH. In order to calculate the capacitor value, first the charge $Q$ is required to be estimated. The total time $\Delta t$ is 20 μs and the value of $Q$ can be expressed as

$$Q = \frac{\Delta t\, I}{2} \tag{10.16}$$

Therefore, with the charge value of $Q = 15$ μC, the capacitor value $C = 37.5$ μF can be calculated according to the following equation:

$$C = \frac{Q}{\Delta V} \tag{10.17}$$

The duty-ratio range of the current control feedback is limited to $D \leq 0.8$. A reduction in the current may occur when the voltage $V_o$ is greater than 80% of the input voltage. Therefore, there would be no problem, since the output voltage is 40 V, so it can still be used in a 48 V circuit. In the concept, it is taken into account that the vehicle could operate using two packs instead of four packs, when two packs are discharged. The input voltage of the IC can vary from 15 V to 450 $V_{DC}$. When this voltage is applied to the $V_{in}$ pin, it maintains a constant 7.5 V level at . This voltage can also be used to power the IC and external circuitry connected to $V_{DD}$. The $V_{DD}$ pin must be bypassed by a low ESR capacitor to provide a low-impedance path for the current pulses of the gate driver. The gate output is used to drive an external MOSFET. For the used IC, the gate charge, $Q_g$, of the switch

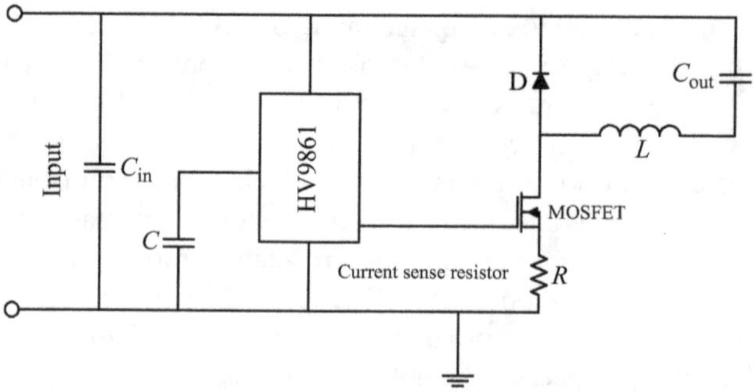


*Figure 10.11 Schematic of the buck converter*

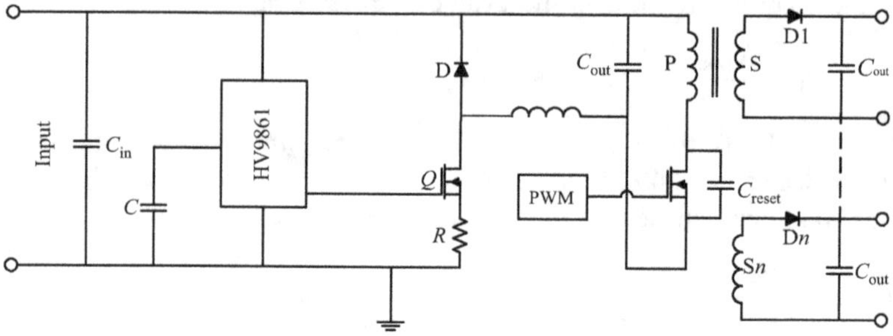


*Figure 10.12 Schematic of the buck–forward converter*

should be less than 25 nC for the switching frequencies $\leq$ 100 kHz. In Figure 10.11, the simplified schematic of the proposed buck converter is presented. The input is fed by the whole total battery pack, which can vary from 105.6 V with all cells at nominal voltage. However, it is designed to operate from 48 V to 150 V. The sensing resistor $R$ can determine the output current. The schematic of the buck-forward converter with one forward converter is shown in Figure 10.12. The output of the buck converter will feed four forward converters in parallel, but for simplicity, one of them is presented in the figure. All the forward converters will be fed by fixed 40 V voltage which is originated from the total battery pack voltage. If some battery cells are below the nominal voltage, the input voltage of the buck will be decreased, but the system has a fixed regulated output voltage; therefore, the input voltage to the forward converters will be a fixed voltage of 40 V.

Figure 10.13 shows the gate-to-source and drain-to-source voltages of the forward circuit. As can be found in the figure, the ZVS is achieved. The red waveform is the gate-to-source voltage, while the blue is drain-to-source voltage. In order to measure the drain-to-source voltage, the oscilloscope probe $\times$ 10 was applied.

The primary and secondary voltages in no-load condition are shown in Figure 10.14(a) and (b), respectively. The oscilloscope probe $\times 10$ and probe $\times 1$ were applied to the primary and secondary side, respectively. In order to verify the

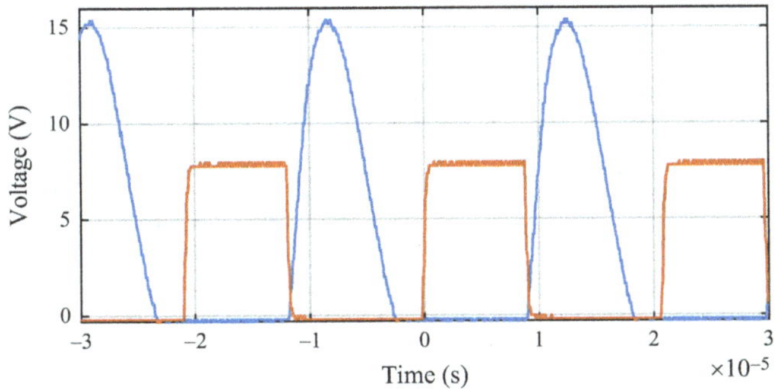


*Figure 10.13 Gate-to-source and drain-to-source voltages of MOSFET Q1*

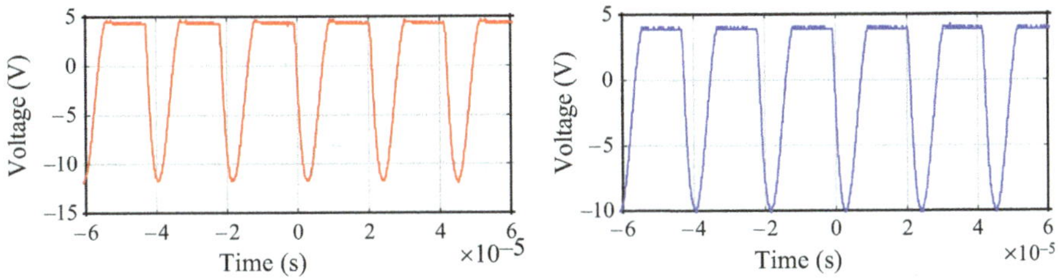


(a) (b)

*Figure 10.14 (a) Primary and (b) secondary voltage waveforms of the forward converter*

buck-forward circuit, the proposed circuit was tested on lithium-ion batteries with a nominal voltage of 3.3 V and 1000 mAh capacity. The cell voltages were measured with the isolated circuit to the Arduino in order to extract the data. To test the system, some conditions were considered to obtain the experimental results. First, the cell charging process without a balancing circuit was tested and the result is shown in Figure 10.15(a). It can be noticed that the cells are starting with different voltage values and show the beginning of the charging process.

In Figure 10.15(b), the discharge process is presented. First, all cells were fully charged and balanced to the nominal voltage. When the discharging is done, it can be seen that the cells are not discharged with the same voltage level, showing an imbalance cells with major voltage differences. In order to test the balancing circuit, the cells with minor voltage differences were charged with the proposed buck–forward balancing circuit. In Figure 10.15(c), the charging with balancing circuit process is shown.

It can be seen that, during charging/balancing process, all cells are at the same voltage level, and that they reached the nominal voltage at the end of the process while converging. Figure 10.15(d) shows the discharging process while the balancing circuit is present. It can be noticed that, at the end of the discharging, the cells are almost of the same value. To verify the speed of balancing circuit, the cells with major voltage differences were tested. The cell voltages were chosen from 2.74 V

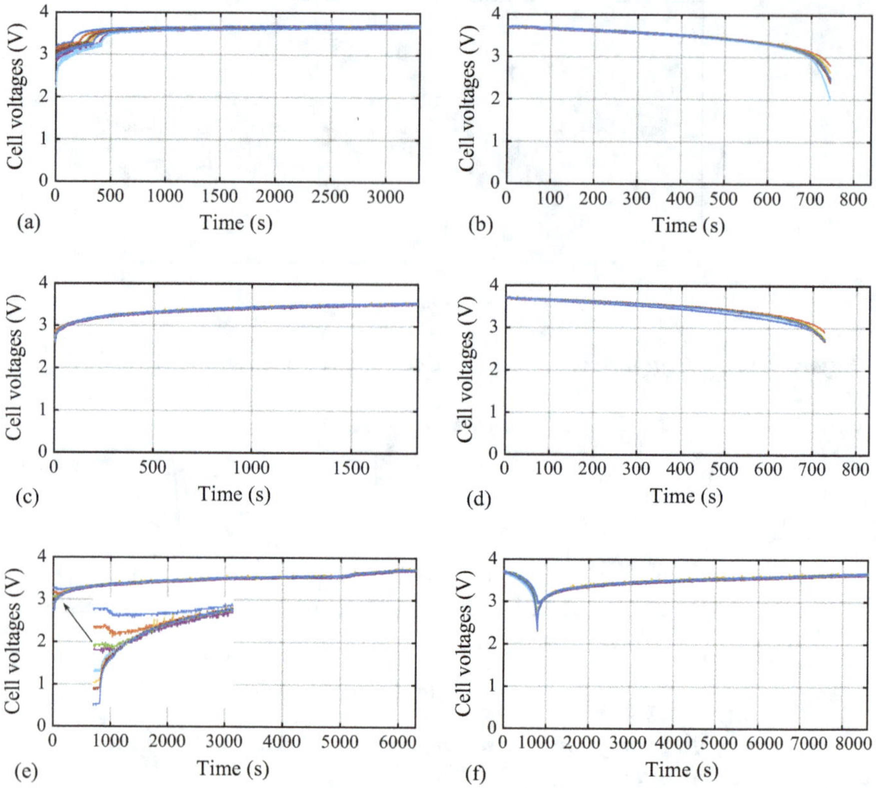


*Figure 10.15 Charging, discharging, balancing with and without balancing circuit: (a) charging without balancing circuit, (b) discharging without balancing circuit, (c) charging with balancing circuit, (d) discharging with balancing circuit, (e) charging (significant voltage differences), and (f) discharging/charging with balancing circuit*

for the lower cell and 3.27 V for the higher one to have sufficient voltage differences, and the results are shown in Figure 10.15(e).

As can be seen in the figure, all cells are fully charged and balanced at the same values with the balancing circuit. Figure 10.15(f) shows the discharging and charging process at the same time. First, all the cells were discharged with the balancing circuit and then, they were charged again with the balancing circuit. As can be found in the figure during discharging and charging, all cells are converging at the beginning. In order to estimate the efficiency of the whole set-up, the efficiency of the two circuits was investigated. The efficiency of the buck-forward combination was calculated and the result is shown in Figure 10.16.

In order to estimate the efficiency of the system, the circuit was tested with different loads. The output current varies from 0.11 A to 0.92 A to achieve the result. It has been found that the highest efficiency of 83.84% can be obtained with

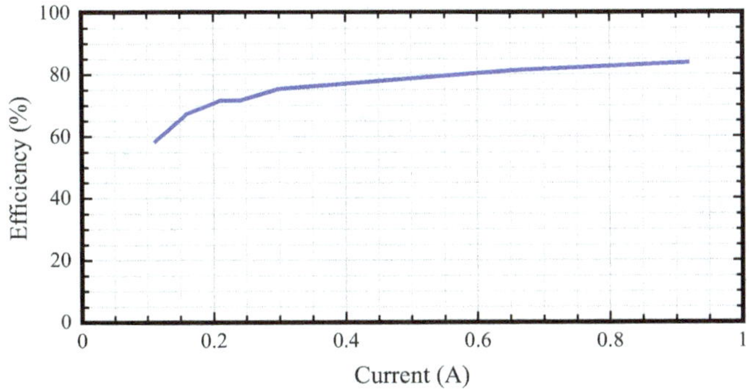


*Figure 10.16 The efficiency of the buck–forward combination*

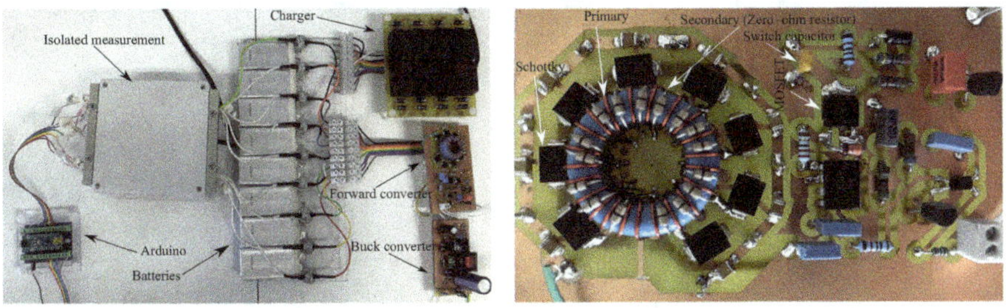


(a) (b)

*Figure 10.17 The photograph of the practical implementation of the forward–buck converter and PCB prototype: (a) testing set-up of the forward–buck and (b) PCB prototype of the forward*

0.92 A output current for the buck-forward converter circuit. Note that the efficiency is mainly limited by the output diode.

The practical implementations of the whole battery balancing system and PCB prototype of the forward converter are presented in Figure 10.17(a) and (b), respectively. As can be seen in Figure 10.17(a), first, the buck converter is fed from total battery packs, which is 105.6 V, if all cells are at nominal voltage. The proposed buck converter can operate from 80 V to 150 V input voltage and delivers a 40 V regulated output.

Then, the forward converters which are connected to each battery pack will be fed by the buck converter. The forward converters will equalize four battery packs in series. The total battery consists of 32 cells in series, 8 cells per each pack. The proposed system is suitable for the applications which consist of several battery cells in series, specially electric vehicles. The system can work with variable input voltages. For example, if one or two packs are discharged, the system can continue to operate with two packs. Figure 10.18 shows the cooperation of balancing and monitoring circuits of multipacks. It is a combination of buck-forward circuit with

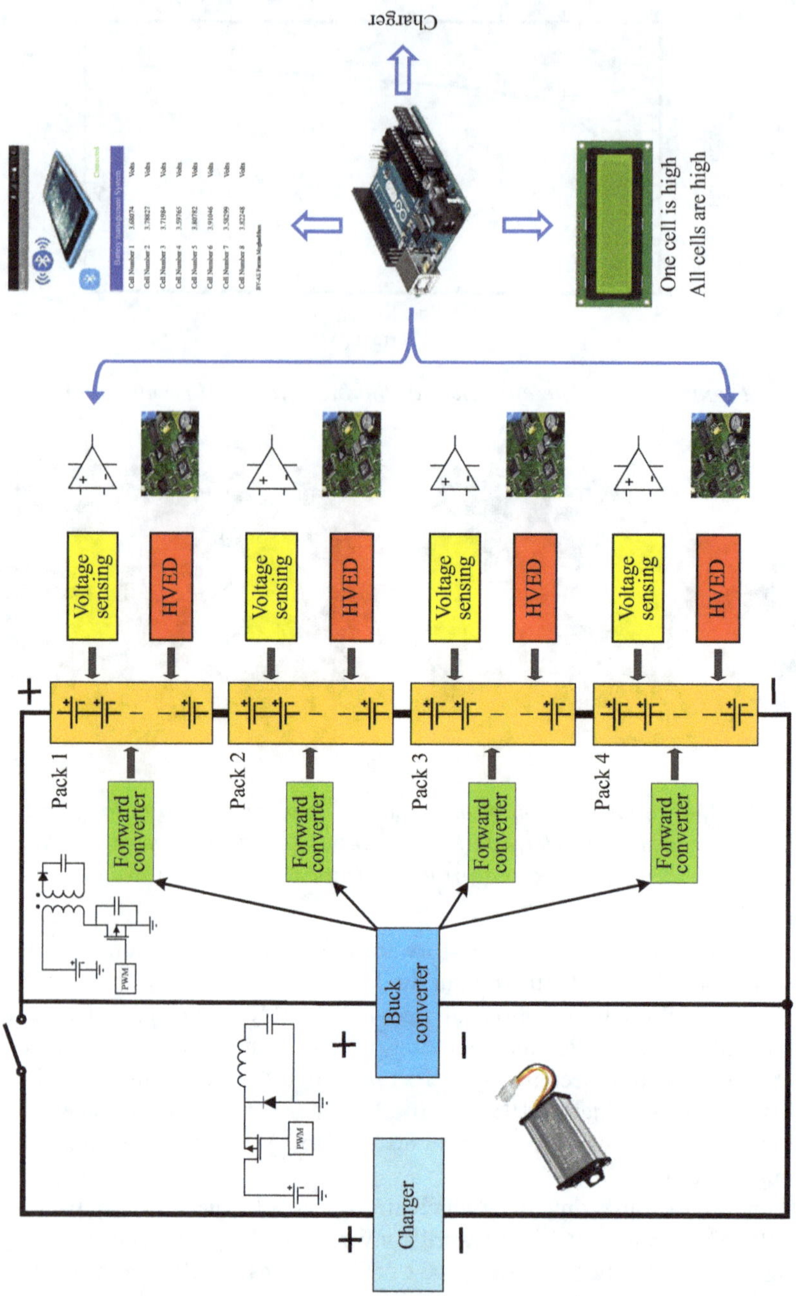


*Figure 10.18 Cooperation of balancing and monitoring circuits of multipacks*

monitoring circuit. Also high-voltage equalizing detection circuit from Chapter 5 is added if a cell reaches its maximum voltage. The monitoring system is explained in detail in Chapter 5.

## Summary

In this chapter, a special multiple-output forward converter equalizing circuit is proposed. It is fed by a buck converter from larger voltage. The case of lithium-based batteries in ultralight electrical vehicles was taken as an inspiration for an example. The special arrangement makes it possible to equalizing eight cells or more in a series in a battery pack. It can be repeated for the other battery packs in the full battery. A common 40 V buck converter is feeding them, so it is a solution for a full battery equalization. Each battery pack consists of eight cells in series and the total battery bank consists of four battery packs in series. The proposed method combines a simple structure with fast balancing time. This method applies a single-transistor zero-voltage switching (ZVS) technique that also avoids a demagnetizing winding. It needs only one transistor and has lower switching losses and does not require a demagnetizing winding. The proposed circuit equalizes each battery pack starting from a common voltage, made by a buck converter. Hence, all cells are equalized in the same way, without sophisticated, or sensitive voltage measurements. The buck converter is used in current mode, with limited voltage. The proposed forward type uses only one diode per cell and one transistor making it cost-effective. The secondaries of the proposed transformer use hairpin windings, using two zero-ohm resistors in series as secondaries. This permits a large number of secondaries and guaranties uniform voltage by symmetry. Each two hairpins form one secondary which are connected in series at the bottom of the PCB. The system can also be used to finalize charging the batteries, or equalizing the cells after a replacement of a cell or a battery set. The number of cells or battery packs can be extended to the desired numbers. The proposed topology can operate with variable input voltages. For instance, if one or two battery packs are discharged, the system can continue to operate with a single packs. In some cases, packs could switch between parallel and series. Experimental and simulation results were performed to show the feasibility of the system.

## References

[1] Gerislioglu B, Ozturk F, Sanli AE, *et al.* The multi-windings forward structure battery balancing. In: 2014 11th International Conference on Electronics, Computer and Computation (ICECCO). IEEE; 2014. p. 1–4.

[2] Lee Y, Yun JJ, and Bae S. Active cell balancing circuit based on the forward converter with an RCD circuit. In: 2015 9th International Conference on Power Electronics and ECCE Asia (ICPE-ECCE Asia). IEEE; 2015. p. 2887–2892.

[3] Sun J, Lu R, Wei G, *et al.* A high efficiency equalizer based on forward converter for series connected battery string. In: 2012 IEEE Vehicle Power and Propulsion Conference. IEEE; 2012. p. 376–379.
[4] Tan FD. The forward converter: from the classic to the contemporary. In: APEC. Seventeenth Annual IEEE Applied Power Electronics Conference and Exposition (Cat. No. 02CH37335). vol. 2. IEEE; 2002. p. 857–863.
[5] Bridge CD. Clamp voltage analysis for RCD forward converters. In: APEC 2000. Fifteenth Annual IEEE Applied Power Electronics Conference and Exposition (Cat. No. 00CH37058). vol. 2. IEEE; 2000. p. 959–965.
[6] Qin X, Hu H, Wu H, *et al.* A series-input forward converter with shared RCD cell for high-reliability and wide input voltage range applications. In: 2013 IEEE Energy Conversion Congress and Exposition. IEEE; 2013. p. 154–158.

*Part IV*

# Battery charging and auxiliary circuits

# *Chapter 11*
# Trickle charging and vehicle start-up

## 11.1 Introduction

This chapter presents the concept of charging of lithium–iron–phosphate (LFP) battery cells in an electric vehicle (EV). An example of an ultralight EV has been taken, but many of the concepts can be enlarged to many other vehicles types or appliances. Charger topologies play an important role in EVs to increase the performance of the batteries.

The battery charging and discharging process should take care to achieve the best performance and long lifetime. In this chapter, the design of a charger with four battery packs, each consisting of eight cells in a series, is investigated. Two battery chargers are performed to feed the whole battery packs.

In the considered case, two smaller chargers were more cost-effective than one bigger, and it could happen in other cases as well. A current controller is proposed to limit the current from the charger. The operational principle is based on a metal–oxide–semiconductor field-effect transistor (MOSFET) constant current supply. The drives should only be connected to the battery when the car needs to drive in order to avoid unnecessary risks and losses.

Also a start-up resistor is needed to avoid high inrush currents; thus, some circuit is required and designed to overcome this problem. The operational principle of the proposed circuit is explained in detail and the experimental results are performed to verify the feasibility of the proposed system. Lithium-ion batteries have an improved energy density and have a long lifetime compared to other types, such as lead–acid [1–3].

Up to now, they have the largest electrochemical potential and provide a high energy per weight. The conventional charging mode uses a rather low current to charge the battery. A conventional charging current is rather low, less than 0.3 C, so the charging time is more than 3 h [4–9]. The advantage of the conventional charging can be expressed as less charging current, less impact on battery life.

However, a disadvantage of this charging method is that it takes longer time. In this chapter, first, the concept of battery charger by using a current limiter circuit is investigated, then the start-up circuit is performed to control the battery pack when it is required to connect to the vehicle. The charger can be easily plugged into the wall. In the considered example, there are four battery packs, each consisting of eight cells.

## 11.2 The proposed current driver circuit

In this section, the charger concept will be discussed. The battery can be charged with 'an external DC source'. It is convenient to have a charger that can be plugged 'into the wall'. How this charger is built can be seen in Figure 11.1. In order to charge the batteries, two AC/DC chargers with the input voltage of 230 V and the output voltage of 58 V are implemented, so the total voltage will be 116 V. Next in line, there is a current driver to limit the current coming from the charger. The current limiter limits the current coming from the charger and will be applied to the batteries in order that the equalizer can still handle it. In smaller vehicles, LFP might be preferred, as the batteries might be close to the driver or passenger and they are safer than the Li-ion types.

LFP has a nominal voltage of 3.3 V and the maximum charge voltage of 3.65 V; thus the total voltage of the whole battery packs can be calculated as

$$V_{\text{Charge}} = 32 \times 3.65 = 116.8 \text{ V} \tag{11.1}$$

Thus, connecting two 58 V chargers in series results in a total voltage of 116 V, a little below the maximum allowed voltage and thus optimal for charging the cells. If it is needed, there is a preferred way to connect the chargers.

The two power plugs need to be connected in parallel, so only one plug remains. At the DC side, the wires need to be connected in series in order to obtain required 116 V. The charge monitor detects if one cell is full, then the current limiter is put into circuit, so that the battery management system (BMS) can manage it. Indeed, BMSs can handle only a limited amount of current. The fact that a linear, non-switching solution is used to face less radio disturbance (less electromagnetic interference), so that it may take shorter time to market. It was also fast designed and made. In fact, current controllers exist for LED lamps, but their operation is not clear if the output voltage almost equals the input voltage; for example a constant frequency, peak-current control is not stable above 50% duty

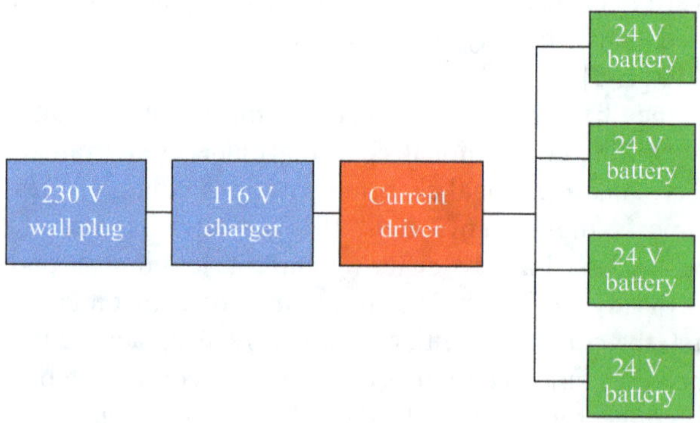


*Figure 11.1 The charging principle overview with four battery packs and two chargers*

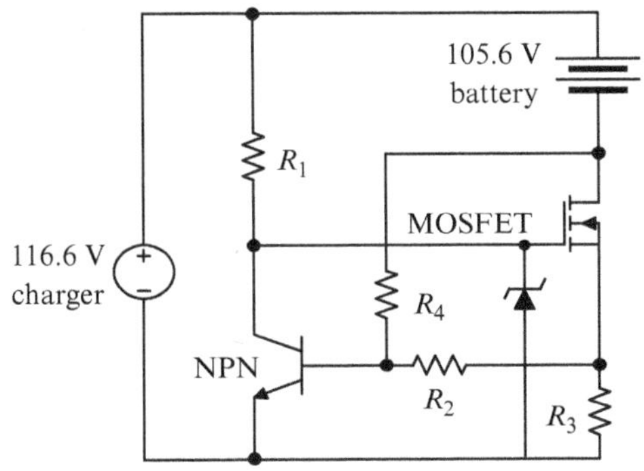


*Figure 11.2 The schematic of the current driver circuit*

ratio. The current driver is used to limit the current to the desired current of 0.3 A. In order to obtain this current, a circuit was designed, shown in Figure 11.2. It consists of one bipolar transistor and one MOSFET. This circuit is working based on MOSFET constant current driver. The circuit is designed with two transistors: one bipolar junction transistor (BJT) and the other, MOSFET. When the charger is plugged in, the gate resistor $R_1$ will gradually turn on the MOSFET, then the current flows through the battery, MOSFET, and resistor $R_3$. Therefore, this current will increase until the voltage over resistor $R_3$ reaches the base-emitter voltage of the transistor, which is 0.7 V. When this voltage is obtained, the transistor will sink the gate of the MOSFET to the ground and it will be turned off; now the current through the battery, MOSFET, and $R_3$ will be decreased until the base-emitter voltage of the transistor reaches below about 0.6 V, then the gate voltage of the MOSFET will not increase any more. The exact value of the constant current can be obtained by the appropriate value of the resistor $R_3$. The constant current is chosen as 0.3 A and the base-emitter voltage of the transistor is found on the data sheet as $V_{BE} = 0.65$ V; thus, the value of the resistor can be calculated as $R_3 = 2.166\ \Omega$ according to the Ohm's law:

$$R_3 = \frac{V_{BE}}{I_{battery}} \tag{11.2}$$

The most important resistor in the circuit is $R_3$. It determines the value of the constant current. For the specific transistor in the circuit, $V_{BE} = 0.55\text{ V} < 0.6\text{ V}$ is found on the data sheet. For the calculations, $V_{BE} = 0.6$ V is chosen. As mentioned earlier, $I_{battery}$ should equal 0.3 A. % This gives the following result Therefore, according to (11.2), $R_3 = 2.2\ \Omega$ delivers a current sufficiently lower than 0.3 A. For 0.3 A battery current, the power dissipation in the resistance is

$$P = R_3 I^2 = 0.198\text{ W} \tag{11.3}$$

A gate resistor $R_1$ is required. This value is not critical, but it is needed to be high enough in order to keep the current low in the circuit. Stability is obtained in

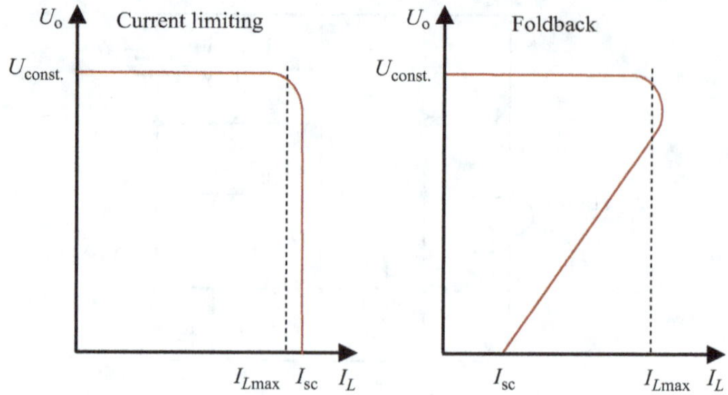


*Figure 11.3 Normal current limiting vs foldback*

the feedback as the gate capacitance is a dominant time constant in the circuit. The resistor $R_4$ acts as foldback resistance. It lowers the current when the voltage drop on the MOSFET is higher and so limits the power dissipation of the MOSFET to limit the amount of required cooling. It could happen if one fully charged cell is inserted between almost discharged cells, and so appearing a quite high voltage on the current limiter. When the battery attempts to draw **too much** current from the charger, the foldback **reduces the output current**. Typically once the output voltage is reduced to zero, the current is a small fraction from the maximum current. The proposed foldback current limiting is to keep the output transistor (MOSFET in this case) within the safe power dissipation. The graphs on Figure 11.3 clarify the concept. Note the difference with normal current limiting. The prime purpose of foldback current limiting is to keep the output transistor (in this case MOSFET) within safe power dissipation limits. Under overload conditions, the voltage over the battery ($V_{out}$) decreases (principle of current limiting) as well as the voltage over the battery ($V_{out}$) decreases. The effect of the power dissipation can be expressed as

$$P_{dissipation} = (V_{in} - V_{out})\, I_{out} \tag{11.4}$$

It is clear that, when $V_{out}$ decreases, the power dissipation will increase. This might lead to an unwanted failure of the MOSFET. Today MOSFETs have a safety operating area with a second breakdown part, so that not the power is limited, but $V^2I$; so a higher voltage is more risky. Foldback prevents this by lowering $I_{out}$ when $V_{out}$ is low. It can even be zero at a battery voltage which is half of the input voltage. It is not good to attempt to charge a battery which is below 50% of its maximal voltage. An N-channel MOSFET is used as it is the cheapest option. The circuit can also be made with a P-channel, allowing one to split the current sources in an upper one and a lower one. Then the mid-point of the chargers can be connected to the mid-point of the battery, which may have some advantages of redundancy. The most important characteristics can be found in Table 11.1.

*Table 11.1 MOSFET characteristics*

| | |
|---|---|
| Drain–source breakdown voltage | 200 V |
| Continuous drain current | 76 A |
| Drain–source resistance $R_{DS(on)}$ | 0.17 Ω |
| Max operating temperature | 175 °C |
| Thermal resistance, junction–case | 0.9 K/W |
| Thermal resistance, junction–ambient | 62.5 K/W |

The drain-to-source voltage can be up to 200 V, which suffices. Still care should be taken not to overload the MOSFET in its second breakdown range (quite high voltage and not so low current). The last point will be the cooling of the MOSFET. It is a non-insulated transistor, which means the thermal resistance (junction to case) is quite low, 0.9 K/W. Next, the heat sink will be designed. A little more information about this process will go first. The formula to calculate the maximum power dissipation can be expressed as

$$P_{\text{max, dissipation}} = \frac{T_{J,\max} - T_{\text{amb}}}{R_{\theta JA}} = \frac{T_{J,\max} - T_{\text{amb}}}{R_{\theta JC} + R_{\theta CH} + R_{\theta HA}} \tag{11.5}$$

with the following definitions: $T_{J,\max}$ is maximum operating temperature of the MOSFET; $T_{\text{amb}}$ is ambient temperature; $R_{\theta JA}$ is the thermal resistance, junction to ambient; $R_{\theta JC}$ is the thermal resistance, junction to case, of the MOSFET; $R_{\theta CH}$ is the thermal resistance of the insulation between the MOSFET and the heat sink; $R_{\theta HA}$ is the thermal resistance of the heat sink.

The actual power through the component may never transcend this $P_{\text{max, dissipation}}$. Now a calculation will be made. In Table 11.1, it can be found that

$$\begin{aligned} T_{J,\max} &= 150\ ^\circ\text{C} \\ R_{\theta JA} &= 62.5\ \text{K/W} \\ R_{\theta JC} &= 0.9\ \text{K/W} \end{aligned} \tag{11.6}$$

where $T_{J,\max} = 150$ °C is the maximum operating temperature of the MOSFET, $T_{\text{amb}} = 21$ °C is the ambient temperature without heat sink, and then $R_{\theta JA} = 62.5$ °C/W is thermal resistance. Therefore, the maximum power dissipation without heat sink can be calculated as

$$P_{\text{max, dissipation}} = \frac{150 - 21}{62.5} = 2.064\ \text{W} \tag{11.7}$$

The maximum power through the MOSFET can be driven as

$$P_{\max} = V_{\max} I_{\max} = 116.6 \times 0.3 = 34.98\ \text{W} \tag{11.8}$$

A heat sink was chosen with a low thermal resistance of $R_{\theta HA} = 1$ °C/W. Together with the internal thermal resistance of the MOSFET and the insulation, a safe condition is obtained. The experimental set-up of the current driver circuit is shown in Figure 11.4.

*Figure 11.4 The experimental set-up of the current driver circuit*

## 11.3 The start-up system circuit

To drive the EV, two brushless DC (BLDC) motors were used. They are mounted on each front wheel. The two BLDC motors are controlled by two inverters which are fed by the whole battery pack. The connection of the battery and the BLDC motor is of great importance. Three main points should be taken in mind:

- The BLDC should only be connected to the battery when the car needs to drive in order to avoid unnecessary loss of battery power.
- A resistor is needed during the start-up to avoid high starting currents, as those could destroy the battery and/or inverter.
- Before the motor is used, the internal capacitors of the inverters need to be charged.
- Only the owner of the vehicle should be able to start the car. This is especially important when the vehicle will get used in public.

To test the car, two DC thermal magnetic circuit breakers of 32 A in series were used for each pole. In the beginning, an external power resistor (80 Ω) was used manually to pre-charge the capacitor. This process is not user-friendly but may be an origin of errors, so the start-up is better implemented with a well-designed circuit. There are a few problems with start-up. First, we do not like heavy relays, and the reason is that they are constant energy loss when they are 'on'. A second problem is that if to implement a manual breaker, you do not know if the driver or operator will operate first the breaker or first the contact switch. The electrical circuit for the start-up system is explained in the next section.

## 11.4 Operational principle of the start-up system

The operational principle of the start-up system circuit will be discussed in this section. The schematic of the proposed circuit is shown in Figure 11.5. This circuit will start up the EV by feeding two inverters which can supply power to the motors of the EV. Two switches are required to turn on in order to start the vehicle, the ignition key, and the circuit breaker.

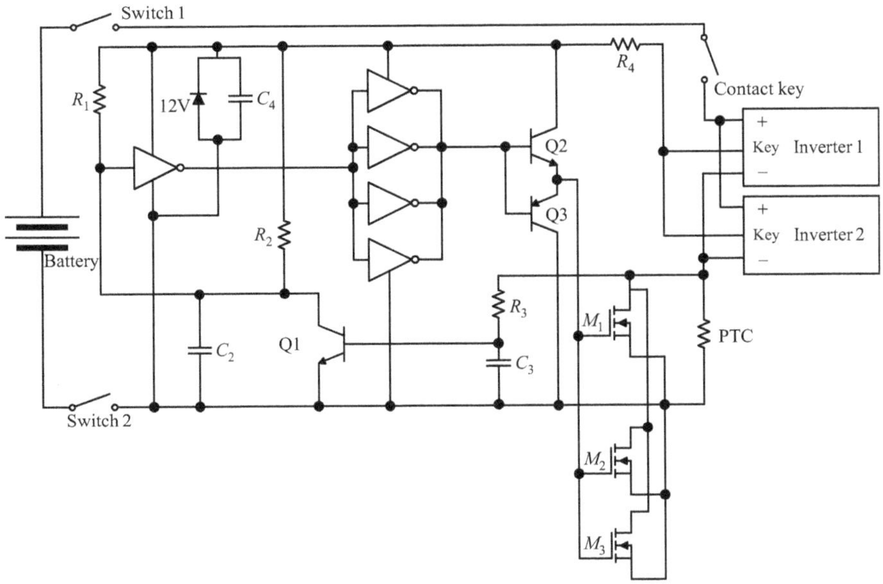


*Figure 11.5 The schematic circuit of the battery start-up system*

The circuit is designed in such a way that it does not matter which one is turned on first. Some people could also try both almost at once. When the circuit breaker is closed, current is allowed to the inverters by a start-up thermistor (positive temperature coefficient (PTC)). It cannot run over the start-up MOSFETs as these are off. If the condition takes too long, the PTC thermistors go in protection and limit the current. This could happen when the drives are defective, or when the throttle is active too early, before the capacitors are charged. Charging an empty capacitor from a DC source needs about the same energy in the charging resistor as energy in the capacitor. Very high peak currents can flow if connected directly. It could damage the circuit breaker, but also MOSFETs that would work in a linear operation, as their safety operating area has a second breakdown area with severe limitations on the current at high $V_{ds}$. The reason is that a high local temperature in the junction results in a hot spot with even more current density at that place due to the typical lower gate threshold at higher temperature. The capacitor is now charging. At a low $V_{ds}$, the Schmitt trigger turns the gate on by the voltage follower Q2 and Q3. Now all the currents run from drain to source over the MOSFETs, and almost nothing through the thermistor. This results in the battery and inverter being directly connected. MOSFETs of a few milli-ohm are cheap now and more reliable than contactors if the right circuit is controlling them. The motors are now fully prepared to start turning. The question may arise what would happen if the ignition key is turned on fast and the throttle pedal directly pushed. There will be no problem because as long as the capacitor is not fully charged, the transistor Q1 has a $V_{BE}>0.7$, which will make it pull down the Schmitt trigger, and the PTCs will protect and limit the current, but will take some time to recover. The system is designed to be 'fail-safe'. When the

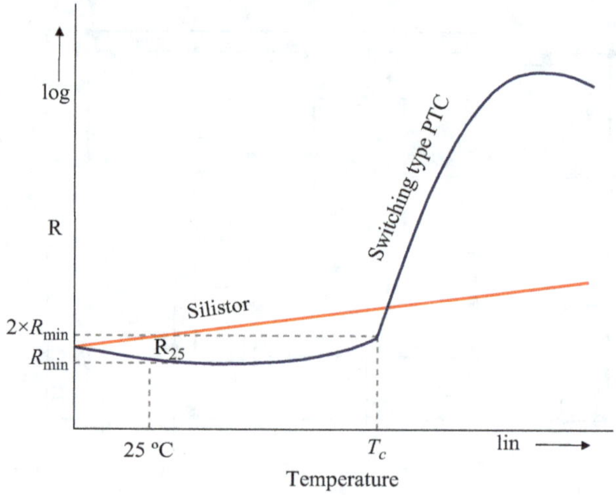


*Figure 11.6 PTC temperature vs resistance*

ignition key is turned on first, nothing will happen as the batteries are not connected to the circuit yet. Once the breakers are closed, the same situation occurs as described before. The most important resistor in the circuit is the PTC thermistor. A little more information about that type of resistor will be given first. A thermistor is a resistor with resistance dependent on its temperature. The relation between the temperature and the resistance of a PTC is shown in Figure 11.6.

The resistance increases as temperature rises 'PTC'. Some resistors (silistors) have a linear PTC but change only about a factor of 2 in 100 °C, are used as temperature sensors, and cannot be used to directly protect circuits. The PTC polymer thermistors (also known as PPTC) can almost 'switch' at a specified temperature. In this way, the resistance increases when more current 'more heat' flows through the component. As a result, the PTC is typically used as a current limiting device. The choice in the resistance of the PTC should balance the maximum current and start-up time. A high resistance results in the lower current, but it takes a longer time to start up. As a result for a suitable resistance, two possibilities are found, '22 Ω PTC and 56 Ω PTC'. Thus in order to make the right choice, both their start-up times were calculated. This can be done by calculating how long it would take to charge the internal capacitor. This can only be done if its capacitance is known. The measurements are performed by an LCR-meter at different frequencies. Based on the results, the capacitor was estimated as 27 mF. The vehicle uses two inverters so for calculations, and the capacitance should be 54 mF. Therefore, multiplying this capacitance with the start-up resistance results in the start-up time. It can be expressed as

$$t_{\text{start, 22 }\Omega} = 0.054 \times 22 = 1.18 \text{ s} \tag{11.9}$$

$$t_{\text{start, 56 }\Omega} = 0.054 \times 56 = 3.02 \text{ s} \tag{11.10}$$

Waiting time of 1.18 s is acceptable; however, 3.024 s is rather long and might be annoying as people are used to wait less. Consequently, a 22 Ω PTC is the only acceptable choice. This will lead to a somewhat higher start-up current, but it is still reasonable. The circuit consists of three transistors and three MOSFETs. The MOSFETs as discussed before will be turned on when the ignition key is turned on, and the internal capacitor of the inverter is charged. This allows the full current to flow towards the BLDC motor. It is clear that this will lead to high currents going from drain to source. The transistor should be able to handle these. Further, there is also the power dissipation which can be calculated as

$$P_{\text{dissipation}} = R_{\text{DS(on)}} I^2 \tag{11.11}$$

The maximum current that the battery is allowed to deliver is 100 A. This will lead to very high power dissipation if $R_{\text{DS(on)}}$ is not very low. Therefore, N-channel MOSFETs are preferred as they have a lower $R_{\text{DS(on)}}$. At last, there is the maximal voltage of 120 V that needs to be handled by the transistor. The most important information from the data sheet is given in Table 11.2. At such currents a heat sink is difficult to avoid. According to the data sheet, the $R_{\text{DS(on)}}$ of the MOSFET is 5.9 mΩ; thus the power dissipation in each transistor can be expressed as

$$P_{\text{dissipation}} = 2 \times 5.9 \times 10^{-3} \times \left(\frac{100}{3}\right)^2 = 13.11 \text{ W} \tag{11.12}$$

The resistance is multiplied by 2 as the $R_{\text{DS(on)}}$ in the data sheet is only for low temperatures. So the actual resistance will increase when the MOSFET gets hot. Multiplying the resistance by 2 anticipates this effect. The current is divided by 3 as it is divided over the three parallel MOSFETs. The obtained 13.1 W is the power dissipation for each MOSFET. As solution, the behaviour of the MOSFETs will be approached as if they were one MOSFET with three times more power dissipation and three times less thermal resistance. The equivalent MOSFET needs to dissipate 39.3 W and has a thermal resistance (junction to case) of 0.183 °C/W. In fact, the three MOSFETs behave like an equivalent MOSFET that needs to dissipate 39.3 W and has a thermal resistance (junction to case) of 0.183 °C/W. An advantage is that the heat is not concentrated on a single point. Even with some added thermal resistivity of electric insulation towards the heat sink of 1.68 K/W, the design is sufficient.

*Table 11.2 MOSFET characteristics*

| | |
|---|---|
| Drain–source breakdown voltage | 150 V |
| Continuous drain current | 112 A |
| Drain–source resistance, $R_{\text{DS(on)}}$ | 5.9 mΩ |
| Max operating temperature | 175 °C |
| Thermal resistance, junction–case | 0.55 K/W |
| Thermal resistance, junction–ambient | 62 K/W |

*Figure 11.7 The experimental set-up of the start-up system circuit*

Next to the MOSFETs, there are three BJTs, two NPN types, and one PNP. Q1 is used to pull down the Schmitt-trigger if needed as a current protection. Q2 and Q3 act as a voltage follower, also called push–pull-emitter follower. A voltage follower is some kind of buffer that creates almost the same voltage at output as at input but can deliver much more current at output. In Figure 11.7, the experimental set-up of the start-up system is shown. The circuit was tested for both switching sequences: circuit breaker closed first and ignition key closed first, which worked flawlessly.

## Summary

In this chapter, the concept of a trickle charging circuit for lithium-based battery is tried out: operate the BMS when one cell is high. If no cell is high, it is normally bypassed. Two battery chargers in series feed the total battery packs, which was convenient in the considered case but may be of help in other cases as well. There are four battery packs, and each battery pack consists of eight cells. The batteries are lithium–iron–phosphate (LFP) with a nominal voltage of 3.3 V. To achieve a best performance of the battery a current driver circuit (trickle charging) is designed to limit the current from the charger to the battery, and its operational principle is based on MOSFET constant current driver. Then a start-up circuit is implemented to connect the charger to the battery when it needs to drive in order to limit peak currents that could damage a manual operated circuit breaker, the battery, or other circuitry in that path, such as a current measurement shunt. The drive will only be connected to the battery if the car needs to drive, which increases safety. The MOSFETs are chosen in such a way to have less $R_{DS(on)}$ to reduce losses. Note that the standby loss of such a circuit is much lower than a contactor. Note that it acts as current protection. The experimental results are performed to verify the feasibility of the proposed system.

## References

[1] Lu L, Han X, Li J, *et al.* A review on the key issues for lithium-ion battery management in electric vehicles. *Journal of Power Sources*. 2013;226:272–288.

[2] Xu J, Cao B, Cao J, *et al.* A comparison study of the model based SOC estimation methods for lithium-ion batteries. In: 2013 IEEE Vehicle Power and Propulsion Conference (VPPC). IEEE; 2013. p. 1–5.

[3] Shafiei A and Williamson SS. Plug-in hybrid electric vehicle charging: current issues and future challenges. In: 2010 IEEE Vehicle Power and Propulsion Conference. IEEE; 2010. p. 1–8.

[4] Castiglia V, Livreri P, Miceli R, *et al.* Design and simulation of a fast DC recharging station for EV. In: 2017 IEEE 6th International Conference on Renewable Energy Research and Applications (ICRERA). IEEE; 2017. p. 1198–1203.

[5] Sakagami T, Shimizu Y, and Kitano H. Exchangeable batteries for micro EVs and renewable energy. In: 2017 IEEE 6th International Conference on Renewable Energy Research and Applications (ICRERA). IEEE; 2017. p. 701–705.

[6] Hattori S, Kajiwara K, Matsui N, *et al.* 72V DC bus output full-bridge converter in solar power system for micro EV. In: 2017 IEEE 6th International Conference on Renewable Energy Research and Applications (ICRERA). IEEE; 2017. p. 544–547.

[7] Fajri P, Heydari S, and Lotfi N. Optimum low speed control of regenerative braking for electric vehicles. In: 2017 IEEE 6th International Conference on Renewable Energy Research and Applications (ICRERA). IEEE; 2017. p. 875–879.

[8] Longo M, Yaïci W, and Foiadelli F. Electric vehicles charged with residential's roof solar photovoltaic system: a case study in Ottawa. In: 2017 IEEE 6th International Conference on Renewable Energy Research and Applications (ICRERA). IEEE; 2017. p. 121–125.

[9] Iyer VM, Gulur S, and Bhattacharya S. Variable DC bus control for a bidirectional on-board electric vehicle charger. In: 2017 IEEE 6th International Conference on Renewable Energy Research and Applications (ICRERA). IEEE; 2017. p. 1041–1046.

# Chapter 12

# Automatic charge selection between normal charge, equalizing and standby

## 12.1 Introduction

As discussed in Chapter 11, a charger was designed to be able to charge the battery using a standard wall socket. Besides that, a current driver circuit was added between the charger and the battery to limit the charging current to a maximum of 300 mA. This limitation on the charging current is required because the equalization current of the battery management system is limited by design. However, equalizing of the battery cells is normally mainly done during the final stage of the charging cycle. Limiting the current during the complete charging cycle is not necessary and would simply result in a very long charging time [1,2]. As the capacity of the battery in the example is 30 Ah, an estimation of the charging time, having a current which is always limited to an equalizing current of 300 mA, could happen on an empty battery if one bad cell is replaced by a new fully charged one. The theoretical time to charge only on equalizing current only would be in the worst case:

$$t_{\text{charge}} = \frac{30\,000 \text{ mAh}}{300 \text{ mA}} = 100 \text{ h} \tag{12.1}$$

In Chapter 5, a smart high-voltage cell equalizing and detecting (HVED) circuit was developed. The battery pack consists of four modules, each composed of eight cells resulting in a total of 32 cells connected in series. In a later development stage, each module will be equipped with an HVED circuit, giving four HVEDs in total. These HVEDs circuits monitor the voltages of the individual cells of the battery and each HVED has two output signals. One output signal is used to indicate when the voltage of one of the eight cells of a module rises above the threshold voltage. The second output gives a signal when all eight cells of a module are high. The rationale of the proposed method is that when all signals are absent, and thus all 32 cells are below their threshold voltage, so the charging current is not limited and the charger can deliver its full current to the battery. When 1 cell of the 32 cells goes above the threshold voltage, the charging current must be reduced to 300 mA such that the balancing circuit is capable of balancing the voltages of the affected module. Once all the four modules indicate that all eight cells are high, the battery must be disconnected automatically from the charger. This can be reduced to two modules at once if the charger is split into two parts.

## 12.2 Working principle of the selection

A circuit is designed to obtain the desired charging modes as discussed before. The simplified schematic of the designed circuit is shown in Figure 12.1. In this figure, the circuit designed in the previous chapter is indicated with the blue dashed lines.

The working principle of this part of the circuit was already discussed in detail. The circuit should satisfy the following requirements:

- maximum current charging (e.g. 6 A),
- equalizing charge current limited to 300 mA,
- charger cut-off.

Before proceeding into detail about the working principle of the upgraded current driver, some extra information about the high-voltage equalizers is needed. As mentioned before, the HVED circuit is able of producing two different signals. To do that two NPN transistors are used in each HVED circuit. One NPN is used to indicate the state one cell high (OCH), the other one is used for the state all cells are high (ACH). When at least one cell exceeds the threshold voltage, the transistor is activated, resulting in the collector being pulled low. The collector of the transistor is directly connected to one of the output pins of the HVED circuit; thus this output pin being low indicates that at least one cell is high. The same scenario is valid for the other transistor. When ACH, the corresponding output pin of the HVED is pulled low.

The newly designed current driver uses the states of the NPNs as logic variables to decide which charging mode the current driver should operate in. Table 12.1 indicates the state combinations of the NPNs with the corresponding charging mode. Note that the combination ACH 'on' and OCH 'off' cannot occur as ACH always results in OCH.

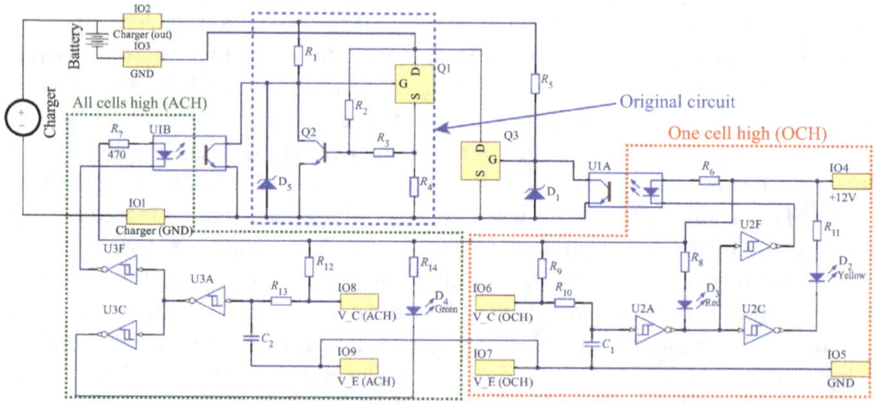


*Figure 12.1 Circuit for automatic charge selection between normal charge, equalizing and standby*

*Table 12.1 Charging modes with corresponding states HVED output*

| NPN OCH | NPN OCH | Charging mode |
|---|---|---|
| off | off | Max current charging (6 A) |
| on | off | Current limited to 300 mA |
| on | on | Charger cut off |

*Table 12.2 Characteristics of the MOSFET*

| Max. continuous drain current ($T_c$ = 25 °C) | 112 A |
|---|---|
| Drain–source breakdown voltage | 150 V |
| Max gate–source voltage | ±20 V |
| Drain–source on-state resistance | 5.9 mΩ |
| Max. operating temperature | 175 °C |
| Thermal resistance, junction–case | 0.7 K/W–0.7 K/W |
| Thermal resistance, junction–ambient (minimal footprint) | 62 K/W |

The charging modes described earlier can be obtained in hardware, by adding components to the existing circuit (indicated with dashed blue lines in Figure 12.1). Note that if a software implementation is done, about the same philosophy is needed, but probably the standby power would be higher. When the full charger current can be used to charge the batteries, the existing current driver has to be bypassed. This is done by placing another MOSFET (Q3) in parallel with it. When both $NPN_{OCH}$ and $NPN_{ACH}$ are off, and thus no signal is present, this MOSFET should be active. This is why the gate is pulled up by $R_5$. Using a Zener diode, the gate–source voltage is limited when the MOSFET is on. The used MOSFET is the same as the one used in the start-up circuit for the drives and its most important characteristics are shown in Table 12.2.

When charging at maximum current, the chargers regulate their output voltage to equal the voltage of the battery plus the small voltage drop due to the MOSFET and the connections. This results in a low $V_{DS}$ and thus it does not cause a problem. When working in another mode, the maximum voltage over the circuit is given by

$$V_{DS,max} = V_{charger,\ max} - 32 \times V_{cell,\ min} = 27\ \text{V} \tag{12.2}$$

where the charger output has a maximum output of 116.8 V and the cells of the battery are assumed to be never discharged lower than 2.8 V. The resulting $V_{DS}$ value is also well below the required limit. The maximum charger current amounts 6 A which is also far below the mentioned limit. The Zener diode $D_1$ guarantees that the gate–source voltage stays limited. The power dissipation in the MOSFET can be calculated as

$$P_{diss} = I_{max}^2 R_{DS(on)} = (6\text{A})^2 \times 5.9\ \text{m}\Omega = 0.21\ \text{W} \tag{12.3}$$

Without a heat sink, the power that can be dissipated by the MOSFET is given by

$$P_{\text{diss, max}} = \frac{T_{J,\text{max}} - T_{\text{amb}}}{R_{\theta JA}} = \frac{175\,^{\circ}\text{C} - 30\,^{\circ}\text{C}}{62\,^{\circ}\text{C/W}} = 2.34\ \text{W} \tag{12.4}$$

As $P_{\text{diss}} \ll P_{\text{diss, max}}$, the MOSFET is capable of dissipating the heat caused by its internal losses without the need for a heat sink. Note that a somewhat higher ambient temperature is used here. This is done because the printed circuit boards (PCBs) will be mounted in a closed space underneath the battery compartment. While driving, ventilation will be provided through the front of the car in the future. However, when charging the battery, the car will be stationary resulting in less ventilation of the storage space. Besides that, due to direct sunlight, the temperature inside the PCB storage space could increase. In case, this ambient temperature would increase too much, forced ventilation could be implemented using a small fan, or charging could be interrupted until lower temperatures are attained. To switch from the first mode, charging at 6 A, to the mode where current is limited, the MOSFET Q3 should be turned off such that the original current driver can take over charging the batteries. The circuit responsible for controlling this mode change is marked by the red dashed line in Figure 12.1 and is named OCH. This part is galvanically isolated from the current driver circuit to prevent problems caused by different ground references and it will be powered from the internal 12 V supply of the car. The isolation is obtained by using an optocoupler. The transistor of the optocoupler is connected in parallel with the Zener diode between gate and source. Activating the light-emitting diode (LED) of the optocoupler thus disables the MOSFET Q3 by pulling its gate low. The OCH part of the circuit is connected to the four OCH NPN transistors from the HVED circuits. These NPNs are all connected in parallel, and the collectors are wired to pin IO6 of the circuit, while the emitters are connected to IO7, which also equals the ground of the internal 12 V supply.

To convert the state of the NPNs into the required action of the optocoupler, the Hex Schmitt trigger integrated circuit with inverting input is used. As long as all battery cells are below the threshold, voltages of the HVED, IO6 and IO7 are not connected. In this situation, capacitor $C_1$ is charged through resistors $R_9$ and $R_{10}$ such that the input voltage of the Schmitt trigger U2A stays above the trigger's threshold voltage. This results in the output of U2A being low, which in turn results in the output of both U2F and U2C being high. Due to the high output of U2F, the optocoupler stays inactive, resulting in Q3 remaining on. The LED $D_3$ will light up to indicate on the dashboard which charging mode is active.

Once the HVED detects a cell voltage above its threshold, one of the four NPN transistors, connected between IO6 and IO7, will turn on, resulting in $C_1$ discharging through $R_{10}$. When the input voltage of U2A drops below the negative going voltage threshold, its output becomes high and the output of both U2F and U2C becomes low. This leads to LED $D_3$ turning off and LED $D_2$ turning on, indicating the mode change on the dashboard. Next to that, the LED of the optocoupler U1A turns on resulting in the gate of Q3 being pulled down and the bypass MOSFET turning off.

It is important to prevent the circuit from switching very fast between two charging modes because this could have a negative impact on the transistors. Imagine all 32 cells are below the HVED threshold voltage, the cells are being charged at a current of 6 A and switching between different modes is instantaneous. When one cell exceeds the threshold voltage of the HVED, the charging mode switches to a limited current of 0.3 A. However, in the meantime, possibly some power is extracted from the battery for the auxiliary equipment in the car and also the balancing of the cells is active. The 'high cell' only just reached the threshold voltage and the power consumption, and balancing would immediately lower the voltage again below the threshold value and thus the mode switches again to charging at 6 A. This switching could repeat itself very fast as the voltage of that one cell would oscillate around the threshold voltage of the HVED. To limit the fast switching to some extent, a small delay is introduced by $C_1$ and $R_{10}$. Their combination gives a time constant

$$\tau = R_{10}\,C_1 = 220\ \text{k}\Omega \times 0.47\ \mu\text{F} = 94\ \text{ms} \tag{12.5}$$

This results in a limit of ten mode switches per second between 'charging at 6 A' and 'charging at 0.3 A' which should be sufficiently low to protect the circuit from transients caused by fast switching back and forth between two modes. The circuit responsible of controlling the switching between 'charging at 0.3 A' and 'charger cut-off' is indicated with green dashed lines in Figure 12.1. The working principle of this part is similarly analogous to the circuit for OCH described earlier. When there is one battery module where an ACH is detected, this circuit activates optocoupler U1B and the gate of MOSFET Q1 is pulled down causing it to turn off. In this case, both MOSFETs of the circuit are turned off and the battery is disconnected from the charger.

## 12.3 Heat sink redesign

When the battery is charged using 0.3 A, MOSFET Q1 is used and it has to be capable of dissipating the internal losses. When limiting the output current of the chargers, both chargers will maintain their maximum output voltage as can be seen from their characteristic, leading to a total charger voltage of 116.8 V. However, this is not the voltage drop across the transistor Q1. In this charging mode, the worst-case scenario for the transistor would be one cell which is high, thus with a voltage of 3.65 V, while all 31 other cells are at their lowest voltage (assume 2.8 V). This situation results in

$$V_{\text{DS, max}} = 116.8\ \text{V} - 3.6\ \text{V} - (31 \times 2.8\ \text{V}) = 26.35\ \text{V} \tag{12.6}$$

With a charging current of 0.3 A, the losses to be dissipated by the MOSFET become

$$P_{\text{loss}} = 26.35\ \text{V} \times 0.3\ \text{A} = 7.9\ \text{W} \tag{12.7}$$

*Table 12.3 Thermal characteristics of MOSFET*

| | |
|---|---|
| $R_{\theta JC}$ | 0.9 °C/W |
| $R_{\theta JA}$ | 62.5 °C/W |

Using the thermal characteristics of MOSFET Q1 as shown in Table 12.3, the maximum power dissipation of the MOSFET without a heat sink is calculated as

$$P_{\text{diss,max}} = \frac{T_{J,\max} - T_{J,\text{amb}}}{R_{\theta JA}} = \frac{150\,^{\circ}\text{C} - 30\,^{\circ}\text{C}}{62.5\,^{\circ}\text{C/W}} = 1.92\ \text{W} \tag{12.8}$$

Again a higher ambient temperature is used, for the same reason mentioned earlier. As $P_{\text{loss}} \gg P_{\text{diss,max}}$, a heat sink is indeed necessary. Note that when looking at the final result, MOSFET Q3 is also mounted onto the heat sink, although it is not really necessary. The polyester film is used as insulation between the MOSFET and the heat sink. The used polyester film has a thermal conductivity $k$ of 0.7 W = (K m) and a thickness $t$ of 0.2 mm. The size of the used film is 10 mm × 30 mm. From this, the thermal resistance is calculated as

$$R_{\theta CH} = \frac{t}{k \cdot A} = \frac{0.0002}{0.7 \cdot 0.01 \cdot 0.03} = 0.95\ \text{K/W} \tag{12.9}$$

The used heat sink has a thermal resistance $R_{\theta HA} = 1$ K/W. The maximum possible power dissipation of the MOSFET together with polyester film and heat sink can be calculated using the following equation:

$$P_{\text{diss, max}} = \frac{T_{J,\max} - T_{J,\text{amb}}}{R_{\theta JC} + R_{\theta CH} + R_{\theta HA}} \tag{12.10}$$

With the values obtained earlier, this gives

$$P_{\text{diss, max}} = \frac{150\,^{\circ}\text{C} - 30\,^{\circ}\text{C}}{(0.9 + 0.95 + 1)\,^{\circ}\text{C/W}} = 42.11\ \text{W} \tag{12.11}$$

This value greatly exceeds the required power dissipation obtained in (12.8) even if large safety margins would be used. Therefore, the size of the available heat sink can be reduced while sufficient power dissipation is still guaranteed. When overestimating the required maximum power dissipation to be 20 W, instead of 7.9 W, and using (12.10) to obtain the maximum allowable value of $R_{\theta HA}$, this gives $R_{\theta HA} \leq 4.15$ K/W.

Figure 12.2 shows the thermal resistance $R_{\theta HA}$ of the heat sink as a function of the heat sink length. However, the obtained maximum limit for the thermal resistance is off-grid of the characteristic. Looking to the trend of the curve corresponding to passive convection, a rather safe length of 45 mm was chosen, which lies still well below the obtained limit. The length of the heat sink was thus reduced from 100 mm to 45 mm as can be seen in Figure 12.3.

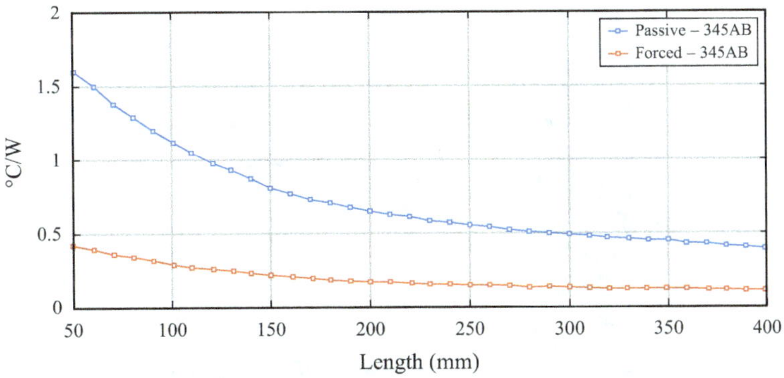


*Figure 12.2 Thermal resistance heat sink as function of length*

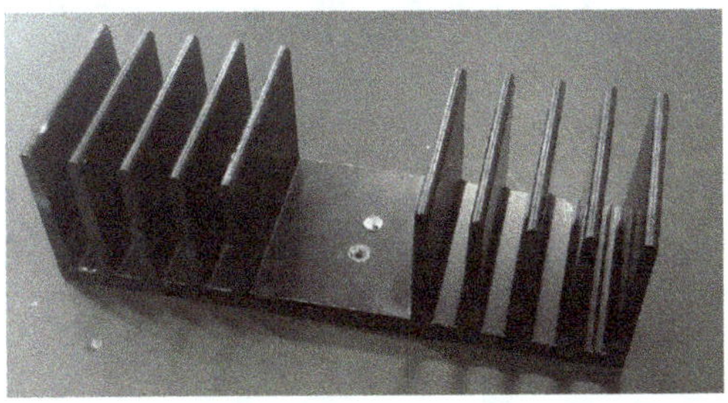

*Figure 12.3 Heat sink*

## 12.4 Result

First of all, the MOSFETs are properly aligned making it possible to attach them both to the heat sink. Second, the copper surfaces for the connection to the grounds of the charger and the battery are made sufficiently large and Q3 is placed directly next to it. When Q3 is on, currents up to 6 A flow through the circuit. The result of the made PCB for testing is shown in Figure 12.4.

The wires are labelled with the number corresponding to the input/output port as shown on the schematic in Figure 12.1. The three blue/purple wires are used to wire the LEDs on the dashboard. The blue wire leads to the anode, purple to the cathode. Table 12.4 shows which LED colour on the dashboard corresponds to which wire and gives the meaning of the LED combinations. Figure 12.5 shows the charging current, battery voltage and charger voltage, respectively, from left to right when the current driver is in maximum current mode. As expected, the charger output voltage is lowered to better match the battery voltage.

A small voltage drop due to the components and the connections is still present. The test results of the charging at 0.3 A were already reported before. When switching to the cut-off mode, the current was interrupted as expected.

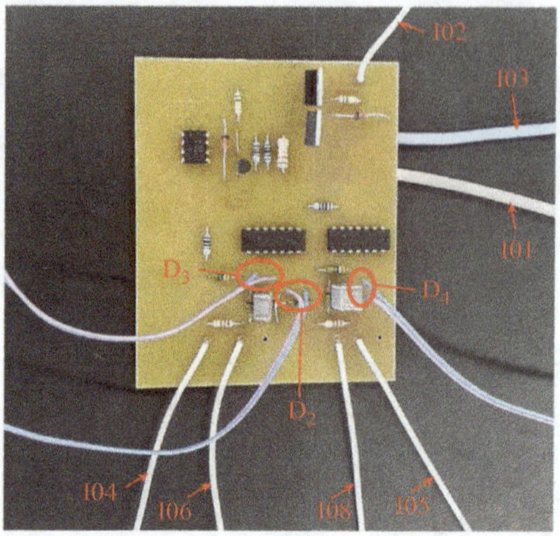


*Figure 12.4 PCB for testing the charge current selection*

*Table 12.4 Charger LEDs and their meaning*

| LED | Colour | Charging mode |
|---|---|---|
| $D_3$ | Red | Charging at 6 A |
| $D_2$ | Yellow | Charging at 300 mA |
| $D_4$ | Green | Charging cut-off |

*Figure 12.5 Charging at full current: (from left to right) charging current, battery voltage, charger output voltage*

The circuit of current selection probably could combined on the same heat sink of the start-up circuit of the first part of this chapter, as the start-up circuit mainly uses its heat sink during driving, whereas the current selection is used during charging from grid or solar. Originally, we estimated that using a switched type circuit could have been preferred but finally, the design, as it is, is simpler, and anyhow the equalizing charge is only a small time at limited current.

## Summary

The section proposed a selection of current between full charge and a current limited trickle charge. If the current is always limited, it would result in a very long charging time. To overcome this problem, two-stage charging is proposed. First, mostly much more than 90% of the battery capacity could be charged using higher currents (6 A). Then the remaining part, much less than 10% of the capacity is then charged using limited current (300 mA). If 1% of the charge has to be equalized, it could take about 1 h. However, very long equalizing times could happen if a repair has been done, which could save the battery by doing this. The proposed circuit is only one possible design, other variants can be imagined, but the circuit is a check if the necessary actions are implemented.

## References

[1] Shafiei A and Williamson SS. Plug-in hybrid electric vehicle charging: current issues and future challenges. In: 2010 IEEE Vehicle Power and Propulsion Conference. IEEE; 2010. p. 1–8.

[2] Longo M, Yaïci W, and Foiadelli F. Electric vehicles charged with residential's roof solar photovoltaic system: a case study in Ottawa. In: 2017 IEEE 6th International Conference on Renewable Energy Research and Applications (ICRERA). IEEE; 2017. p. 121–125.

# Chapter 13

# Push–pull converter battery charger fed by PV for electric vehicles

## 13.1 Introduction

In this chapter, the charging concept of lithium-based batteries for electric vehicles (EV) by using photovoltaic (PV) panel and push–pull converter is investigated, directly from DC to DC, as simple as possible. The aim of this chapter is to charge the batteries with PV panels on the vehicle roof, or panels independent from grid, or on the car port. Today, most of DC charging is designed for massive power, for fast charging. However, the tiny power from PV may finally be the most cost-effective one. The maximum power point voltage of this PV panel is much lower than the battery packs; thus a DC–DC converter is needed to overcome this problem. Therefore, a push–pull converter is designed and implemented to be connected to the solar panel. As the trial was at a small vehicle, the main aim was simplicity. The batteries are lithium–iron–phosphate (LFP) battery. The batteries are connected in series with a total number of 32 cells. The battery bank includes four battery packs and each pack consists of eight cells in series. The operational principle of the circuit is explained, and the experimental results are performed to verify the feasibility of the proposed system.

The electricity directly from solar panels is often much cheaper than from the grid or charging poles. There is also less discussion if it is really 100% renewable. Indeed some 'green' electricity may need more $CO_2$ for the production unit, conversion, transportation, grid infrastructure, and storage. In sunny areas, vehicles could simply be operated on it. But if they are often in the shadow in sunny weather, tandem cells would perform better, as they also use the scattered blue light well, where the silicon types do harvest better red and infrared wavelengths so are poor for blue light. In northern countries, it is also realistic for 50% of the year to have a significant contribution in vehicle charging. In addition, global warming and depletion of fuels cause EVs to be important transportation in the future. The most expense in EVs is the cost of the batteries; thus an alternative to always more batteries is the use of some PV panels on the roof. LFP batteries are widely used in many application [1–5] due to their advantages such as a higher safety, low self-discharge rate, lower weight, high density, no memory effect [6–8]. In Li-ion-based batteries, voltage differences exist between cells due to charging and discharging process. These unavoidable differences occur due to the chemical and electrical characteristics, such as asymmetrical degradation with ageing, production tolerance, slightly different temperatures. They reduce the life cycle of the battery [9, 10]. To overcome the problem of imbalances, battery management system is introduced to monitor battery modules and it provides early warning protection. A part of the task is equalizing the cells, but during that process much less power is needed, which can be supplied by PV. So equalizing can be done without occupying a charging pole and so liberate it for other vehicles. It also extends the driving-range in sunny periods or avoids self-discharge if not plugged in. PV panels permit also lowering the total $CO_2$ emissions of EV and not just to be 'zero tailpipe emission'. The chapter proposes a battery charging circuit by a push–pull converter, without the need for grid connection. By doing this, the battery cells will be fully charged and balanced and the driving range will be extended.

## 13.2 Charging principle with PV panel

When it comes to the green energy, solar panels play an important role. A PV cell converts light into electricity by using semi-conducting materials. A typical PV system is a solar panel built out of solar cells. When using a PV panel for electrical vehicles, they have some advantages. The most important benefit of a solar panel is that solar energy is a renewable energy source and sunlight will be available for at least 5 billion years according to the scientists. For vehicles, lighter PV panels are chosen, which can be bent easily. In Figure 13.1, the photograph of the PV panel is shown.

A little more information about PV panels will be given first. If PV panels are used at home, it is good to oversize them, as their electricity cost is less than the grid. It is expected to have an excess for a big part of the year, which can be used in vehicles. Vehicles could even help to supply electricity to houses in evening, as their batteries are often bigger than what are used in power-walls. In domestic applications, a low customer price of naked panels is about 300–350 euro/kWp. Second hand prices are even much lower. In addition, other costs are also low, such as installation, cabling, rules for compliance, and converters. One of the major advantages of PV panels is their low maintenance cost, and their life expectancy largely exceeds the vehicle and battery life. This chapter helps for the

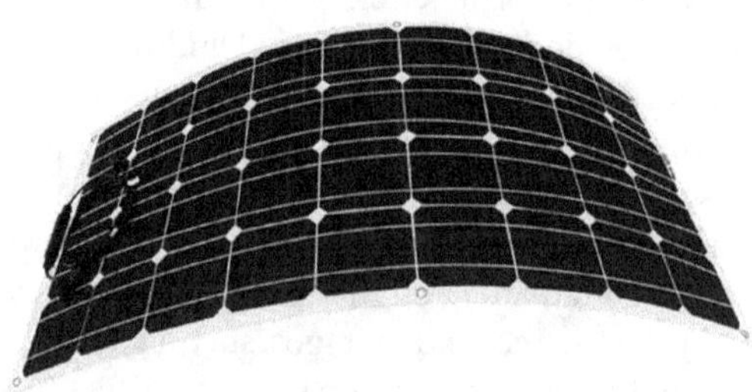

*Figure 13.1 Photograph of the used PV panel*

converter part. DC–DC converters could be very simple if the right design is chosen, even about five times less expensive than DC–AC followed by AC–DC. Knowing that, we expect a growing interest for DC–DC. It can also have a longer life than the AC types as they can be made without electrolytic capacitors and are less sophisticated.

A PV panel generates no pollution and no greenhouse emissions. These advantages could make solar panel suitable for electrical vehicles by charging the vehicle while driving to increase the range. When using a PV panel for household purposes, they have the following advantages and disadvantages.

**Advantages:**

- Solar energy is a truly renewable energy and sunlight will be available as long as the sun does not stop burning.
- They have low maintenance costs. Often the inverter has a much shorter life than the panels.
- Solar energy is free. All the energy needs that can be fulfilled with the sun will reduce the electricity bill.
- The technology is constantly advancing and improvements will intensify in the future.
- A PV panel generates no local pollution and nor greenhouse gas emissions. The amount of $CO_2$ for manufacturing is in the order of 10–20 times lower than it may produce in its life.

**Disadvantages:**

- The sun is free, but solar panels certainly are not. The initial cost of purchasing a PV panel is rather high.
- Although solar energy can still be collected on cloudy and rainy days, and shadowed cells, the output drops to 5%–25%. This is definitely a problem in the Belgian climate. On top of that, solar energy can obviously not be collected at night.
- Just as every type of electrical energy, solar energy needs to be used right away. Taking into account the variability of the sun, this raises quite some problems. Relying only on PV panels would require a form of energy storage to store energy at day to use at night, for example a battery.
- PV panels act as a DC voltage source. Consequently, for most applications, an inverter is needed as AC voltage is more widely used.
- Using a PV panel does not pollute, but producing and transporting them does.

These advantages and disadvantages mentioned earlier are more general. The solar panel could be used for the following purposes:

1. Charging the vehicle while driving to increase the range: nice in summer holiday.
2. 'Helping' the net with charging the vehicle whilst not driving in order to lower the electricity bill. It is in the aim of increasing 'self-consumption'.
3. Avoid bottle necks in the grid at noon. The purpose is that PV panels are available for charging both at home and at the job.

4. It can act as an emergency solution. Suppose you get lost and suddenly run out of battery. Just take some rest and the solar panels will give you enough power to drive to the closest wall plug or charging pole. In emergency situation at home, little people know that a lot of equipment works from 120 Vdc on, and some LED lamps intended for 115 or 230 Vac work even down to 70 Vdc [11].

Therefore, choosing for a PV panel seems about the only right solution. However, this cannot be done without first taking the disadvantages into account. An electric car runs entirely on DC voltage; thus no inverter is needed, as for the energy storage, a battery is already present in the vehicle. Concerning the price, it should be noted that only one panel is used, and on a small vehicle, much more would not fit well on the roof. At last, the biggest problem is, 'Will the PV panel pay of its energy debt?'. It is known that household solar panels work for about 30 years. For this type of panels, the energy debt is paid back many times. However, for smaller electric products, for example a smartphone, which are only supposed to 'live' about 2–3 years, this is not always the case. Adding a solar panel to these products is more polluting than just using the electrical grid. The average lifetime of a vehicle needs to be determined first. Next, the periods concerning the energy debt of a solar panel should be examined. Figure 13.2 shows the global energy payback period of multicrystalline silicon PV panels [12].

Based on the graph, it seems a good estimate to take an energy payback period of 2.1 years as this graph is specifically made for rooftop panels which work at all time. Estimating that the car will be inside or in the shadow about 50% of the day is a cautious estimate. Consequently, 4.2 years is taken as payback period for the PV panel. Knowing that the vehicle should at least drive around 10 years, it can be concluded that the panel will pay back its energy debt at least two times. However, a less amount of batteries may be needed, so it can be better. Designers may have a challenge to incorporate it, but having panels on the roof or other places of a vehicle may look futuristic. Panels bent in one direction only are affordable. Instead of an exception, the authors estimate that PV panels on vehicles will get standard. It

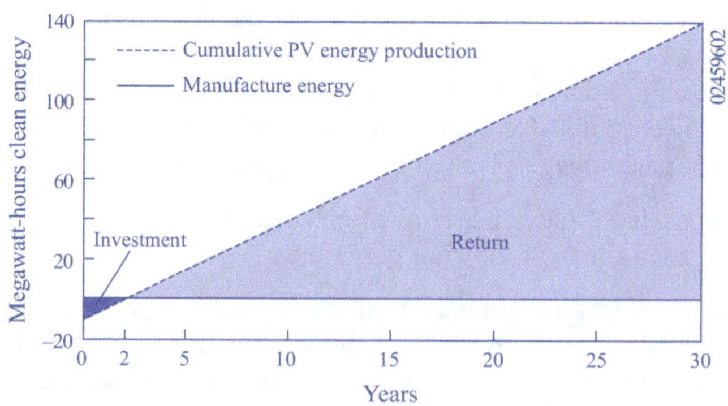


*Figure 13.2 Global energy payback period of multicrystalline PV panels*

*Table 13.1 Electrical information of the PV panel*

| **Parameter** | **Values** |
|---|---|
| Peak power | 100 W |
| Solar cell efficiency | 17%–18% |
| Maximum power voltage | 18 V |
| Maximum power current | 5.56 A |
| Open circuit voltage | 21.24 V |
| Short circuit current | 6.11 A |
| Power allowance range | ±3% |
| Maximum system voltage | 500 V |

will even be the case for fuel operated vehicles, as the efficiency from tank to electricity of the Lundell alternator is about two times lower than to the wheels. From fuel to electricity in a fuel operated vehicle, the efficiency is about 10%, at 10 kWh/L, and it results in an electricity price of a litre fuel/kWh. True, it may be expensive or not, it depends on the country. For a small vehicle, a 100 W Flexible Lensun Solar Panel was chosen. The choice was especially based on the weight and flexibility. The stand-alone PV panel was tested and compared to the main specifications of the panel as summarized in Table 13.1.

To achieve the maximum possible power output in different irradiation situations, a more advanced control strategy such as maximum power point tracking (MPPT) should be implemented. However, for simplicity, a maximal power point (MPP) control has been omitted, but if that control or an additional converter to do the job is implemented, it will also add to losses, so some 2%–5% deviation from MPP could be still accepted.

By looking at data of residential solar PV, a module with 100 W output results in a yearly production of 85 kWh 'ideally oriented'. Ultralight vehicles can drive with 3–4 kWh/100 km, 2833–2125 km, and in practice it will be a fraction of it but not negligible.

## 13.3 Push–pull converter circuit

In the previous section a desired solar panel was discussed. An overview of its specifications shows that the PV panel has a maximum power voltage of 18 V. This voltage is considerably lower than the desired voltage to charge the batteries, which is approximately 105.6 V according to (13.1). Therefore, a DC–DC converter will be necessary.

$$32 \times 3.3 = 105.6 \text{ V} \tag{13.1}$$

The choice will have to make between a regulated or an unregulated converter. The regulated type has a main advantage that fixed output voltage is available. As a disadvantage, the converter is more complex, bigger, and less efficient. The

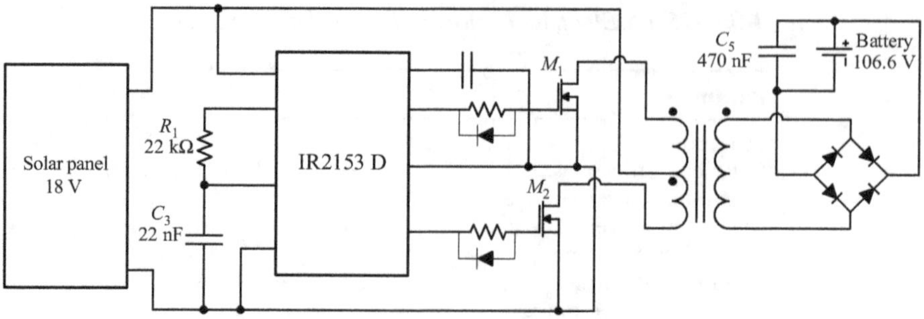


*Figure 13.3 The schematic of the proposed push–pull converter*

problem with an unregulated converter is the uncertain output voltage. However, a well-chosen constant voltage load for the PV can get the system quite close to the MPP. If a sophisticated controller needs 1 W to work, one can accept 1 W deviation while using a simple circuit. Thus it can be more beneficial to make the voltage proportional to the PV and possibly make adjustments at the secondary side, where currents are lower. As a type of DC–DC converter, a push–pull converter was implemented. The simplified schematic of the proposed push–pull converter is presented in Figure 13.3.

The transformer in a push–pull converter is supplied by two transistors, typically high-power MOSFETs, with an ON-state resistance of a few milli-ohms. The two switches are controlled to be ON or OFF by a pair of complementary signals in a synchronous trigger pattern generated by a MOSFET driver. This periodically reverses the EMF in the transformer which leads to current being drawn during both halves of the switching cycles. So it draws less current ripple compared to a boost or a flyback converter and has a higher efficiency. The push–pull converter is designed around the IC that creates the switching signal. The MOSFET driver is a self-oscillating half-bridge driver (IRS2153D). It has a high-voltage half-bridge gate driver with a front-end oscillator similar to the industry standard CMOS TLC 555. The typical connection diagram, as found in the data sheet, is shown on Figure 13.4.

The IC is originally intended for a half-bridge use but is also compatible for a push–pull type of converter. The price is not expensive and it incorporates two transistor drivers, an oscillator and undervoltage lock-out (UVLO) function, and has only eight pins. Pins RT and CT determine the oscillating frequency with their, respectively, named resistor RT and capacitor CT. Pin 4 is connected to the ground. Pins HO and LO send the periodic signals towards the two MOSFETs. In Table 13.2, pins of the gate driver are explained. At the beginning, the chip is in UVLO mode. This mode occurs when $V_{cc}$ has not reached the threshold voltage yet. It is designed to assure that the IC is fully functional before the MOSFETs get activated. It is important that the transistors are always ON or OFF, nothing in between, to prevent unwanted high-power dissipation. The UVLO mode assures that both MOSFETs are off when the IC is not fully functional. Once the threshold voltage is obtained at $V_{cc}$, the IC starts working. The voltage at the RT-pin at the

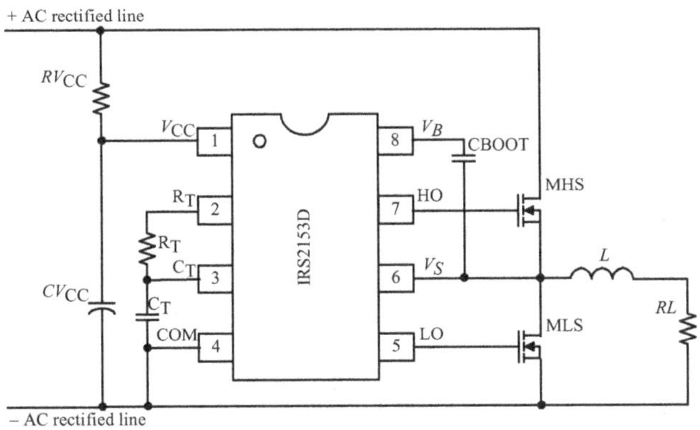


*Figure 13.4 Typical connection diagram for the IRS2153D*

*Table 13.2 Pins description of the IR2153D*

| Symbol | Description |
|---|---|
| $V_{cc}$ | Logic and internal gate drive supply voltage |
| RT | Oscillator timing resistor input |
| CT | Oscillator timing capacitor input |
| COM | IC power and signal ground |
| LO | Low-side gate driver output |
| $V_S$ | High-voltage floating supply return |
| HO | High-side gate driver output |
| $V_B$ | High-side gate driver floating supply |

internal enable increases to approximately $V_{cc}$ and the external capacitor CT starts charging. Once it reaches 1/3 $V_{cc}$, LO turns on with a delay, equivalent to the dead time. This delay is important as the two MOSFETs may never be ON at the same time. Then when the capacitor CT reaches 2/3 $V_{cc}$, LO will be low and RT will be down approximately to the ground. Then CT starts discharging and the dead time circuit is activated. When the dead time is passed, HO will be high. The signal stays high until CT reaches 1/3 $V_{cc}$.

Again HO becomes low, RT becomes high, the dead time circuit is activated and the capacitor starts charging again. After the dead time, LO becomes high and the cycle starts over again. The oscillating frequency can be determined by RT and CT by the following equation:

$$f = \frac{1}{1.435 \times \text{RT} \times \text{CT}} \tag{13.2}$$

With the knowledge of a push–pull converter and the gate driver IR2153D, the proposed circuit as shown in Figure 13.3 was implemented. To explain the working principle, consider that the threshold voltage of the IC is reached, and the driver can start working. It will give an oscillating signal to Q1 and Q2 as extensively explained before. This will alternately turn Q1 and Q2 ON with a dead time in between, which is essential to prevent transistor shoot through. They indeed may never both be ON. At the secondary side, this will result in current running through the winding clockwise and counter clockwise. The full diode bridge at the secondary side acts as a rectifier towards the battery. Due to the dead time, there are very small gaps at which no current flows. This switching noise will be filtered by capacitor $C_5$. A choke coil could smooth more, but as this would also increase the cost and complexity, it is not added to the circuit. The oscillating frequency is determined by values of the resistor $R_1$ and capacitor $C_3$. As mentioned before, the relation can be presented with (13.2) but the actual relation is better determined by the graph given in the data sheet. The graph is shown in Figure 13.5. To determine the switching frequency a high resistance for RT is preferred as this lowers energy consumption. Thus, 22 kΩ is chosen and CT = 1 nF is preferred. This results in 31.28 kHz of switching frequency.

$$f = \frac{1}{1.435 \times 22 \times 10^3 \times 1 \times 10^{-9}} = 31.28\ \text{kHz} \qquad (13.3)$$

The power MOSFETs have a high current switching capability and they have low ON resistance '$R_{DS(on)}$'. These qualities are required as the transistor will be constantly switching and high-power dissipation should be avoided. In order to choose a MOSFET, some conditions were considered.

- The PV panel output voltage and current are 18 V and 5.56 A, respectively, so a drain–source voltage of at least 2×18 V + 50% headroom is chosen and a low loss of a current of at least 5.56 A should be required.

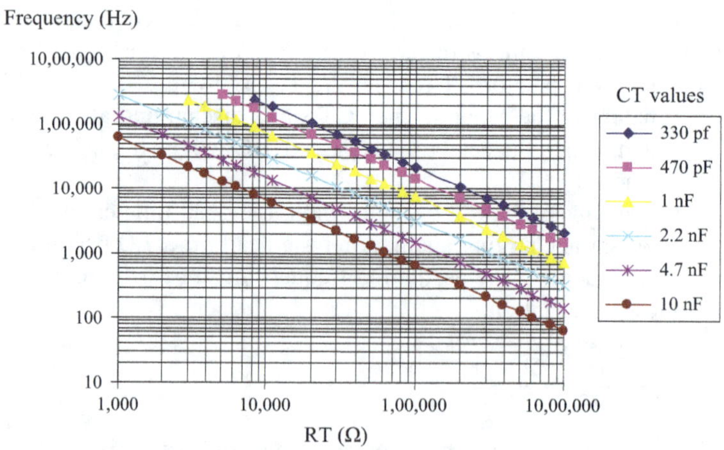


*Figure 13.5 Frequency vs RT for different values of CT*

*Table 13.3 MOSFET characteristics*

| Parameter | Values |
|---|---|
| Drain–source breakdown voltage | 60 V |
| Continuous drain current | 43 A |
| Drain–source resistance | 12.2 mΩ |
| Max operation temperature | 150 °C |
| Thermal resistance, junction–case | 5 °C/W |
| Thermal resistance, junction–ambient | 62.5 °C/W |

- The switching time of the MOSFET should be negligible in comparison with the oscillating period imposed by the chip. This period equals ±30 μs as the frequency is 33 kHz.

Eventually, the N-channel MOSFET TK30A06N1 was chosen. In Table 13.3, some of the MOSFET characteristics are presented. It has an $R_{DS(on)}$ of 12.2 mΩ, which is quite low. The voltage rating for drain–source is 60 V is sufficient. The voltage drop in the ON-resistance is low and the current rating is 43 A, even if the $R_{DS(on)}$ increases by a factor 2 with temperature. The only issue is, if a heat sink will be necessary or not. The current flows always in one of both transistors. The conduction loss for both transistors together, for a two times higher resistance (taking in account a large temperature increase), is

$$\begin{aligned} P_{\text{dissipation}} &= R_{DS(on)} \times (I_{DS})^2 \\ &= 2 \times 0.0122 \times 5.56^2 = 0.754 \text{ W} \end{aligned} \tag{13.4}$$

The maximum acceptable power dissipation, without heat sink, for each transistor can be calculated as

$$\begin{aligned} P_{\text{dissipation}} &= \frac{T_{J,\max} - T_{\text{amb}}}{R_{\theta JA}} \\ P_{\text{dissipation}} &= \frac{150 - 21}{62.5} = 2.064 \text{ W} \end{aligned} \tag{13.5}$$

In the data sheet, it can be found that $T_{J,\max} = 150$ °C and $R_{\theta JA} = 62.5$ °C/W. Therefore, as 0.754 W $\ll 2 \times 2.064$ W, no heat sink will be required for the MOSFETs. The purpose of the transformer is to increase the voltage from the solar panel to the battery. This will be obtained by choosing an appropriate winding ratio. A nominal charging value for the battery is about 3.3 V/cell. As 32 cells are in series, a voltage of 105.6 V is required. The maximal-power voltage of the solar panel is 18 V. The expected turns ratio is calculated as

$$\frac{N_s}{N_p} = \frac{V_s}{V_p} = \frac{105.6}{18} = 5.86 \tag{13.6}$$

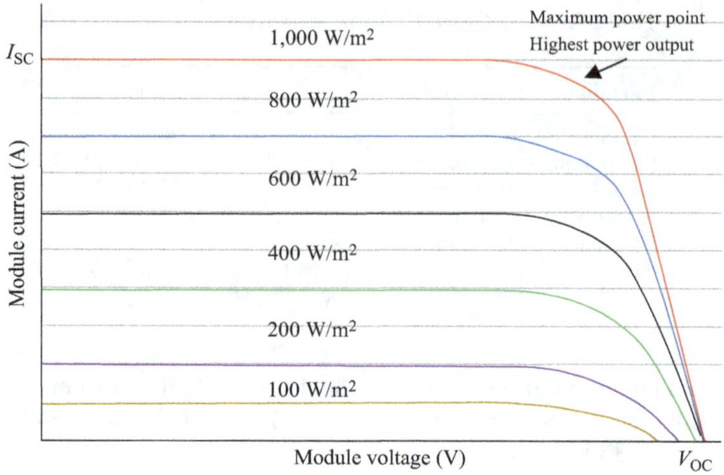


*Figure 13.6 Maximal power point for a solar panel*

When the previous winding ratio is chosen, the battery will force the wanted 18 V on the solar panel, independent of the amount of lightening. This is close to the MPP.

The graph in Figure 13.6 proves this [13]. For each level of lightening, the MPP of a PV panel is reached when the voltage is at the given maximal power voltage, 18 V in this case. It should be noted that evidently the current will lower when the irradiance is lower. The MPP point only guarantees the maximum power belonging to each specific irradiance level.

For the transformer, a coil former and core are needed. The PQ32 × 20N95 cores and accompanying coil former are selected for the circuit. Next, the number of turns at the primary winding needs to be calculated. This can be obtained by the equation for voltage per turn as

$$N_t = \frac{V_{\text{dc}}}{4f\,A_e B_m} \tag{13.7}$$

where $N_t$ is the number of turns, $B_f$ is the operating frequency, $Bm$ is the magnetic flux density [0.22 T], $A$ is effective cross sectional area [$127.5 \times 10^6$ m$^2$], and Vdc is DC input voltage (V).

$$N_t = \frac{18}{4\;32.28\;127.5\;0.22} = 4.97 \text{ turns} \tag{13.8}$$

For 18 V, this results in 5.08 turns, which is rounded to 5. Now the turn ratio and number of turn in the primary are known, and the number of turns in the secondary can be calculated as

$$N_s = \frac{V_s}{V_p} \times N_p = 29.35 \text{ turns} \tag{13.9}$$

It can be rounded up to 30 turns. However, this will be increased to 33 turns to avoid that the voltage is not optimal at lower sun, but also for some voltage drop due to primary–secondary leakage. Now, the width of each winding is required. The coil former has 18.5 mm height for wires. Thus, 33 turns can have a maximum diameter of $d = 18.5/33 = 0.56$ mm, which is rounded to 0.5 mm. As the primary has only five turns, the height will not be a problem. Therefore, 1.5 mm wire is chosen as a good balance between wire resistance and AC component. The smaller secondary windings are placed first to have a smooth surface for the primary windings. If it were to be done the other way around, the bigger radius of the primary wires would result in a bumpy surface for the secondary winding. The primary windings are bifilar wound. Between primary and secondary, some insulation sheet is put, rather for mechanical reasons. The proportions of the core and windings on the figure are correct. This gives the opportunity to see if the coils are not too wide or high. As the 'grey block' on the left represents the outer core, the gap is wide enough to hold the entire core.

Finally, at the output a diode bridge is used as a rectifier for the circuit. At the secondary of the transformer, a block wave is present that alternates between + and – 105.6 V (and 0 V during dead time). The diode bridge turns thus wave into a steady signal of 105.6 V. The ripple is filtered by a single, large enough, capacitor. The diodes make sure that the battery is never used to 'charge' the solar panel. Now a specific type of diode should be selected. The most important performance to look upon is the reverse recovery time $t_{rr}$. It represents how long the diode needs to 'recover' after it has been put in non-conducting mode. During the recovery, current runs in the opposite direction as it is meant to be. As a maximum battery life and a good circuit performance is wanted, the used diodes should have a fast recovery. Ultra-fast diodes were chosen. The diodes' reverse recovery time is 50 ns. It should be mentioned that the problems caused by the recovery are anyhow limited as they occur during the dead time of the IC, which is much longer than the recovery time, and the current has the time to decay.

## 13.4 Experimental results

In this section, the experimental results and efficiency will be discussed. In Figure 13.7, the photograph of the push–pull converter is presented. The circuit was tested on a sunny day and was able to charge the batteries. The voltage measured over the panel was 19 V. The output voltage of the push–pull converter is presented in Figure 13.8. The oscilloscope probes × 10 are selected as well as oscilloscope scale of 50 V. As can be seen, the output voltage reaches 104 V which is close to 105.6 V. In the hypothetic case that 100 W of the data sheet of the solar panel could be obtained, we can calculate the efficiency as

$$\eta = \frac{P_{\text{tot}} - P_{\text{loss}}}{P_{\text{tot}}} \tag{13.10}$$

where $P_{\text{loss}}$ can be calculated by adding different power losses. The transformer has losses in the core and coil. The ferrite losses in the core will be discussed first. The

Figure 13.7 Photograph of the push–pull converter circuit

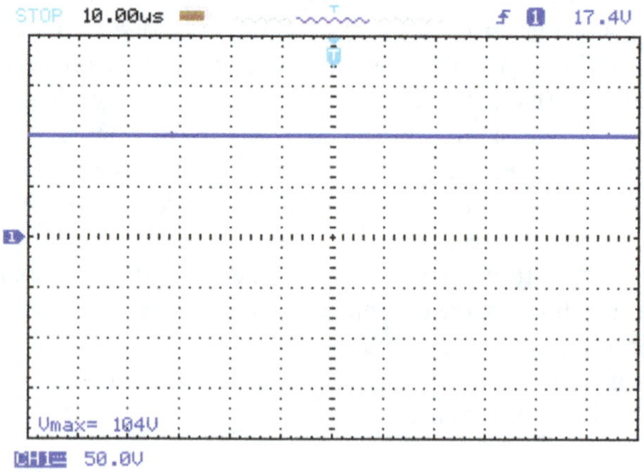


Figure 13.8 Output voltage of the push–pull converter circuit

data sheet mentions a maximal loss <4.7 W at 0.2 T in a temperature range of 25 °C–100 °C at 100 kHz; note that it is better to start from the data of the core than the material (measured on R34 toroids) as larger dimensions generate larger E-field losses. As the oscillating frequency is 31.28 kHz, the losses will be smaller. In order to calculate the actual losses, the Steinmetz equation (logarithmic interpolation) can be used. From the material data sheet, we obtain about $\alpha = 1.2$ and $\beta = 2$.

$$P_v = k f^{\alpha} B^{\beta} \tag{13.11}$$

This results in the following equation can be expressed as

$$P_v = 4.7\,(31.28/100)^{1.2}\,(0.22/0.2)^2 = 1.41\text{ W} \tag{13.12}$$

Further, there are losses in the windings, which can be calculated if the resistance of the winding is known. This can be calculated with Pouillet's law as

$$R_{\text{winding}} = \frac{l \times N \times \rho}{A} \tag{13.13}$$

where $l$ is the mean length of one turn, $N$ is the number of turns, $\rho$ is the electrical resistivity, and $A$ is the cross section of the wire.

The secondary winding will be calculated first. The electrical resistivity for copper is $23 \times 10^9\ \Omega\text{m}$. The mean length of one turn can be calculated according to the dimension of the core as

$$l_s = 2 \times \pi(0.0159/2 + 0.0002/2) = 0.0515\ \text{m} \tag{13.14}$$

This results in the following power loss as

$$R_{\text{winding}} = \frac{0.0515 \times 33 \times 23 \times 10^9}{\pi \times 0.00025^2} = 0.199\ \Omega \tag{13.15}$$

At 1 A, this gives a power loss of 0.2 W. At 33 kHz, the penetration depth is a bit smaller than the wire diameter or the equivalent square side [14]. The losses will be somewhat higher, about 0.3 W. The primary winding is wound above the secondary. These windings carry both a part DC and a part AC. The AC part has an amplitude of about six larger than the secondary but is carried by both windings that have three times more diameter, so the losses by skin-effect are of the same order of magnitude. The DC current component flows through the whole section and has a much lower loss. The order of magnitude:

$$P_{\text{loss, transformer}} = 1.41 + 0.3 + 0.4 = 2.11\ \text{W} \tag{13.16}$$

We could indeed calculate much more precise, but for a first trial it is enough. In the diodes, there is about 1.6 V drop, and this result is about 2 W power loss, switching losses included, so $P_{\text{loss, diode}} = 2$ W. The transistor losses are already discussed for the heat sink, which are 0.754 W. At last, there are some losses in the printed circuit board, capacitors, which can be safely estimated at 0.3 W.

All above-mentioned losses result in the following total loss as

$$P_{\text{loss, total}} = 2.11 + 2 + 0.754 + 0.3 = 5.16\ \text{W} \tag{13.17}$$

Now the efficiency of the circuit at full load can be calculated as

$$\eta = \frac{100 - 5.16}{100} \times 100 = 94.8\% \tag{13.18}$$

In practice, a bit higher as it is not likely that the transistors will be at the maximal temperature, as this corresponds to a very high ambient temperature as well. As a test, the circuit was connected to a voltage source and loaded with a variable power resistor. To simulate realistic situations, which could occur when the converter is

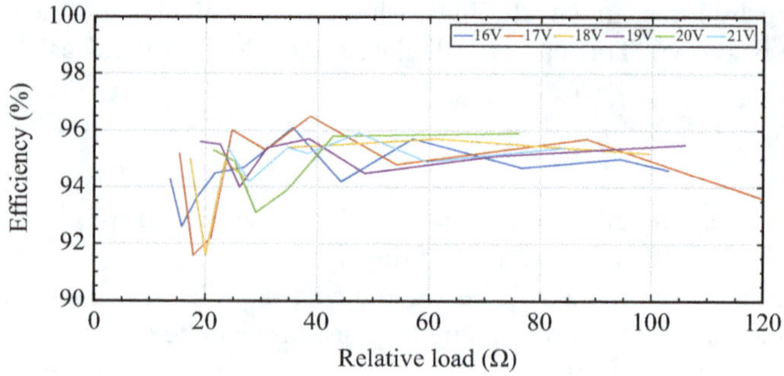


*Figure 13.9 Converter efficiency as a function of the load at different input voltages*

installed in the car, an input voltage from 16 V to 21 V with increments of 1 V was applied. For each fixed input voltage, the load was varied while measuring input and output current and voltage. From this, the converter's efficiency for different input voltages and loads could be calculated. The results of these measurements are shown in Figure 13.9. The load is given relative to the maximum power output of the PV panel, which is 100 W. From the measurements follows an efficiency of about 95% while no clear correlation between the efficiency and the input voltage can be observed. At lower loads, the efficiency is somewhat lower due to the increase of the no-load losses of the transformer, IC, and driving of the transistors, but is still sufficiently high because in cloudy weather not all generated power is lost due to these no-load losses. For the simplicity, no current protection was implemented, If the diode bridge survives, no big short circuit currents are expected from the output, also short-circuiting the PV panel is maximally some 7 A, and the voltage is low enough not to arc.

## Summary

In this chapter, the charging concept of lithium-based batteries for electric vehicles from PV is investigated. For a small power, we looked at a very simple circuit, with its shortcomings, but also its merits. A simple push–pull converter was implemented. More control, or a boost or buck converter could be added to improve the MPPT effect, but it is not sure that finally more output power would be obtained. The maximum voltage of the PV panel is much lower than the battery; therefore, a DC–DC converter was required. Also in larger vehicles, it might be needed, as the designers limit the PV voltage in the case of a crash and for the safety of fireman. In the considered case, the battery bank includes four battery packs and each pack consists of eight cells in series. The aim is to charge the batteries with PV panel on the vehicle roof, but also a fixed panel source could be considered. The PV panel and push–pull converter are implemented and tested. The experimental results show the feasibility of the proposed circuit. At the end, the maximal efficiency of the circuit is approximated as 95%. The calculation is only approximate as this would be beyond the aim of the book. Note that PV panels make also sense on cars with combustion engines, as the electricity comes finally from fossil fuel at a low efficiency of 8–15% (15–30% for the engine and a low efficient Lundell alternator (about 50%), so about 1 liter fuel is needed for 1 kWh, which is expensive electricity.

## References

[1] Moghaddam AF and Van den Bossche A. An active cell equalization technique for lithium ion batteries based on inductor balancing. In: 2018 9th International Conference on Mechanical and Aerospace Engineering (ICMAE). IEEE; 2018. p. 274–278.

[2] Moghaddam AF and Van den Bossche A. A cell equalization method based on resonant switched capacitor balancing for lithium ion batteries. In: 2018 9th International Conference on Mechanical and Aerospace Engineering (IC-MAE). IEEE; 2018. p. 337–341.

[3] Mi C and Masrur MA. *Hybrid electric vehicles: principles and applications with practical perspectives*. Chichester, UK: John Wiley & Sons; 2017.

[4] Park YH, Choi SY, Choi YJ, *et al.* A novel charge equalizer with auxiliary circuit to control the allowable charging and discharging current of the lithium-ion battery. In: 2017 IEEE 3rd International Future Energy Electronics Conference and ECCE Asia (IFEEC 2017-ECCE Asia). IEEE; 2017. p. 500–505.

[5] Williamson M. Air power the rise of electric aircraft. *Engineering & Technology*. 2014;9(10):77–79.

[6] Shang Y, Xia B, Zhang C, *et al.* An automatic battery equalizer based on forward and flyback conversion for series-connected battery strings. In: 2017 IEEE Applied Power Electronics Conference and Exposition (APEC). IEEE; 2017. p. 3218–3222.

[7] Lu L, Han X, Li J, *et al.* A review on the key issues for lithium-ion battery management in electric vehicles. *Journal of Power Sources*. 2013;226: 272–288.

[8] Uno M and Tanaka K. Single-switch cell voltage equalizer using multi-stacked buck-boost converters operating in discontinuous conduction mode for series-connected energy storage cells. *IEEE Transactions on Vehicular Technology*. 2011;60(8):3635–3645.

[9] Kim Y, Samad NA, Oh KY, *et al.* Estimating state-of-charge imbalance of batteries using force measurements. In: 2016 American Control Conference (ACC). IEEE; 2016. p. 1500–1505.

[10] Lin X, Stefanopoulou AG, Li Y, *et al.* State of charge imbalance estimation for battery strings under reduced voltage sensing. *IEEE Transactions on Control Systems Technology.* 2014;23(3):1052–1062.

[11] Alex vdB. Direct use of PV panels [database on the Internet]. www.researchgate.net. 2019. Available from: https://www.researchgate.net/publication/331907430_Direct_use_of_PV_panels.

[12] U.S. Department of Energy *What is the energy payback for PV?* Washington, DC: The National Renewable Energy Laboratory; 2004.

[13] Beaudet A. *How do I read the solar panel specifications?* The altE Blog; 2016.

[14] Van den Bossche A and Valchev VC. *Inductors and Transformers for Power Electronics*. Boca Raton, FL: CRC Press (Taylor and Francis Group); 2005.

# Chapter 14
# Auxiliary DC/DC converters

## 14.1 Introduction

In the example, a battery pack with 32 cells in series was considered. The battery could control brushless DC motors, as they can operate using the full battery voltage without any problem. However, in the electric vehicles, also some auxiliary equipment (lighting, dashboard, on-board computer, etc.) exists. Thus to power the auxiliary equipment in the car, converters had to be implemented to convert the battery voltage to lower voltage levels, and this should be done in an efficient way. Lowering the voltage from battery voltage immediately to lower voltage using only one converter would lead to higher voltage levels across the internal components and a less efficient voltage conversion. Because of that, the voltage converter was split into two separate stages. The main purpose of doing this 'two-stage DC/DC converter' is to equalize all battery cells from the total battery packs. In the first stage, the first converter, which is a buck converter, will be fed by the total battery pack consisting of four packs in series. Also, each pack includes 8 cells in series, thus a total of 32 cells in series. The nominal voltage of each cell is 3.2 V in this case; therefore, the total pack results in 102.4 V. The first converter 'buck converter' [1] will decrease the total battery packs voltage to a lower voltage, for example 40 V, which is suitable to feed the second-stage converter to equalize eight cells in series. The second converter could be, a forward converter, a flyback converter, or a multi-secondary transformer balancing circuit. In fact, in the range of 20 Vdc–60 Vdc, plenty of DC/DC converters exist, mainly made for IT. Own-made equalizing converters can equalize each battery pack '8 cells in series'. Therefore, for total battery packs, 4 converters are required to equalize 32 cells. The number of cells or battery packs can be extended to the desired numbers. Also the vehicle can be operated by two times two packs in parallel. However, if the number of cells is doubled, 64 cells, it results in a voltage of 204.8 V. An intermediate voltage of 160 V would then comply with most of larger cars. The voltage of 160 Vdc corresponds to a rectified 110 V–140 V grid, so many switched-mode power supplies are available. In fact, between 60 Vdc and 140 Vdc, not much standard converters are offered. The principle of the two-stage converter is illustrated in Figure 14.1.

## 14.2 Buck converter

Most converters available on the market have quite complicated internal circuits, resulting in stable and precise characteristics, which can even be tuned in some cases. However, each component added to a circuit may introduce extra losses and reduce the efficiency. It also increases the cost. A circuit for a buck converter was made while trying to limit the number of used components to a minimum. The idea of this converter 'first stage' is to drive a buck converter using the integrated circuit (IC) HV9861A. The HV9861A is actually a light-emitting diode (LED) driver which operates using Average-Mode Constant Current Control [2]. The typical application circuit of the LED driver is shown in Figure 14.2. In our case, the 'LED load' of the circuit is replaced by the output terminals of the buck converter. To obtain a fixed output voltage a feedback from the output to the LED driver is provided, which will be discussed later.

The HV9861A has an internal voltage regulator allowing a wide range of external supply voltages. The pin $V_{DD}$ is kept constant at 7.5 V by this voltage regulator. The pins PWMD and LD can be used for the 'dimming of LEDs' driven by the IC. The functions corresponding to these pins are not needed for our implementation and are disabled by connecting them to $V_{DD}$. The MOSFET of the buck converter is driven by the GATE output of the LED driver. The state of the pins current sensing (CS) and $R_T$ determines the waveform used to drive the gate of the MOSFET. The CS pin measures the voltage across an external sense resistor in

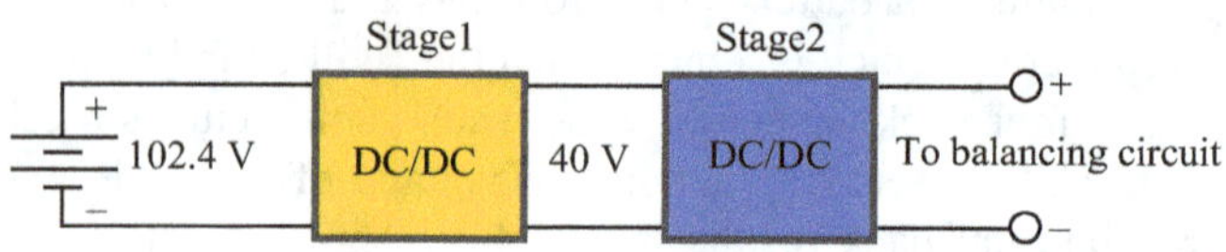


*Figure 14.1 Two-stage voltage conversion*

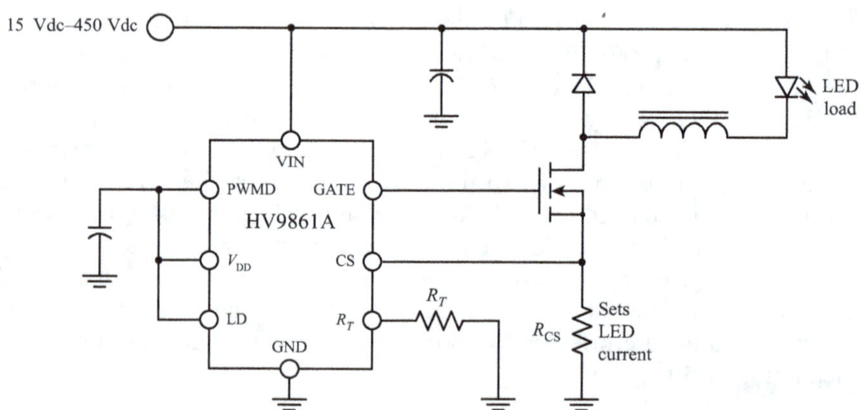


*Figure 14.2 Typical application circuit of HV9861A*

order to measure the current through the MOSFET. The driver operates using a constant off-time for the MOSFET. The duration of this off-time is determined by the value of the timing resistor connected between ground and the RT-pin.

## 14.3 Component selection

A previous version of the circuit was designed to produce a constant output voltage of 40 V. Now let us assume that the maximum output power equals 56 W, which results in a maximum output current of 1.4 A. When $V_{LD} \geq 1.5$ V, the following formula from the data sheet is valid. Because LD is directly connected to the internal voltage regulator of 7.5 V, the condition $V_{LD} \geq 1.5$ V is always satisfied.

$$I_{out} = \frac{0.27\ \text{V}}{R_{CS}} \tag{14.1}$$

As $I_{out} = 1.4$ A, this gives $R_{CS} = 0.193\ \Omega$. To better approximate this value of $R_{CS}$, two resistors of 0.39 Ω were placed in parallel. The value of $R_T$ was chosen equal to 500 kΩ. By using the formula extracted from the data sheet which is valid when $30\ \text{k}\Omega \leq R_T \leq 1.0\ \text{M}\Omega$, the off-time can be calculated

$$T_{off}(\mu\text{s}) = \frac{R_T(\text{k}\Omega)}{25} + 0.3 = 20.3\ \mu\text{s} \tag{14.2}$$

The schematic of the final circuit, including the feedback to obtain a constant output voltage, is shown in Figure 14.3. The calculations were done assuming an output voltage of 40 V. Later on, this voltage was fine-tuned to better match the optimal input voltage of the second-stage converter.

This fine-tuning can be achieved by changing the resistance values of the voltage divider created by $R_1$–$R_2$–$R_3$. For an output voltage of 40 V, the duty ratio of the MOSFET equals

$$\delta = \frac{V_{out}}{V_{in}} = \frac{40\ \text{V}}{102.4\ \text{V}} = 0.39 \tag{14.3}$$

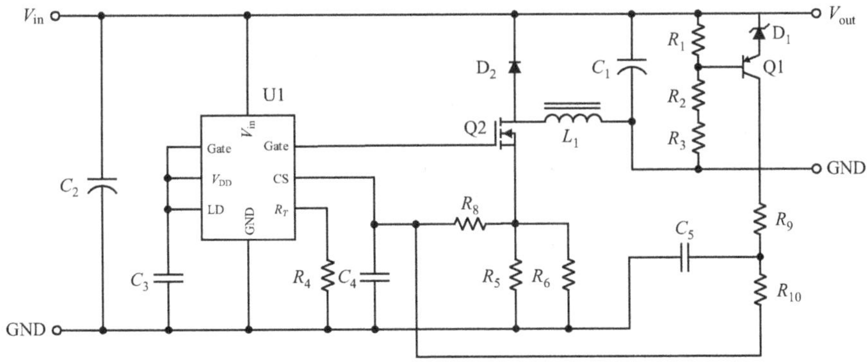


*Figure 14.3 Buck converter circuit*

As the off-time is kept constant by the driver, the period $T$ of one cycle amounts

$$T = \frac{T_{\text{off}}}{1-\delta} = \frac{20.3\ \mu}{1-0.39} = 33.31\ \mu\text{s} \tag{14.4}$$

and this corresponds to a switching frequency of 30 kHz. This switching frequency is much higher than the human hearing range (20 Hz to 20 kHz) and thus will not generate audible noise.

## 14.4 Inductor design

To sufficiently limit the ripple in the output current, a well-chosen inductor has to be used. In the data sheet of HV9861A, it is advised to limit the peak-to-peak value of the ripple current to 30%–40% of the full load average DC. To limit the size of the inductor, the upper limit will be used. As it has been known, the full-load output current $I_{\text{out}}$ is 1.4 A, which results in

$$I_{L,\text{pp}} = 0.4 \times I_{\text{out}} = 0.56\ \text{A} \tag{14.5}$$

From the peak-to-peak current, the required inductance can be calculated as

$$L = \frac{V_{\text{out}} \times T_{\text{off}}}{I_{L,\text{pp}}} = \frac{40\ \text{V} \times 20.3\ \mu\text{s}}{0.56\ \text{A}} = 1.45\ \text{mH} \tag{14.6}$$

To keep the full design freedom, the inductor was made by ourselves. For the inductor, a PQ32/20 core with a corresponding coil former is used. The selected material is N87. For the design, the required inductance is rounded up to 1500 μH. The expected peak current is given by

$$I_{\text{peak}} = I_{\text{out}} + \frac{I_{L,\text{pp}}}{2} = 1.4\ \text{A} + \frac{0.56\ \text{A}}{2} = 1.68\ \text{A} \tag{14.7}$$

A maximum current of 1.8 A is used to have a certain margin and to take imperfections of the control into account. With this peak current, the corresponding peak flux linkage is

$$\Phi_{\text{peak}} = L\,I_{\text{peak}} = 1.5\ \text{mH} \times 1.8\ \text{A} = 2.70\ \text{mVs} \tag{14.8}$$

The data sheet of N87 shows a saturation induction of 490 mT at 25 °C and 390 mT at 100 °C. To be sure the core does not saturate, a saturation induction $B_{\text{sat}} = 350$ mT is used. The minimum core cross section $A_{\text{min}}$ of PQ32/20 is 127.5 mm$^2$. From this, the required number of turns can be calculated

$$n = \frac{\Phi_{\text{peak}}}{B_{\text{sat}}\,A_{\text{min}}} = \frac{2.70\ \text{mVs}}{350\ \text{mT} \times 127.5\ \text{mm}^2} = 60.5\ \text{turns} \tag{14.9}$$

This value can be rounded down to 60 turns because the chosen safety margins on previous selected limit values were quite large. For the winding, 0.71 mm copper wire is used. Taking the insulation into account results in an outer diameter

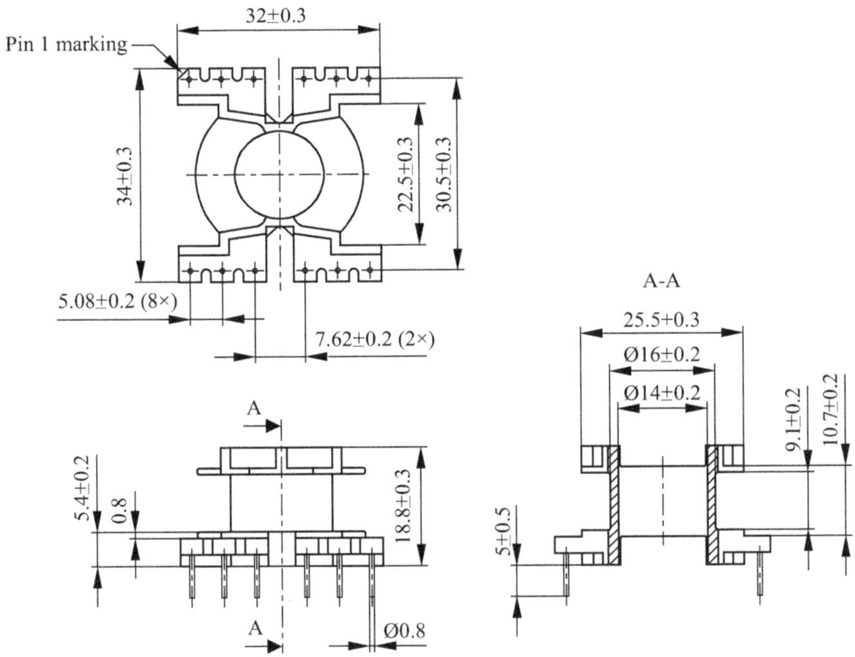


*Figure 14.4 Dimensions of coil former for PQ32/20 core*

of around 0.79 mm. The maximum winding width of the used coil former is 9.1 mm (see Figure 14.4). This gives a maximum of 9.1 mm/0.79 mm = 11.5 turns per layer. Choosing 10 turns per layer results in six needed layers. Assuming a square fitting of the wires gives a winding height of 6 × 0.79 mm = 4.74 mm. The maximum winding height is till which the coil former supports the turns equals

$$\frac{25.5\text{ mm} - 16\text{ mm}}{2} = 4.75\text{ mm} \tag{14.10}$$

The required winding height thus seems to be at the limit of the maximum winding height. However, assuming a square fitting of the wires gives the maximum value of the winding height one can expect and also the winding diameter of the upper layer can be somewhat larger than 25.5 mm as the core opening has a diameter of 27.5 mm, Figure 14.5. The losses in the inductor consist of joule losses in the wire and core losses, due to the hysteresis and eddy currents in the core. The mean radius of the turns is 10.37 mm. This gives for the mean length of the turns (MLT)

$$l_{\text{MLT}} = 2\pi \times 10.37\text{ mm} = 65.2\text{ mm} \tag{14.11}$$

From this the resistance of the wire can be calculated

$$R_{\text{wire}} = \frac{\rho\, l_{\text{MLT}}}{A}\, n \tag{14.12}$$

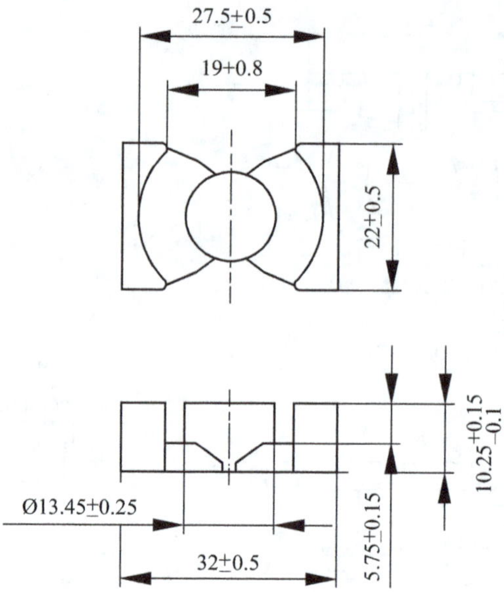


*Figure 14.5 Dimensions of PQ32/20 core*

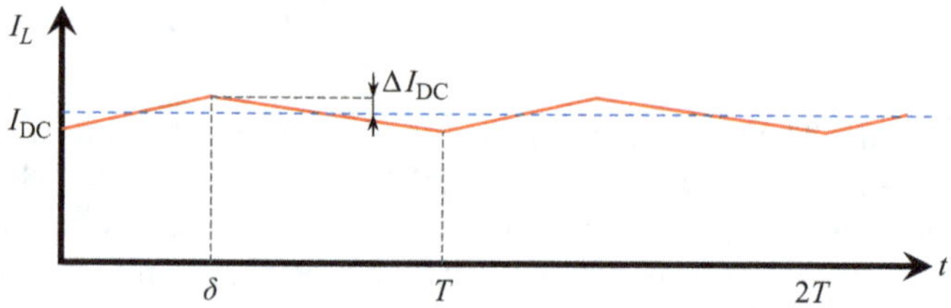


*Figure 14.6 Inductor current waveform*

where $\rho$ is resistivity of copper, $\rho = 2.3 \cdot 10^{-8}$ Ωm; $A$ is wire cross section, $A = \pi((0.71\ \text{mm})^2/4) = 0.396\ \text{mm}^2$.

This gives a resistance of 230 mΩ for the wire. The rms value of the current through the conductor can be calculated as

$$I_{rms} = \sqrt{I_{DC}^2 + \frac{(\Delta I_{DC})^2}{3}} \tag{14.13}$$

The waveform of the current through the inductor is illustrated in Figure 14.6. Here $I_{DC}$ equals 1.4 A and $\Delta I_{DC} = \dfrac{0.4 \times 1.4\ \text{A}}{2} = 0.28$ A. Using these values, (14.13) gives 1.409 A. The joule losses in the copper wire are

$$P_{j,loss} = R\, I_{rms}^2 = 0.23\ \Omega \times (1.409\ \text{A})^2 = 0.457\ \text{W} \tag{14.14}$$

Next the core losses are estimated. The data sheet of the inductor core only gives the value of 5.5 W/set as a relative loss at 100 kHz and 200 mT at 100 °C. To estimate the core losses at another frequency and induction, Steinmetz's equation is used

$$P_{v,\text{loss}} = k\, f^{\alpha}\, B^{\beta} \tag{14.15}$$

The frequency was already calculated to be 30 kHz. The amplitude of the AC component of the magnetic induction can be calculated from the peak induction in the core. The peak-to-peak ripple current amounts 40% of the average current and the same holds for the magnetic induction. This gives for the peak induction $B_{\text{peak}} = 1.2\, B_{\text{avg}}$. The amplitude of the AC component equals 0.2 times the average value, resulting in

$$\Delta B = \frac{0.2}{1.2} \times B_{\text{peak}} = \frac{0.2}{1.2} \times 350\ \text{mT} = 58.3\ \text{mT} \tag{14.16}$$

In the data sheet of the material N87 relative core losses at different frequencies and magnetic induction amplitudes are given. These losses are given in Table 14.1.

Fitting (14.15) to the losses given in Table 14.1 gives $\alpha = 1.4$ and $\beta = 2.1$. Using these values, the loss of the core can be estimated starting from the value given in the core's data sheet

$$P_{v,\text{loss}} = 5.5\ \text{W} \times \left(\frac{30}{100}\right)^{1.4} \times \left(\frac{58.3}{200}\right)^{2.1} = 0.08\ \text{W} \tag{14.17}$$

Using a rule of the thumb [3], the heat-transfer capability of the core can be estimated as

$$P_{\text{th}} = a \times b \times 0.2\ \text{W/cm}^2 \tag{14.18}$$

where $a$ and $b$ are the two largest dimensions of the surfaces participating to the heat transfer. It is somewhat less as the major part of the losses are concentrated in the copper and that the ferrite will be at a lower temperature and so participates less in heat removal. Also the eddy current losses were neglected, but the DC is much larger than the AC rms component. From Figure 14.5 follows that these two dimensions are 3.2 cm and 2.1 cm, respectively. Using these values (14.18) gives 1.34 W. As the heat-transfer capability exceeds the losses of the inductor, no overheating is expected. The result of the wound inductor core is shown in

*Table 14.1 Relative core losses N87 (at 100 °C)*

| Frequency (kHz) | Magnetic induction (mT) | Relative core losses (kW/m$^3$) |
|---|---|---|
| 25 | 200 | 57 |
| 100 | 200 | 375 |
| 300 | 100 | 390 |

*Figure 14.7 Wound PQ32/20 inductor core*

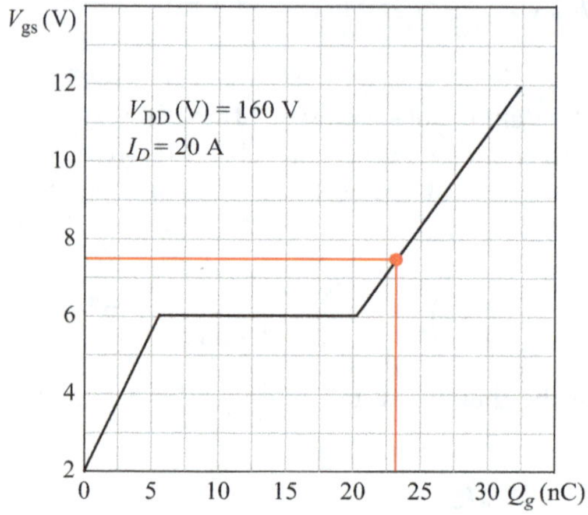


*Figure 14.8 Gate charge vs gate–source voltage*

Figure 14.7. The inductance was tuned by adapting the air gap. Finally an inductance of 1.75 mH was obtained.

When looking for a MOSFET suited to use in the buck converter, the MOSFET STD25NF20 was chosen. The MOSFET has to fulfil certain requirements to guarantee a good operation of the converter. A very important requirement for this application is the total gate charge of the MOSFET. For the LED driver to be able to control the MOSFET, the total gate charge should be less than 25 nC for operating frequencies below 100 kHz and less than 15 nC for higher frequencies. The operating frequency of our converter is 30 kHz, setting the limit for the total gate charge to 25 nC. Figure 14.8 shows the relation between the total gate charge and $V_{gs}$ of the MOSFET. The gate voltage is limited by the internal voltage regulator of the IC which is 7.5 V and thus, neglecting the small voltage drop across the sensing resistors, $V_{gs}$ will be limited to 7.5 V. To this gate–source voltage corresponds a gate charge of 23 nC, which is just below the limit of 25 nC.

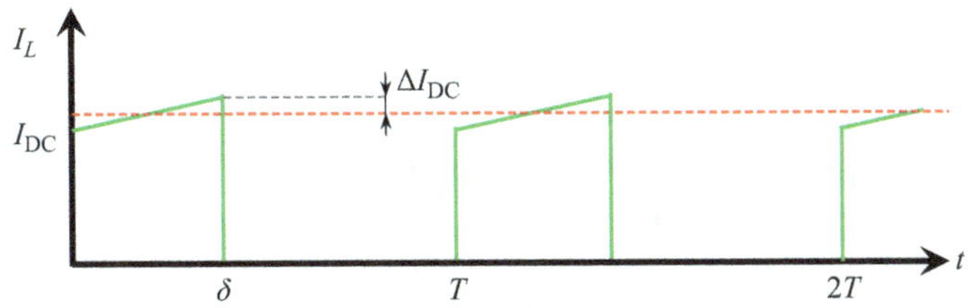


*Figure 14.9 Inductor current waveform*

The MOSFET also meets the requirements regarding applied voltages and expected currents. To determine the losses in the MOSFET, first the rms value of the drain–source current, as illustrated in Figure 14.9, has to be calculated as

$$I_{rms} = I_{DC}\sqrt{\delta}\sqrt{1 + \frac{1}{3}\left(\frac{\Delta I_{DC}}{I_{DC}}\right)^2} \tag{14.19}$$

The DC equals 1.4 A obtained earlier and $\delta = 0.39$. $\Delta I_{DC}$ can be calculated as the ripple of the inductor current is known to be 40% of the DC

$$\Delta I_{DC} = \frac{0.4 \times 1.4\ \text{A}}{2} = 0.28\ \text{A} \tag{14.20}$$

Using these values (14.19) gives $I_{rms} = 0.88$ A. From this, the losses in the MOSFET can be calculated from

$$P_{loss} = R_{DC(on)} I_{rms}^2 = 0.1\ \Omega \times (0.88\ \text{A})^2 = 0.08\ \text{W} \tag{14.21}$$

Mainly switching losses could dominate, they are estimated at 0.2% of power. $40 \times 1.4 \times 0.002 = 0.112$ W. The freewheel diode would take some (about 60% duty ratio, at 0.9 V) $0.9 \times 1.4 \times 0.6 = 0.756$ W. We can give a loss estimate for gate charging, coming from the battery: 23 nC, 30 kHz, and 102.4 V, so $23 \times 30 \times 102.4 = 71$ mW. The internal current consumption of the IC is about 1 mA. The summation of losses at full load (without capacitor, PCB losses, which are low) $0.457 + 0.08 + 0.08 + 0.112 + 0.756 + 0.071 + 0.1024 = 1.66$ W.

## 14.5 Constant voltage feedback working principle

As mentioned in the beginning, the IC driving the circuit is designed to obtain a *constant current* for LEDs. However for our application, a constant output voltage is desired. To obtain this, feedback from the output voltage to the IC was added using a PNP transistor (Q1), a Zener diode ($D_1$), and some resistors. Properly choosing their values results in a constant voltage output of 47.6 V. The working principle is based on the way the driver IC senses the load current through the MOSFET. By forcing another voltage at the CS pin, the IC is tricked and it will change its current control accordingly. Imagine the circuit does not have to provide the full power output it was designed for. This would result in a higher output voltage if the circuit still tried to control the current to equal 1.4 A. The voltage across $D_1$ remains the same, but the voltage across $R_1$ will increase. This causes the PNP's $V_{BE}$ to drop below the value for which it will become active resulting in a collector current $I_C$ flowing of the PNP. This current will flow through resistors $R_8$, $R_5$, and $R_6$ and the voltage across the resistors due to this current adds up to the voltage due to the real load current through the MOSFET. Due to the voltage increase at the CS pin of the IC, the IC thinks the current is too high and it will control the MOSFET in such a way that the load current is reduced, causing the output voltage to reduce back to its nominal value. When trying to increase the output power above the designed value, however, the PNP will not become active. Thus output load current is limited to its designed value of 1.4 A and the output voltage will drop when trying to increase the power output.

## 14.6 Result

The circuit shown in Figure 14.3 was converted into a PCB layout and the final result is shown in Figure 14.10. All calculations were done assuming an output voltage of 40 V. When used in the vehicle, a second converter is connected to this converter to further lower the voltage to the desired voltage. To test the circuit, it was connected to the battery of the vehicle and a power resistor was used as a load. Input voltage and current were measured, as were output voltage and current. From these measurements the *V–I* output characteristic shown in Figure 14.11 was constructed. The addition of the voltage feedback to the circuit results in a fixed output voltage close to the nominal input voltage of the second converter. Once the current reaches 1.4 A, the output voltage starts to drop and the output current remains limited to 1.46 A. The efficiency is determined for several load set points by calculating the input and output power using the measured voltages and currents.

Figure 14.12 shows the efficiency as the function of the relative load. The maximum obtainable output power of 65.5 W is used here as a reference. For low loads, the efficiency is still somewhat lower, due to the higher relative importance

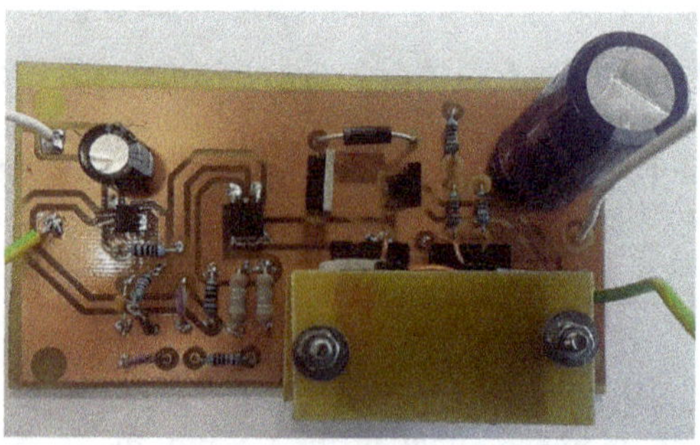

*Figure 14.10 Buck converter final result*

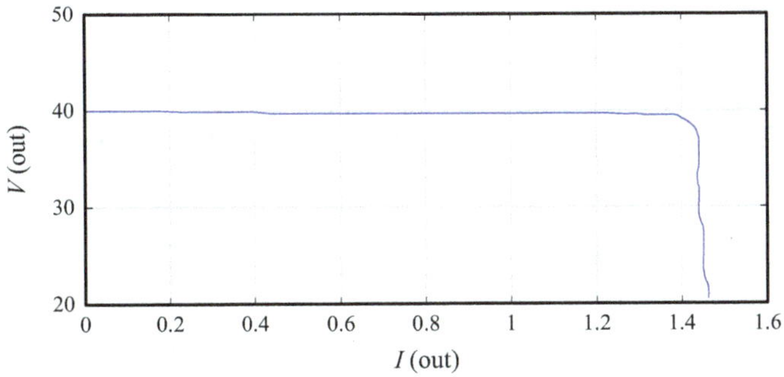


*Figure 14.11 V–I characteristic converter*

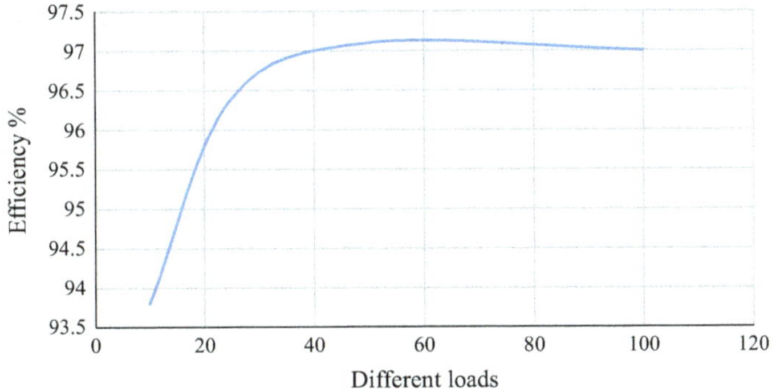


*Figure 14.12 Converter efficiency at different loads*

of the no-load losses of the components. By the transistor feedback, the losses associated with an optocoupler's feedback are avoided, e.g. 5 mA at 40 V would mean 200 mW, which would lower the efficiency a lot at low load. Once the load is above 15% of the maximum load, an efficiency between 96% and 98% is reached. The deviations seen in the curve are caused by the limited accuracy of the current measurements, resulting in rounding errors.

## Summary

Using a common intermediate voltage seems to be a good way to power the voltage equalizers from. For smaller vehicles, it could be 40 V, for larger 160 V. Starting from these also, other supplies can be generated. In this chapter, a 'minimal' circuit design has been tried out. The use of an LED-driver IC results in a much better current control in overload than usual converter ICs. By the fact, no optocoupler is used in the feedback, and a much lower loss close to no-load is obtained. True that this style of design can have many applications, not only for BMS. In battery applications, lots of auxiliary supplies may be needed, which all like a high efficiency and a low no-load loss. The biggest losses seem to be at the freewheel diode, but at that voltage level, it still permits good efficiencies.

## References

[1] Buck converter. USA: https://en.wikipedia.org. Available from: https://en.wikipedia.org/wiki/Buck_converter.
[2] LED driver/average-mode constant current control. USA: https://www.microchip.com/en-us/product/HV9861A.
[3] Van den Bossche A and Valchev VC. *Inductors and Transformers for Power Electronics*. Boca Raton, FL: CRC Press (Taylor and Francis Group); 2005.

# Conclusions

The general purpose of the book is to give an overview of the battery problem setting in combination with battery management systems (BMSs). It provides building blocks to compose a BMS, where blocks are compared, and also less known inductive equalizing circuits are proposed.

Chapter 1 started in the Mesopotamian time, but after that it took almost two millennia to batteries to be rediscovered. Different types of batteries are studied. Furthermore, lithium-ion battery characteristics, types, and shapes are investigated and gradually come to the problem setting why some batteries have no or only a reduced BMS and why the high performance batteries of today need it.

In Chapter 2, discussion about batteries, several points of batteries are tackled. Batteries may not be a good solution for every energy storage problem. The manufacturing of the batteries also releases $CO_2$ and other pollutants. It makes no sense to store electricity first in a battery to just heat water with it afterwards; in fact the whole storage problem is concerned: gravity, pressure, flywheel, hydrogen, energy fluids as methanol, and ammonia.

In Chapter 3 an overview is given over different battery types, in groups, depending on the electrolyte: water, organic solvent, molten salt, fuel cell, and flow battery, where at the end the chemical potential is given, to understand why some elements are chosen.

Chapter 4 describes why a BMS is needed, and the state of charge and state of health are explained. Then the battery configuration is described. In the end, several battery balancing topologies are studied and examples are given.

In Chapter 5, Section 1, a combined passive, equalizing and detecting circuit is built up to equalize battery cells. It will detect if the individual cell has reached the maximum fully charged voltage, then it will dissipate the extra energy through a shunting resistor. The circuit will detect if one cell is high, or all cells are high, and the results can be seen on an LCD screen. The circuit will top each battery cell to a maximum fully charged voltage. A Bluetooth model is added to command the microcontroller by voice or text message to cut off the charger if all cells are at a high voltage level or when desired. A major feature of the system is to draw a very low standby current. Therefore, the system does not contribute significantly to the self-discharge of the battery. In Section 2, a smart monitoring cell voltage measurement circuit is proposed. It will measure each individual cell voltage by an operational amplifier used as differential amplifier. The measured voltage will be stored in a microcontroller and will be transmitted via Bluetooth to the smartphone. The cell voltages can be displayed on the mobile app, which is generated with a software 'MIT App Inventor 2'.

Chapter 6 proposes a resonant switched capacitor converter (RSCC) balancing method. In switched-capacitor converter types, the advantages are 'easy to implement' and 'simple circuit', while a longer balancing time is a disadvantage, also more transistors are required. The RSCC improves the equalizing time by the resonant effect and achieves lower losses due to zero current turn-on and off.

In Chapter 7, Section 1, an inductor balancing circuit has been studied. Inductor balancing decreases the balancing time, but it requires a more complex magnetic core. The proposed circuit requires fewer switches, resulting in correspondingly less switching losses. In Section 2 of Chapter 6, a Ćuk converter balancing method is studied. The Ćuk converter balancing circuit has a high efficiency with non-dissipative currents and a bidirectional energy transferring capability. The traditional Ćuk balancing method requires $2N-2$ switches for $N$ cells; however, the suggested coupled method uses $N$ switches and $N+1$ inductors for $N$ number of cells, resulting in fewer components and corresponding losses.

In Chapter 8, Section 1, a coupled inductor equalizing is proposed. It only requires one magnetic core and four windings to equalize eight cells; thus the number of components is reduced as compared to the conventional separate inductor based balancing method. Indeed, the traditional inductor based equalizer needs $N-1$ inductors and magnetic circuits for $N$ number of cells and more switches. In Section 2, a Ćuk converter balancing method using a coupled inductor is investigated. The proposed circuit equalizes eight battery cells in series. The conventional Ćuk converter necessitates $2N-2$ switches to balance $N$ cells, while the proposed circuit requires only $N$ switches for $N$ cells in series. Also, the proposed developed topology uses coupled inductors instead of uncoupled inductors. The coupled inductor shares the same magnetizing inductance. It can be done by using five wires 'in one hand'. The circuit contains five windings, one magnetic core, for eight cells, and one capacitor for two cells.

In Chapter 9, Section 1, a transformer balancing circuit is investigated. The transformer balancing is an active method to move energy from a cell to another. It has fast equalization time compared to other methods. The circuit requires $N$ secondaries per $N$ cells. In the primary side, it has a full-bridge inverter which is powered from the total battery pack. In Section 2, investigates a transformer balancing circuit based on a double forward converter. It controls two times more cells than secondaries, and it equalizes eight cells in a series. This means a low pin count on the transformer. Conventional schemes using a multi-winding transformer require an equal number of secondaries per cell. The proposed scheme requires only one secondary for two adjacent cells. Section 3, a flyback converter balancing is investigated. A flyback equalizer is also suitable for multi-battery cells because of the possibility of having multiple secondaries. The flyback converter is widely used in power supplies and so a lot of control ICs are offered, which has a low cost, a simple design, and reasonable efficiency. A drawback is the presence of an air gap that may result in an unequal back EMF.

In Chapter 10, a forward–buck converter balancing is proposed. The topology equalizes 32 cells in series in a battery pack, by using 4 forward converters, one for each battery module and one for the whole battery pack using a buck converter.

The battery pack consists of four battery modules which are in series. Therefore, the proposed system equalizes 32 cells in series. The proposed circuit employs a single transistor used in zero-voltage switch for the forward converter. To overcome the problem of many pins on a coil former, the transformer secondary windings are made by using hairpin turns on a ring core. It permits, e.g. having eight secondaries and uniform output voltages. The proposed topology has fewer components and circuitry and it can equalize multiple battery packs by using a single-buck converter and several forward converters for each battery pack.

In Chapter 11, the concept of battery charging for lithium–iron–phosphate cells in electric vehicle (EV) applications is investigated. It is a design of a charger with four battery packs, each consists of eight cells in series. Two battery chargers are implemented to feed the whole battery packs. A current driver is proposed to limit the current from the charger. The operational principle is based on a MOSFET constant current driver. The battery should only be connected when the car needs to drive to avoid unnecessary losses. Moreover, a resistor is required to avoid high starting currents; hence the start-up system is required and designed to overcome this problem.

In Chapter 12, an automatic change from charging, equalizing to standby is presented. Power-loss calculation and heat sink design are investigated. As in Chapter 2, a high-voltage equalizing detection circuit was developed. The rationale of the proposed method is that when all signals are absent, and thus all 32 cells are below their threshold voltage, the charging current is not limited and the charger can deliver its full current to the battery. When one cell of the 32 cells goes above the threshold voltage, the charging current is reduced to 300 mA such that the balancing circuit is capable of balancing the voltages of the affected module. Once one of the four modules indicates that all eight cells are high, the battery is disconnected automatically from the charger.

In Chapter 13, the charging concept of lithium-based batteries for EVs by using photovoltaic (PV) panel and push–pull converter is investigated. This chapter aims to charge the batteries with a PV panel. The maximum power voltage of this PV panel is much lower than the battery packs; therefore, a DC–DC converter is required to overcome this problem. Hence, a push–pull converter is designed and implemented to be combined with the solar panel. The battery bank includes four battery packs and each pack consists of eight cells in series, so a total of 32 cells in series will be charged. To do this, a 100 W flexible Lensun Solar Panel was preferred to be implemented on the roof of the vehicle.

In Chapter 14, the concept of DC/DC converter to supply the BLDC motor or auxiliary equipment such as lighting, dashboard, on-board computer in the car is investigated. To power the supplemental equipment in the car, converters are required to convert the battery voltage to lower voltage levels. Lowering the voltage from battery voltage immediately to lower voltage using a single converter would lead to higher voltage levels across the internal components and also less efficient voltage conversion. Thus the voltage converter was divided into two separate stages. The main purpose of doing this 'two-stage DC/DC converter' is to equalize all battery cells from the total battery packs.

Unfortunately, the area of batteries and BMSs got so large that a book has to focus on some aspects, which in this case is more axed to hardware and inductive solutions. Also a wide application area is aimed and not the specific of cars with often CAN bus controlled systems. Batteries will have to be recycled but may also get a second life, where a non-dedicated BMS could be interesting. Too much sophistication might reduce the BMS life expectancy, and may in this way also damage the battery.

# Index

www.ingramcontent.com/pod-product-compliance
Lightning Source LLC
LaVergne TN
LVHW011848200926
848484LV00002B/421